T0314001

Individuality and Entanglement

Individuality and Entanglement: The Moral and Material Bases of Social Life

Herbert Gintis

Princeton University Press
Princeton and Oxford

Copyright ©2017 by Princeton University Press

Published by Princeton University Press, 41 William Street, Princeton, New Jersey 08540

In the United Kingdom: Princeton University Press, 6 Oxford Street, Woodstock, Oxfordshire OX20 1TR

press.princeton.edu

Jacket art courtesy of Shutterstock

Library of Congress Cataloging-in-Publication Data

Names: Gintis, Herbert, author.

Title: Individuality and entanglement : the moral and material bases of social life / Herbert Gintis.

Description: Princeton : Princeton University Press, [2016] — Includes bibliographical references and index.

Identifiers: LCCN 2016020687 — ISBN 9780691172910 (hardcover : alk. paper)

Subjects: LCSH: Economics–Psychological aspects. — Economics–Sociological aspects. — Social psychology. — Rational choice theory–Social aspects. — Social sciences.

Classification: LCC HB74.P8 G46 2016 — DDC 306.3–dc23 LC record available at https://lccn.loc.gov/2016020687

British Library Cataloging-in-Publication Data is available

The publisher would like to acknowledge the author of this volume for providing the camera-ready copy from which this book was printed

This book has been composed in Times and Mathtime by the author

Printed on acid-free paper.

Printed in the United States of America

10 9 8 7 6 5 4 3 2 1

Dedicated to Gary Becker, Jack Hirshleifer and Edward O. Wilson for their ability to explore deep connections across the various regions of human social life. I love these guys.

> Play is a uniquely adaptive act, not subordinate to some other adaptive act, but with a special function of its own in human experience.
>
> Johan Huizinga

> How selfish soever man may be supposed, there are evidently some principles in his nature, which interest him in the fortune of others, and render their happiness necessary to him, though he derives nothing from it except the pleasure of seeing it.
>
> Adam Smith

Contents

Overview

The eternal mystery of the world
is its comprehensibility.

Albert Einstein

Physicists may one day have found the answer
to all physical questions,
But not all questions are physical questions.

Gilbert Ryle

HG: Who are you?
C: I am the Choreographer.
HG: What are you?
C: An emergent property of social systems.
HG: What are you here for?
C: To ask you some questions.

Choreographer interview

I develop several related themes in this book. The first theme is that *society is a game with rules, people are players in this game, and politics is the arena in which we affirm and change these rules*. Unlike the rules in standard game theory, however, social rules are continually *contested* by players allying to scrap old rules and create new rules to serve their purposes.

Rule contestation is conflictual. People do not always agree on what rules should govern their lives. Societies usually have rules concerning how the rules themselves get changed, but people do not always play by the rules when they want to change the rules. *Individuality* is central to our species because the rules do not change through inexorable macrosocial forces or through a biochemical dynamic of gene substitution. Rather, individuals band together to change the rules.

Everything distinctive about human social life flows from the fact that we construct and then play social games. Other animals are playful, but they do not make up the games they play. They do not change the rules of the game to suit their purposes. Similarly, other animals live in societies. But the rules of the game for nonhuman societies are inscribed in the genome of the species, while ours is not. Other animals do not change these rules to suit their purposes. We do.

Second theme: Playing games with socially constructed rules requires a *moral sense*. Humans treat some rules as purely *conventional*, such as what side of the road to drive on. But many social rules are *morally binding*. Especially when there is general agreement about the rules, people gain satisfaction by playing by the rules, are ashamed when they break the rules, and are offended when others break the rules. Indeed, individuals often reward others who play by the rules and punish others who break the rules even at considerable personal cost and without a prospect of personal return in the future.

Even societies that lack government, judges, juries, and jails effectively reward and punish behavior. This is why Thomas Hobbes was profoundly incorrect in saying that before the emergence of modern institutions, life was "solitary, poor, nasty, brutish, and short" (Hobbes 1968[1651]). Our moral sense was developed long before there were courts, jails, and teachers to lecture us on morality. Despots would love to be able to determine the moral sense of their subjects, but they can never rest securely in their beds knowing that they have done so. Morality has a dynamic that, at least up to now, cannot be controlled by states. New technologies may change this. Hopefully not.

Third theme: Our minds are socially *entangled*. Cognition (what people know and believe) does not reside in individual minds. Minds are *socially networked* with cognition distributed *across* social networks. An individual in making a decision draws not only on his own information and personally ascertained beliefs, but on information and beliefs located in the minds with which he is networked as well. Entangled minds produce social behavior that is *rational*, as we shall see, although it does not always conform to the standard axioms of rational choice in contemporary decision theory. We extend rational choice theory to deal with entangled minds.

Fourth theme: Human morality has an important *nonconsequentialist* dimension: individuals generally do the right thing, when they do the right thing, not because of the *personal or social consequences* of their actions, but simply because they believe it is the right thing to do. I am honest, when I am honest, not simply because I care about the people with whom I daily interact (although I may care about them), and not simply because I am afraid to be caught cheating (although I may indeed be afraid), but because *honesty is good*. Being dishonest dirties me, like not brushing my teeth. I vote not because my vote can possibly alter the outcome of an election, but because contributing to the election of good leaders is the right thing to do, even if one makes no difference to the outcome. I vote for the candidate I want to win because this is my contribution to a collective effort, even though I know my vote will not affect the outcome of the election. This nonconsequentialist moral sense is akin to Aristotle's notion of *virtue* (Aristotle 2002[350BC]), and even more closely to Immanuel Kant's (2012[1797]) notion of the *categorical imperative*.

The nonconsequentialist dimension of morality is connected to the entanglement of minds. Our minds work with a form of causal efficacy that is *collectively rational* as opposed to being *instrumentally rational*. I call this *distributed effectivity*, a principle analyzed in Chapter 3. An example is the belief that one has helped a candidate for political office to win the election, despite the obvious and well understood fact that the candidate would have won even if one had not voted, or even if one had voted for a different candidate.

Fifth theme: Careful inspection of human behavior reveals that there are three distinct dimensions of human preferences: *self-regarding*, *other-regarding*, and *universal*. Self-regarding preferences deal with what we want for our personal selves, including consumption, leisure, wealth, love, health, and the respect of others. Other-regarding preferences deal with our care about other people's well-being. Wanting to help someone for whom we feel compassion, or hurt someone who has offended or annoyed us, are other-regarding preferences. Universal morality is neither self- nor other-regarding. Universal moral principles can have consequences, as when I help a stranger in need, but include Aristotle's nonconsequentialist *character virtues*, such as courage, truthfulness, and loyalty. Nonconsequentialist moral principles correspond to Kant's notion of acting on purely moral grounds.

As a scientist rather than a philosopher, I am not concerned with what people *should* value, but rather with what they *do* value. I am particularly not concerned with the relative moral coherence of utilitarian, deontological, and virtue theories. Real human beings mix and match. So will I.

Sixth theme: Individuals *trade off* among these three sorts of values. Someone, for instance, may sacrifice personal reward (self-regard) on behalf of the well-being of others (other-regard), and/or on behalf of universal moral principles. Similarly, an individual may lie (violation of virtue) out of regard for the feelings of others (other-regard) or because he can gain materially therefrom (self-regard). Aristotle understood this idea by championing the *mean* over both *excess* and *deficiency*, whereas Kant went off track by asserting a rigorous nonconsequentialism in which perfectly moral behavior does not take self-interest or the welfare of others into account.

We discover, through game-theoretic experiments in the laboratory (Fehr and Gintis 2007) and in the field (Herbst and Mas 2015), not only that people trade off, but these trade-offs can be modeled in terms of *rational choice theory*. Indeed, unless an individual satisfies some basic principles of rational choice, it is virtually meaningless to say he is moral or immoral.

There are whole academic disciplines in which rational choice is actively rejected by most researchers. These include psychology, sociology, and anthropology. The grounds for rejection are uniformly unpersuasive and ill-informed. Even great thinkers have offered embarrassingly shoddy reasons for rejecting rational choice. Rejection of rational choice theory accounts in part for these fields' lack of a coherent analytical core.

The current vogue of behavioral economists also include many who spend considerable time arguing that people are not rational (Ariely 2010). While there are certainly systematic violations of the rational actor model (see Chapter 5), rationality is generally strongly empirically supported. There is absolutely no way to do serious social theory without recognizing that human behavior is purposive and can generally be modeled as rational choice. Denying this is like asserting that people cannot see what is really there because people sometimes fall down the steps and laboratory experiments reveal that there are optical illusions.

Seventh theme: The above aspects of human behavior can be understood only through *transdisciplinary research*. Many social scientists take pride in their deep knowledge of one discipline uncontaminated by "foreign" ideas. Some of these researchers do fine work, but the above themes cannot be effectively handled in this way. There is therefore a pressing need for a thorough cleansing of the behavioral disciplines of the glaring incompatibilities among them. For it is precisely these incompatibilities that prevent effective transdisciplinary research. I stressed this in *The Bounds of Reason* (2009), and the theme is reflected in virtually every chapter of this book.

The central fact of the behavioral sciences is that *social species live in complex social systems* that must be studied through the lens of evolutionary theory. The first common background of all the behavioral disciplines is the scientific method, mathematical modeling, statistical testing, and where formal analytical methods fail, and fail they must at some point in conceptualizing complex systems, historical insight, ethnographic description, and agent-based modeling become critical.

Chapter 1 ("Gene-Culture Coevolution") explores the evolutionary dynamic that applies to any species for which epigenetic information takes the form of *culture* that accumulates reliably and long-term across generations. That applies strongly only to *Homo sapiens*. This chapter is based on a paper that appeared in a special issue of the *Philosophical Transactions of the Royal Society B (Biological Sciences)* in 2011. The main point is that culture includes techniques and social norms that determine which genes will be evolutionarily rewarded in a given society. Therefore human genes are the product of human culture as much as the reverse (Boyd and Richerson 1985). I give as an extended example the evolution of the physiology of communication in humans (Section 1.3). This example is important because it involves the development of key human genetic capacities that are the product of *social structure* as opposed to simple *individual adaptation*. Some researchers have called this group selection, but I avoid this term, being rather fed up with the arcane and futile arguments that have arisen over its use. The physiology of communication in humans simply is what it is—an emergent property of human social evolution.

The natural habitats for gene-cultural coevolution are the fields of sociology and anthropology. It is curious, then, that the most powerful models of cultural structure and evolution come from biology. It is equally curious that economic theory has no concept of culture whatever, and as I show in my book *The Bounds of Reason* (2009), deploys an implausible form of methodological individualism that attempts, quite unsuccessfully, to explain human behavior on the basis of rational choice and game theory alone. Rational choice theory and game theory are simply mathematical constructs. Standing alone, like vector spaces or semi-simple algebras, they say nothing.

Chapter 2 ("Zoon Politikon: The Evolutionary Origins of Human Sociopolitical Systems") addresses the evolutionary roots of human sociality. The argument is a direct application of gene-culture coevolutionary theory. The material is taken from a collaboration with coauthors Carel van Schaik and Christopher Boehm, primatologists and anthropologists, that appeared as a target article in *Current Anthropology* in 2015. The inspiration for this analysis was two-fold. First, studying animal behavior reveals that *Homo sapiens* is also *Homo ludens*—Man, the game player. This idea lies at the analytical core of political theory, where politics is viewed as a process of creating, maintaining, and transforming the social rules of the game. Second, an exciting genetic study by Shultz et al. (2011) in *Nature* using coalescent theory identified the most likely socio-political structure of our last common ancestor with the primates. This allowed my coauthors and I to chart a plausible evolutionary cascade that led from this common ancestor, through the hominin line, to *Homo sapiens*.

Chapter 3 ("Distributed Effectivity: Political Theory and Rational Choice") explores the implications of the moral predispositions, both personal and political, that have emerged from the coevolutionary process described in Chapters 1 and 2. This chapter is based on a paper that appeared in the *Journal of Economic Behavior and Organization* in 2015. The value of the rational actor model becomes clear in this chapter as it allows one to explore in some depth the interplay between self-regarding, other-regarding, and universal preferences.

In developing the rational actor model throughout this book I argue that the notion of the *subjective prior* that is standard in rational decision theory (Savage 1954) must be expanded to include entanglement: *human minds are networked* and *cognition is distributed across minds*. Among the themes explored in this chapter is that in dealing with political participation, individuals reason not using standard consequentialist logic, but rather a logic of *distributed effectivity*. From this understanding flows the insight that such phenomena as voting in mass elections and participating in large collective actions is both rational and has a deeply moral dimension. For instance, each individual supporter believes he helped elect the winner even if the outcome would have been no different had he not voted. And each is of course right. The alternative, which is that no one helped the winner get elected, is just silly.

Chapter 4 ("Power and Trust in Competitive Markets") forges links among economic, political, and sociological theory by showing how incomplete contracts, which are ubiquitous and unavoidable in market economies, lead to the *power relations* studied in political theory and the *moral behaviors* characteristic of highly productive advanced economic systems, involving legitimacy, integrity, trust, and trustworthiness. The notion that economic theory can provide the conditions for economic success without specifying the distribution of political power in the economy, or without specifying the moral commitments of market participants, is simply and dramatically incorrect. The material in this chapter draws on many years of research with Samuel Bowles (Bowles and Gintis 1988, 1990, 1993, 1999).

Chapter 5 ("Rational Choice Revealed and Defended") is an explication and a defense of the rational actor model as foundational in analyzing social behavior. Rational decision theory is *de rigeur* in economics and biology but rather strongly spurned by the majority of researchers in the other behavioral disciplines. This chapter explains why this is a serious error. The critics of rational choice invariably—and embarrassingly—misrepresent the theory. In particular, the rational actor model does not assert that rational actors are egoists, or that they maximize pleasure, or in fact, that they maximize anything. It is useful to keep in mind at all times that the rational choice model is a key tool of animal behavior theory (Maynard Smith 1982; Alcock 1993). It is difficult to consider a creature lacking nociceptors (e.g., most insects) as a happiness maximizer, and yet the rational actor model is very illuminating even for such creatures. This chapter also explores extensions of the rational actor model that take account of *social rationality* and the character of beliefs as products of social networking. The most important extension replaces the subjective prior assumption with the notion that minds are socially networked and cognition is distributed across networks.

Chapter 6 ("An Analytical Core for Sociology") is based on a target article that appeared in the *Review of Behavioral Economics* in 2015, coauthored with sociologist Dirk Helbing. I have long been a fan of sociological theory and cannot imagine that a serious behavioral scientist could get away with ignoring this immense body of social knowledge. Yet, it is not taken seriously by most researchers outside sociology. The reason is that sociology is not properly theoretically grounded, so its results are difficult to interpret and integrate with the other behavioral disciplines. I think of this chapter as an extension of Talcott Parsons' abandoned quest for an analytical core for sociology (Parsons 1937). An ambitious aspect of this chapter is the suggestion that the general equilibrium model of economic theory be extended to a general social equilibrium model based on the actor-role model of social participation. Also very important is the idea of treating the game-theoretic notion of a social norm as *Choreographer*, developed at greater length in Gintis (2009a).

The most important innovation, however, is perhaps our incorporating the socialization phenomcnon explored so brilliantly by Durkheim, Linton, and Parson into a model of the individual as situated with a social network of minds. This replaces the oft-criticized, and justly criticized, Durkheim-Parsons notion that society has a single dominant culture that individuals internalize through socialization processes, and those who fail to internalize the dominant culture are deviant and pathological. Adherence to the dominant culture/deviance theory leads rather directly to structural functionalism, to which Parsons retreated after abandoning his theory of action. As we shall see, the notion that individuals are situated in a variety of consonant and conflicting social networks, and that these networks are the major forces conditioning their preferences, values, and beliefs, provides the basis for a social dynamic quite at odds with structural-functionalism.

My suggestion for an analytical core for sociology will be called economic imperialism by some because it suggests that all the behavioral disciplines, including sociology, social and cognitive psychology, and anthropology, be organized theoretically using the analytical tools of rational choice and game theory. My retort is that these fields are in such serious need of a unifying theoretical framework that a little imperialism from more successful fields should be welcome. In fact, however, I am equally happy to apply sociological and sociobiological imperialism to the reform of economic theory. This book thoroughly rejects neoclassical economic theory's *methodological individualism.* By embracing an evolutionary perspective, we can understand why applications of economic theory outside its traditional domain often have an outlandish and bizarre character.

The great sociologist Talcott Parsons attempted as a young man to develop an analytical core for sociological theory (Parsons 1937), but he abandoned the attempt for reasons discussed in Chapter 7 ("The Theory of Action Reclaimed"). The most important reasons, I believe, were his lack of mathematical training, the fact that game theory was at the time in an infant stage, and the strong fear that sociologists had of being swallowed up by economic theory. Moreover, his lamentable embrace of the dominant culture/deviance model of socialization afforded him an easy path to structural-functionalism, from which he never recovered.

I never met Talcott Parsons, though he had an office nearby when I began teaching at Harvard. I dedicated my economics Ph.D. dissertation to Karl Marx and Talcott Parsons. Perhaps Parsons was uncomfortable being approved of by a raving anti-Vietnam War dissident. At any rate, he wrote an article in the *Quarterly Journal of Economics* (Parsons 1975) criticizing some of my work (his criticisms were remarkably off-track). I took the occasion of a reply to return the favor (Gintis 1975). Looking back, I find my criticisms of Parsons valid but unconstructive. This chapter says what I should have said forty years ago in the *Quarterly Journal of Economics.*

Chapter 8 ("The Evolution of Property"), based on Gintis (2007b), argues that there is a biologically rooted concept of property that is *prehistorical* in the sense of not depending on formal social institutions, yet *ubiquitous* in daily life. We can understand the social norm of rights of ownership by virtue of its close relationship to *territoriality* in birds and animals. A natural property right is enforced endogenously within the animal society itself: the incumbent is prepared to fight harder to keep his property than an intruder is willing to fight to take it from him, and the intruder behaves as if he knows that this is the case. Without secure property rights, many social species would simply not exist. Take nesting birds: a pair of robins would not spend two days building a nest if they were likely to be confronted with another pair, equally capable and willing to fight to take the completed nest from them.

In their boisterous attack on inclusive fitness theory, analyzed in Chapter 9, Nowak et al. (2010) wisely note that when one actually models the social behavior of a species, issues concerning levels of selection simply do not arise. Interpreting territoriality as a natural property right is a wonderful example of their point. It is impossible to *understand* territoriality—when it occurs and when it does not—in terms of either genes, or individuals. Territoriality is an emergent property of social systems. Nevertheless, all of territorial dynamics can be *described* using gene-level fitness parameters.

Moreover, the territoriality example violates one of the cherished tenets of group selection theory. This is the notion that high-level social behavior necessarily involves altruistic behavior. There is no altruism in the models developed in this chapter, yet respect for property emerges through biological evolution at the level of the population itself.

Chapter 9 ("The Sociology of the Genome") is an exposition of sociological theory as it applies not just to humans but to all social species. By a social species I mean one to which the general model of social equilibrium outlined in Chapter 6 applies: there are social roles, individuals are actors who fill these roles, and the interaction among roles defines the life history of the organism. Virtually every sexually reproducing species is a social species in this broad sense.

Because gene-culture coevolution is a special case of social evolution, a purely logical presentation of the material in this book would place this foundational chapter on social evolution even prior to the coevolution Chapter 1. However, so as not to discourage the many readers who have little taste for mathematically sophisticated analyses, I have deferred the presentation of this critical issue until late in *Individuality and Entanglement*. If you like math, read this chapter before the gene-culture coevolution chapter.

I argue that a social species is defined by its *core genome*, the subset of genes that is shared by all individuals and that specifies the social structure and pattern of social interactions characteristic of the species. My approach appears to conflict with the population biology approach to sociobiology (Dawkins 1976; Bourke 2011). This approach asserts that the gene, not the genome, is the key actor in the evolution of social behavior. The reason offered for this view is that in a sexually reproducing species, gene combinations are broken up by meiosis in each generation, so only individual genes maintain their identity across time (Dawkins 1982b). I show that this view is incorrect, and that the core genome, like individual genes, maintains its integrity across generations.

The population biologist's gene's-eye view of social evolution also holds that the maximization of inclusive fitness is sufficient to explain social organization. This view is inspired by William Hamilton's famous proof that individual genes are utterly selfish maximizers of their total (inclusive) fitness in the population (Hamilton 1964a). Despite the many attempts to criticize Hamilton's argument it is, as we show in this chapter, perfectly correct when properly formulated. However, the fact that each individual gene maximizes its inclusive fitness does not imply that individuals maximize their inclusive fitness. This completely obvious and transparent point is curiously conveniently ignored by many proponents of the gene's-eye view, including the 137 signers of a letter to *Nature* defending inclusive fitness theory against its detractors (Abbot 2011).

The reason individuals in a social species do not maximize their inclusive fitness is that a successful mutant gene is as likely to be antisocial (reducing the fitness of the reproductive population) as to be prosocial (increasing population fitness). When a gene is antisocial, it is in the inclusive fitness interest of other genes to *suppress* it, and when a gene is prosocial, it is in the interests of the other genes in the genome to *enhance* it. Enhancement can occur either on the intragenomic (biochemical) level or the social level, but suppression, we show, cannot occur within the genome, and hence is a primordial process of social sanctioning. It is the resulting complex of gene regulatory networks that promote the evolutionary success of a social species, not individual-level inclusive fitness maximization. This gives us the proper balance between the *gene's-eye view* that has seized the hearts of so many evolutionary biologists and the *multilevel selection* view that has captured the minds of so many others. The correct view flows from applying standard sociological principles to biological species. The result includes the valid insights of both camps.

Chapter 10 ("Gene-Culture Coevolution and the Internalization of Norms") forges a strong link between standard sociology and sociobiology by modeling the evolutionary emergence of the human capacity to internalize social norms. This chapter is based on a paper published in the *Journal of Theoretical Biology* in 2003. An *internal norm* is a pattern of behavior enforced in part by *internal* psychological self-sanctions, such as shame or guilt, as opposed to purely *external* sanctions, such as material rewards and punishment. The ability to internalize norms is widespread among humans, although in some so-called "sociopaths," this capacity is diminished or lacking. Our model shows that if an internal norm is fitness-enhancing, then for plausible patterns of socialization, an allele that promotes the internalization of norms is evolutionarily stable.

This framework shows that social norms adherence to which is costly to the individual can "hitchhike" on the general tendency of respecting social norms to be personally fitness-enhancing. An important implication of this analysis is that the standard argument that Darwinian fitness maximization implies that organisms maximize inclusive fitness is incorrect. But of course we already knew that from Chapter 9. The models in this chapter, like those of Chapter 8, depict complex social structure, yet do not involve group selection.

Chapter 11 ("The Economy as Complex Dynamical System") applies complexity theory and evolutionary game theory to general equilibrium theory. The Walrasian general equilibrium model of competitive market systems is in fact a *complex, dynamical, nonlinear, adaptive system* that simply fails to fit into the mold of classical dynamical systems theory. When the market economy is modeled as a complex system, problems that traditional economic theory has been unable to penetrate for nearly three score years simply melt away. In particular, it becomes relatively straightforward to prove stability of general equilibrium, and using agent-based simulations of complex market dynamics, we can begin to address the question of what institutions protect the economy against extreme sensitivity to exogenous shocks.

Chapter 12 ("The Future of the Behavioral Sciences") began as a target article in *Behavioral and Brain Sciences* (Gintis 2007c), and my further ideas on the subject of the unification of the behavioral sciences were presented in the final chapter of Gintis (2009a). I recapitulate here the main ideas of these earlier contributions, with a summary of some of the main themes of this book.

1

Gene-Culture Coevolution

The eye and the animal to which it belongs...are only so many out of the possible varieties and combinations of being, which the lapse of infinite ages has brought into existence...Millions of other bodily forms and other species having perished, being by the defect of their constitution.

William Paley

Man is nothing else but what he makes of himself.

Jean-Paul Sartre

He who understands baboons would do more towards metaphysics than Locke.

Charles Darwin

C: How are genes and human culture similar?
HG: Both genes and human culture consist of information passed across generations and subject to mutation and selection.
C: How have humans managed to evolve their extraordinary capacity for speech?
HG: Persuasive and informative speakers were rewarded with higher quality mates and increased reproductive opportunities. Their offspring inherit their cognitive and physiological communicative powers.
This is gene-culture coevolution.

Choreographer interview

The acclaimed archeologist V. Gordon Childe called his account of human evolution *Man Makes Himself* (Childe 1936). Nothing could be more true, and it is true of our species alone. The reason is that at some point in the remote past our ancestors invented a new form of culture, one that is *transmissible across generations* and therefore capable of *accumulation across time and space*. This means that man makes culture, but how does man thereby make himself? This chapter provides the answer. In brief, the answer is that human culture affects the fitness payoffs to alternative social behaviors, rewarding some and penalizing others. The genes that predispose individuals to behave prosocially, according to the cultural rules of the group, are rewarded by having more copies in the next generation while correspondingly antisocial genes are disciplined by having fewer.

1

We are the species that we are because genes make culture and culture makes genes. Or, more accurately, genes provide individuals with the capacities and incentives to transform culture, and culture guides the transformation of the gene pool from generation to generation. We call this *gene-culture coevolution*.

The critical point in understanding this dynamic is that DNA is simply a *library of information* passed from parents to offspring, and genes are the books in this library (Noble 2011). Each of our cells contains a complete copy of this library, along with specialized information as to what books to read with the instructions to make the specific cell in question. A liver cell and a neuron are constructed by reading different books in the library and following the directions therein. Brain cells, for instance, open DNA books that provide information through which our mental lives are constructed.

Culture is also a library of information passed across generations. This library includes technical information, such as how to build and maintain fire, how to construct and maintain tools and weapons, how to speak and understand language, and which conventions govern social interaction within the group.

Both culture and genes are subject to the forces of evolution (Dawkins 1976; Mesoudi et al. 2006). Both are transmitted, but with a significant level of mutation, and mutants are selected and incorporated in to the library according to their relative fitness, by which we mean the average number of copies that appear in the next generation. Successful genetic mutations are generally *adaptive*, meaning they improve the long-term success of the population. Successful cultural changes are often maladaptive (Edgerton 1992), but so far, and in the long run, human culture has been extremely adaptive. Whether this will continue in the face of the proliferation of nuclear weapons, climate change, and reduction in biodiversity remains to be seen.

I grew up with the notion that man makes himself not only in the sense of V. Gordon Childe, but also in the sense expressed by Jean-Paul Sartre in the chapter head quote above. Sartre's is the notion that there is no such thing as human nature. Rather, we are the complete product of the choices we make. Our morality is a purely personally and socially constructed morality. This is profoundly mistaken (Cosmides et al. 1992). Just as ducks have duck nature and mosquitoes have mosquito nature, so do humans have human nature. This nature was forged through gene-culture coevolution over hundreds of thousands of years (Brown 1991). This book is an analysis of human nature and a tribute to its wonders.

1.1   Culture Determines Biological Fitness

Because of the importance of culture and complex social organization to the evolutionary success of *Homo sapiens*, individual fitness in humans depends on the structure of social life. Those who are successful according to social norms are differentially rewarded with more and healthier offspring, and violators of social rules are likely to be ostracized or killed.

Because culture is both constrained and promoted by the human genome, human cognitive, affective, and moral capacities are the product of an evolutionary dynamic involving the interaction of genes and culture. Whence *gene-culture coevolution* (Lumsden and Wilson 1981; Cavalli-Sforza and Feldman 1982; Boyd and Richerson 1985; Dunbar 1993; Richerson and Boyd 2004).

This coevolutionary process has endowed us with preferences that go beyond the self-regarding concerns emphasized in traditional economic and biological theory, and with a social epistemology that facilitates the sharing of intentionality across minds. Gene-culture coevolution is responsible for the salience of such other-regarding values as a taste for cooperation, fairness and retribution, the capacity to empathize, and the ability to value such character virtues as honesty, hard work, piety, and loyalty (Bowles and Gintis 2011; Wilson 2012; Tomasello 2014).

Gene-culture coevolution is the application of *sociobiology*, the general study of the social organization of biological species, to humans—a species that transmits culture in a manner that leads to quantitative growth across generations. Gene-culture coevolution is a special case of *niche construction*, which applies to species that transform their natural environment so as to facilitate social interaction and collective behavior (Odling-Smee et al. 2003). Examples are the beaver's dam and the bee's hive. In the case of gene-culture coevolution, the environmental change is that of *the social structure within which individuals live out their lives*. The natural environment may be involved as well, as when settled agriculture alters the ecology of disease-carrying insects, and hence selects for individuals who are relatively immune to these diseases (Laland et al. 2000).

The genome encodes information that is used both to construct a new organism and to endow it with instructions for transforming sensory inputs into decision outputs. Because learning is costly and time-consuming, efficient information transmission will ensure that the genome encodes those aspects of the organism's environment that are constant, or that change only very slowly through time and space, as compared with an individual lifetime. By contrast, environmental conditions that vary rapidly can be dealt with by providing the organism with *phenotypic plasticity* in the form of the capacity to learn. For instance, suppose the environment provides an organism with the most nutrients where ambient temperature is highest. An organism may learn this by trial and error over many periods, or it can be hard-wired to seek the highest ambient temperature when feeding. By contrast, suppose the optimal feeding temperature varies over an individual's lifetime. Then there is no benefit to encoding this information in the individual's genome, but a flexible learning mechanism will enhance the organism's fitness.

There is an intermediate case, however, that is efficiently handled neither by genetic encoding nor learning. When environmental conditions are positively but imperfectly correlated across generations, each generation acquires valuable information through learning that it cannot transmit genetically to the succeeding generation, because such information is not encoded in the germ line. In the context of such environments, there is a fitness benefit to the *epigenetic transmission* of information concerning the current state of the environment; i.e., transmission through non-genetic channels. This is called *cultural transmission*.

Several epigenetic transmission mechanisms have been identified (Jablonka and Lamb 1995), but cultural transmission in humans and to a lesser extent in other animals (Bonner 1984; Richerson and Boyd 1998) is a distinct and extremely flexible form. Cultural transmission takes the form of vertical (parents to children), horizontal (peer to peer) and oblique (elder to younger), as in Cavalli-Sforza and Feldman (1981), prestige (higher influencing lower status), as in Henrich and Gil-White (2001), popularity-related as in Newman et al. (2006), and even random population-dynamic transmission, as in Shennan (1997) and Skibo and Bentley (2003). The parallel between cultural and biological evolution goes back to Huxley (1955), Popper (1979), and James (1880)—see Mesoudi et al. (2006) for details.

The idea of treating culture as a form of epigenetic transmission was pioneered by Dawkins (1976), who coined the term "meme" in *The Selfish Gene* to represent an integral unit of information that could be transmitted phenotypically. There quickly followed several major contributions to a biological approach to culture, all based on the notion that culture, like genes, could evolve through replication (intergenerational transmission), mutation, and selection. Cultural elements reproduce themselves from brain to brain and across time, mutate and are subject to selection according to their effects on the fitness of their carriers (Parsons 1964; Cavalli-Sforza and Feldman 1982). Moreover, there are strong interactions between genetic and epigenetic elements in human evolution, ranging from basic physiology (e.g., the transformation of the organs of speech with the evolution of language) to sophisticated social emotions, including empathy, shame, guilt, and revenge-seeking (Ihara 2011; Zajonc 1980,1984).

Because of their common informational and evolutionary character, there are strong parallels between models of genetic and cultural evolution (Mesoudi et al. 2006). Like biological transmission, culture is transmitted from parents to offspring, and like cultural transmission, which is transmitted horizontally to unrelated individuals, so in microbes and many plant species, genes are regularly transferred across lineage boundaries (Jablonka and Lamb 1995; Abbott et al. 2003; Rivera and Lake 2004). Moreover, anthropologists reconstruct the history of social groups by analyzing homologous and analogous cultural traits, much as biologists reconstruct the evolution of species by the analysis of shared characters and homologous DNA (Mace and Pagel 1994). Indeed, the same computer software developed by biological systematists is used by cultural anthropologists (Holden 2002; Holden and Mace 2003). In addition, archeologists who study cultural evolution have a similar *modus operandi* as paleobiologists who study genetic evolution (Mesoudi et al. 2006). Both attempt to reconstruct lineages of artifacts and their carriers. Like paleobiology, archaeology assumes that when analogy can be ruled out, similarity implies causal connection by inheritance (O'Brien and Lyman 2000). Like biogeography's study of the spatial distribution of organisms (Brown and Lomolino 1998), behavioral ecology studies the interaction of ecological, historical, and geographical factors that determine distribution of cultural forms across space and time (Winterhalder and Smith 1992).

Perhaps the most common criticism of the analogy between genetic and cultural evolution is that the gene is a well-defined, discrete, independently reproducing and mutating entity, whereas the boundaries of the unit of culture are ill-defined and overlapping. In fact, however, this view of the gene is outdated. We now know that overlapping, nested, and movable genes have some of the fluidity of cultural units, whereas quite often the boundaries of a cultural unit (a belief, icon, word, technique, stylistic convention) are quite delimited and specific. Similarly, alternative splicing, nuclear and messenger RNA editing, cellular protein modification, and genomic imprinting, which are quite common, undermine the standard view of the insular gene producing a single protein, and support the notion of genes having variable boundaries and having strongly context-dependent effects. Moreover, natural selection requires heritable variation and selection, but does not require discretely transmitted units.

In *The Extended Phenotype* Dawkins (1982a) added a second fundamental mechanism of epigenetic information transmission, noting that organisms can directly transmit environmental artifacts to the next generation, in the form of such constructs as ant nests, tree galls, and even social structures, such as mating and hunting practices. A species creating an important aspect of its environment and stably transmitting this environment across generations, known as *niche construction*, is a widespread form of epigenetic transmission (Odling-Smee et al. 2003). Niche construction includes gene-environment coevolution, because a genetically induced environmental regularity becomes the basis for genetic selection, and gene mutations that give rise to novel niche elements will survive if they are fitness-enhancing for their constructors.

An excellent example of gene-environment coevolution is the honeybee, in which the origin of its eusociality probably lay in a high degree of relatedness, but which persists in modern species despite the fact that relatedness in the hive is generally quite low, due to multiple queen matings, multiple queens, queen deaths, and the like (Gadagkar 1991; Seeley 1997; Wilson and Hölldobler 2005). The social structure of the hive, a classic example of niche construction, is transmitted epigenetically across generations, and the honeybee genome is an adaptation to the social structure laid down in the distant past.

Gene-culture coevolution in humans is a special case of gene-environment coevolution in which the environment is culturally constituted and transmitted (Feldman and Zhivotovsky 1992). The key to the success of our species in the framework of the hunter-gatherer social structure in which we evolved is the capacity of unrelated, or only loosely related, individuals to cooperate in relatively large egalitarian groups in hunting and territorial acquisition and defense (Richerson and Boyd 2004; Boehm 1999). While some contemporary biological and economic theorists have attempted to show that such cooperation can be supported by self-regarding rational agents (Alexander 1987; Fudenberg et al. 1994; Trivers 1971), the conditions under which their models work are implausible even for small groups (Boyd and Richerson 1988; Gintis 2009a). Rather, the social environment of early humans was conducive to the development of prosocial traits, such as empathy, shame, pride, embarrassment, and reciprocity, without which social cooperation would be impossible (Sterelny 2011).

Neuroscientific studies exhibit clearly the genetic basis for moral behavior. Brain regions involved in moral judgments and behavior include the prefrontal cortex, the orbitalfrontal cortex, and the superior temporal sulcus (Moll et al. 2005). These brain structures are virtually unique to or most highly developed in humans and are doubtless evolutionary adaptations (Schulkin 2000). The evolution of the human prefrontal cortex is closely tied to the emergence of human morality (Allman et al. 2002). Patients with focal damage to one or more of these areas exhibit a variety of antisocial behaviors, including the absence of embarrassment, pride and regret (Beer et al. 2003; Camille 2004), and sociopathic behavior (Miller et al. 1997). There is a probable genetic predisposition underlying sociopathy, and sociopaths comprise about 4% of the male population, but they account for between 33% and 80% of the population of chronic criminal offenders in the United States (Mednick et al. 1977). It is clear from this body of empirical information that culture is directly encoded into the human brain with symbolic representations in the form of cultural artifacts. This, of course, is the central claim of gene-culture coevolutionary theory.

1.2 Reciprocal Causality

Gene-culture coevolution is an empirical fact, not a theory. However, it is a complex and variegated process that takes many forms. Modeling gene-culture coevolution began with Feldman and Cavalli-Sforza (1976), followed by their book (Cavalli-Sforza and Feldman 1981), in which they modeled vertical (parent to child), oblique (non-parental elders to youngers), and horizontal (peer to peer) cultural transmission. Lumsden and Wilson (1981) presented an alternative model, as did Boyd and Richerson (1985). For enlightening contemporary reviews of these pioneers, see Lewontin (1981) and Maynard Smith and Warren (1982).

It might be thought that the complex and intimate interaction of genes and culture outlined above is overdrawn, and that human genetic evolution is the effect of genetic inclusive fitness maximization, culture being an effect of genes that can be factored out in the long run. For instance, the eminent evolutionary psychologist David Buss holds that "culture is not an autonomous casual process in competition with biology for explanatory power" (Buss 1999, p. 407). This denial of gene-culture coevolution can be shown to be *prima facie* untenable. To see this, suppose we have a vector g of genetic variables, a vector c of cultural variables, and a vector e of environmental variables, including the prevalence of predators and prey, weather and the like. In an evolutionary model, the rate of change of variables is a function of the variables, so we have

$$\dot{g} = F(g,c,e) \tag{1.1}$$

$$\dot{c} = G(g,c,e) \tag{1.2}$$

$$\dot{e} = H(e) \tag{1.3}$$

Note that it is plausible for c to affect the nature and pace of environmental change, in which case it should be included in the third equation above. We abstract from this causal path in order to strengthen the case for Buss' argument. The contention that culture is an effect of genetic fitness maximization in this framework is the assertion that c can be eliminated from these equations. Under what conditions can this occur? Taking the derivative of equation (1.1), and substituting equations (1.2) and (1.3) into equation (1.1), we get

$$\ddot{g} = F_g(g,c,e)F(g,c,e) + F_c(g,c,e)G(g,c,e) + F_e(g,c,e)H(e). \tag{1.4}$$

If c is to be absent from this second order differential equation, the derivative of the right-hand side of equation (1.4) with respect to c must be identically zero. Thus, we have

$$0 \equiv F_{gc}F + F_gF_c + F_{cc}G + F_cG_c + F_{ec}H. \tag{1.5}$$

All five of the above terms must then be identically zero, so $Fc = 0$, implying that c does not enter on the right-hand side of the defining equations (1.1)–(1.3); i.e., genes are not a function of culture. This is obviously not appropriate for humans, since both genes and culture are functions of culture. Note that as long as there is high fidelity cultural transmission over multiple generations (signified by the middle row of horizontal arrows), genetic and cultural evolution are inextricably intertwined. By contrast, for species that do not have cumulative learning, these arrows are absent, and despite the fact that genes affect culture in every period, there is no cumulative interrelatedness of genes and culture.

There are many obvious examples of culture affecting genes. For instance, tribes that raise cattle tend to develop lactose tolerance in place of the default condition for humans, which is lactose intolerance (Gerbault et al. 2011). This development is easy to understand because the ability to digest milk is individually fitness-enhancing, so if the genes that permit lactose processing exist in the population or can be created through high probability mutations, lactose tolerance will evolve. Similar arguments apply to the evolution of the human gut after the control of fire for cooking (Gowlett and Wrangham 2013), the structural transformation of the human hand when social life moved from the trees to the ground (Marzke 1997), and the role of culture in creating a genetic predisposition for cooperative activity in humans (Gintis 2003a). We will use gene-culture evolution to illuminate especially complex issues of this and other physiological changes facilitating linguistic communication in humans (Deacon 1998).

1.3   The Physiology of Human Communication

The evolution of the physiology of speech and facial communication presents a theoretical challenge. It is easy to explain, if everyone else is gabbing away and you can only grunt and pant, why you might be handicapped in finding a spouse and teaching your children. But how could the use of complex phonics begin? When everyone is grunting and panting, what is the fitness benefit of being able to make more complex varieties of sounds? What is the fitness benefit of being able to interpret complex sounds? The answers are far from obvious and go far beyond simple individual fitness, or even inclusive fitness, maximization.

For this reason, the evolution of the physiology of speech is a dramatically complex example of gene-culture coevolution. A most common error in the literature is to consider language as a purely *mental* phenomenon that can be explained simply as a byproduct of brain size and intelligence. In fact, the ability to communicate through facial sign and speech has required major changes in human physiology. These could only have come about because individuals with superior communication capacities were rewarded with more and healthier children. Why might this have occurred?

The increased social importance of communication in human society rewarded genetic changes that facilitate speech. Regions in the motor cortex expanded in early humans to facilitate speech production. Concurrently, nerves and muscles to the mouth, larynx, and tongue became more numerous to handle the complexities of speech (Jurmain et al. 1997). Parts of the cerebral cortex, Broca's and Wernicke's areas, which do not exist or are relatively small in other primates, are large in humans and permit grammatical speech and comprehension (Binder et al. 1997; Belin et al. 2000).

Adult modern humans have a larynx low in the throat, a position that allows the throat to serve as a resonating chamber capable of a great number of sounds (Relethford 2007). The first hominids that have skeletal structures supporting this laryngeal placement are the *Homo heidelbergensis*, who lived from 800,000 to 100,000 years ago. In addition, the production of consonants requires a short oral cavity, whereas our nearest primate relatives have much too long an oral cavity for this purpose. The position of the hyoid bone, which is a point of attachment for a tongue muscle, developed in *Homo sapiens* in a manner permitting highly precise and flexible tongue movements.

Another indication that the tongue has evolved hominids to facilitate speech is the size of the hypoglossal canal, an aperture that permits the hypoglossal nerve to reach the tongue muscles. This aperture is much larger in Neanderthals and humans than in early hominids and nonhuman primates (Dunbar 2005). Human facial nerves and musculature have also evolved to facilitate communication. This musculature is present in all vertebrates, but except in mammals it serves feeding and respiratory functions alone (Burrows 2008). In mammals, this mimetic musculature attaches to the skin of the face, thus permitting the facial communication of such emotions as fear, surprise, disgust, and anger. In most mammals, however, a few wide sheet-like muscles are involved, rendering fine information differentiation impossible, whereas in primates, this musculature divides into many independent muscles with distinct points of attachment to the epidermis, thus permitting higher bandwidth facial communication. Humans have the most highly developed facial musculature by far of any primate species, with a degree of involvement of lips and eyes that is not present in any other species.

In short, humans have evolved a highly specialized and very costly complex of physiological characteristics that both presuppose and facilitate sophisticated aural and visual communication, whereas communication in other primates, lacking as they are in cumulative culture, goes little beyond simple calling and gesturing capacities. This example is quite a dramatic and concrete illustration of the intimate interaction of genes and culture in the evolution of our species.

2

Zoon Politikon: The Evolutionary Origins of Human Socio-political Systems

> We are caught in an inescapable network of mutuality, tied in a single garment of destiny.
>
> Martin Luther King

> There is no such thing as society. There are individual men and women, and there are families.
>
> Margaret Thatcher

> **C:** On what is power based in chimpanzee society?
> **HG:** The physical prowess of the alpha male.
> **C:** On what was power based in the societies of our human ancestors?
> **HG:** On the power to persuade and to lead creatively.
> **C:** How do you know this?
> **HG:** The short answer is that humans are extremely gracile and delicate, with only a small fraction of the physical power of other primate species. For the long answer, read on.
>
> Choreographer interview

2.1 Accounting for Human Exceptionalism

This chapter deploys the most up-to-date evidence available in various behavioral fields in support of the following hypothesis: The emergence of bipedalism and cooperative breeding in the hominin line, together with environmental developments that made a diet of meat from large animals adaptive, as well as cultural innovations in the form of fire, cooking, and lethal weapons, created a niche for hominins in which there was a significant advantage to individuals with the ability to communicate and persuade. These forces added a unique political dimension to human social life which, through gene-culture coevolution, became *Homo ludens*—Man, the game player—with the power to conserve and transform the social order. *Homo sapiens* became, in the words of Aristotle's *Nicomachean Ethics*, a *zoon politikon*.

Strong social interdependence plus the availability of lethal weapons in early hominin society undermined the standard social dominance hierarchy, based on pure physical prowess, of multi-male/multi-female primate groups, characteristic, for instance, of chimpanzees. The successful political structure that ultimately replaced the ancestral social dominance hierarchy was an egalitarian political system in which the *group controlled its leaders*. Group success depended both on the ability of leaders to persuade and motivate, and of followers to submit to a consensual decision process. The heightened social value of non-authoritarian leadership entailed enhanced biological fitness for such traits as linguistic facility, political ability, and indeed for human hypercognition itself.

This egalitarian political system persisted until cultural changes in the Holocene fostered the accumulation of material wealth, through which it became possible again to sustain a social dominance hierarchy with strong authoritarian leaders atop.

2.2 Models of Political Power

The behavioral sciences during the second half of the twentieth century were dominated by two highly contrasting models of human political behavior. In biology, political science, and economics, a *Homo economicus* self-interest model held sway. In this model, individuals are rational self-regarding maximizers (Downs 1957a; Alexander 1987; Mas-Colell et al. 1995). Sociology, social psychology, and anthropology, by contrast, embraced a *cultural hegemony* model. In this model, individuals internalize the cultural principles of the society in which they operate. In this view, a dominant culture supplies the norms and values associated with role-performance, and individual behavior meets the requirements of the various roles individuals are called upon to play in daily life (Durkheim 1902; Mead 1963; Parsons 1967). Contemporary research has been kind to neither model.

Contra cultural hegemony theory, daily life provides countless examples of the fragility of dominant cultures. African Americans in the era of the civil rights movement, for instance, rejected a powerful ideology justifying segregation, American women in the 1960s rejected a deep-rooted patriarchal culture, and gay Americans rejected traditional Judeo-Christian treatments of homosexuality. In succeeding years, each of these minority counter-cultures was adopted by the American public at large. In the Soviet

Union, Communist leaders attempted to forge a dominant culture of socialist morality by subjecting two generations of citizens to intensive indoctrination. This effort was unsuccessful, and was rejected rather decisively, immediately following the fall of the Soviet regime. Similar examples can be given from political experience in many other societies.

There has always been an undercurrent of objection to the cultural hegemony model, which Dennis Wrong (1961) aptly called the "oversocialized conception of man." Konrad Lorenz (1963), Robert Ardrey (1997[1966]), and Desmond Morris (1999[1967]) offered behavioral ecology alternatives, a line of thought culminating in Edward O. Wilson's *Sociobiology: The New Synthesis* (1975), the resurrection of human nature by Donald Brown (1991), and Leda Cosmides and John Tooby's withering attack in *The Adapted Mind* on the so-called "standard social science model" of cultural hegemony (Barkow et al. 1992). Meanwhile, the analytical foundations of an alternative model, that of *gene-culture coevolution* (see Chapter 1), were laid by Geertz (1962), Dobzhansky (1963), Wallace (1970), Lumsden and Wilson (1981), Cavalli-Sforza and Feldman (1973, 1981), and Boyd and Richerson (1985). This gene-culture coevolution model informs our analysis of the evolution of human socio-political systems.

Undermining the self-interest model began in economics with the ultimatum game experiments of Güth et al. (1982) and Roth et al. (1991). In the ultimatum game, one subject, called the "proposer," is presented with a sum of money, say $10, and is instructed to offer any portion of this, from nothing to the full $10, to a second subject, called the "responder." The two subjects never learn each other's identity, and the game is played only once. The responder, who knows that the total amount to be shared is $10, can either accept the offer or reject it. If the responder accepts the offer, the money is shared accordingly. If the responder rejects the offer, both players receive nothing. If the players care only about their own payoffs and have no concern for fairness (i.e., they are self-interested), a rational responder will always accept any positive amount of money. Knowing this, a rational proposer will offer $1, and this will be accepted.

When the ultimatum game is actually played, however, this self-interested outcome is almost never observed and rarely even approximated. In many replications of this experiment in more than 30 countries, under varying conditions and in some cases with substantial amounts of money at stake, proposers routinely offer responders very generous shares, 50% of the total generally being the modal offer. Responders frequently reject offers below

25% (Roth et al. 1991; Camerer and Thaler 1995; Camerer 2003; Ooster-beek et al. 2004).

In post-game debriefings, responders who have rejected low offers often express anger at the proposer's greed and a desire to penalize unfair behavior. The fact that positive offers are commonly rejected shows that responders have fairness concerns, and the fact that most proposers offer between 40% and 50% of the pie shows that proposers too have fairness concerns themselves, or at least understand that responders' fairness concerns would motivate them to reject low offers. Of special interest are those who reject positive offers. The explanation most consistent with the data is that they are motivated by a desire to punish the proposer for being unfair, even though it means giving up some money to do so. While initially considered odd, these and other experimental results violating the self-interest axiom are now commonplace.

These and related findings have led in recent years to a revision of the received wisdom in biology and economics towards the appreciation of the central importance of *other-regarding preferences* and *character virtues* in biological and economic theory (Gintis et al. 2005; Henrich et al. 2005; Okasha and Binmore 2012). It might reasonably be thought, however, that these behaviors are the product of the culture of advanced complex societies. To assess this possibility, a team of anthropologists ran ultimatum game experiments in which the subject pool consisted of members of fifteen small-scale societies with little contact with markets, governments, or modern institutions (Henrich et al. 2004). The fifteen societies included hunter-gatherers, herders, and low technology farmers.

This study found that many small-scale societies mirror the results of the advanced economies, but others did not. Among the Au and Gnau people in Papua New Guinea, ultimatum game offers of more than half the pie were common. Moreover, while even splits were commonly accepted, both higher and lower offers were rejected with about equal frequency. This behavior is not surprising in light of the widespread practice of *competitive gift giving* as a means of establishing status and subordinacy in these and many other New Guinea societies. By contrast, among the Machiguenga in Amazonian Peru, almost three-quarters of the offers were a quarter of the pie or less and yet there was just a single rejection among 70 offers. This pattern was strikingly different from the standard experiments in advanced economies. However, even among the Machiguenga, the mean offer was

27.5%, far more than would have maximized the proposer's payoffs given the scant likelihood of a rejection.

Analysis of the experiments led to the following conclusions: (a) behaviors are highly variable across groups; (b) not a single group conformed to or even approximated the model of self-interested agents; and (c) despite the anonymous and asocial setting of the experiments, between-group differences in behavior reflected differences in the kinds of social interaction experienced in everyday life; i.e., people generally conform to cultural rules of their societies *even when there is no chance a deviation will be punished*.

The evidence for this latter conclusion is compelling. For example, the Aché in Paraguay share equally among all group members some kinds of food (meat and honey) acquired through hunting and gathering. In our experiment, most Aché proposers contributed half the pie or more. Similarly, among the Lamalera whale hunters of Indonesia, who hunt in large crews and divide their catch according to strict sharing rules, the proposer's average allocation to the responder was 58% of the pie. Moreover, the Indonesian whale hunters played the game very differently from the Indonesian university students who were the subjects in another set of experiments (Cameron 1999). Indeed, where voluntary public goods provision was customary in real life (for example, the *Harambee* system among the Orma herders in Kenya, whereby individuals contribute resources to build a school or repair a road), contributions in the experimental public goods game were patterned after actual contributions in the actual Harambee system. Those with more cattle contributed more. By contrast, in the ultimatum game, for which there apparently was no everyday life analogue, the wealthy and non-wealthy Orma behaved similarly.

2.3   The Moral Basis of Modern Political Systems

The untenability of the self-interest model of human action is also clear from everyday experience. Political activity in modern democratic societies provides unambiguous evidence. I here preview the extended argument of Chapter 3, noting that in large elections, the rational self-regarding agent will not vote because the costs of voting are positive and significant, but the probability that one vote will alter the outcome of the election is vanishingly small, and adding a single vote to the total of a winning candidate enhances the winner's political efficacy at best an infinitesimal amount (Riker and Ordeshook 1968). Thus the personal gain from voting is too small to motivate behavior. For similar reasons, if one chooses to vote, there is no plausible reason to vote on the basis of the impact of the outcome of the election on one's personal material gains. It follows also that the voter, if rational, self-regarding, and incapable of personally influencing the opinions of more than a few others, will not bother to form opinions on political issues, because these opinions cannot affect the outcome of elections. Yet people do vote, and many do expend time and energy in forming political opinions. Although voters do appear to behave strategically (Fedderson and Sandroni 2006), their behavior does not conform either to the self-interest model (Edlin et al. 2007) or the rational actor model of contemporary decision theory (Savage 1954).

It also follows from the logic of self-regarding political behavior that rational self-regarding individuals will not participate in the sort of collective actions that are responsible for the growth in the world of representative and democratic governance, the respect for civil liberties, the rights of minorities and gender equality in public life, and the like. In the self-interest model, only small groups aspiring for social dominance will act politically. Yet modern egalitarian political institutions are the result of such collective actions (Bowles and Gintis 1986; Giugni et al. 1998). This behavior cannot be explained by the self-interest model.

Except for professional politicians and socially influential individuals, electoral politics is a vast morality play to which models of the rational self-regarding actor are not only a poor fit, but are conceptually bizarre. It took Mancur Olson's *The Logic of Collective Action* (1965) to make this clear to many behavioral scientists, because virtually all students of social life had assumed without reflection the faulty logic that rational self-regarding individuals will vote, and will "vote their interests" (Downs 1957a).

Defenders of the *Homo economicus* model may respond that voters *believe* their votes make a difference, however untenable this belief might be under logical scrutiny. Indeed, when asked why they vote, voters' common response is that they are trying to help get one or another party elected to office. When appraised of the illogical character of that response, the common reply is that there are in fact close elections, where the balance is tipped in one direction or another by only a few hundred votes. When confronted with the fact that one vote will not affect even such close elections, the common repost is, "Well, if everyone thought like that, we couldn't run a democracy."

Politically active and informed citizens appear to operate on the principle that voting is a prerogative of citizenship, an altruistic act that is governed by the categorical imperative: act in conformance with the morally correct behavior for individuals in one's position, without regard to personal costs and benefits. Such mental reasoning, which is built on our urge to conform and our shared intentionality, is implicated in many uniquely human cognitive characteristics, including cumulative culture and language (Sugden 2003; Bacharach 2006). Shared intentionality rests on a fundamentally prosocial disposition (Gilbert 1987; Bratman 1993; Tomasello and Carpenter 2007; Hrdy 2009).

2.4   The Socio-political Structure of Primate Societies

Humans are one of more than two hundred extant species belonging to the Primate order. All primates have socio-political systems for regulating social life within their communities. Understanding human socio-political organization involves specifying how and why humans are similar to and differ from other social species in general, and other primate species in particular.

Concerning the latter, there are two major sources of information. First, some traits are distributed widely and linked to other well-known traits, and thus were almost certainly already present before humans evolved. For instance, many primate species, including humans and our closest living relatives, seek to dominate others and are adept at forming coalitions. It is thus likely that their most recent common ancestor also possessed these traits. Dominance-seeking and coalition-formation in humans, then, are not purely cultural. Rather, humans are endowed with the genetic prerequisites for this behavior, as are numerous other primate species (Wrangham and Peterson 1996).

A second source is similarity with our close relatives, the great apes, and especially the genus *Pan* (chimpanzees and bonobos). Most nonhuman primate species have great trouble in acting collectively in conflict with neighboring groups (Willems et al. 2013). Chimpanzees are a major exception: they engage in war-like raids where larger parties cooperate closely to target and destroy much smaller ones (Goodall 1986; Wilson 2012). War among human hunter-gatherers likewise largely consists of such a raiding strategy (Keeley 1996), suggesting a shared predisposition to engage in this type of warfare (Wrangham and Glowacki 2012). Obviously, the dramatic changes in human social organization accompanying the origin of defensible wealth (discussed below) produced major changes in the nature of warfare, linked to additional genetic predispositions, such as insider favoritism (LeVine and Campbell 1972; Otterbein 2004; Bowles 2006, 2007, 2009; Bowles and Gintis 2011).

Using this logic, we can examine the social structure of multi-male/multi-female primate societies (de Waal 1997; Maestripieri 2007) to identify the elements of human socio-political organization that were already likely present among the first hominins.

Primates live in groups to reduce the risk of predation (Alexander 1974; van Schaik 1983), exchange information about food location (Eisenberg et al. 1972; Clutton-Brock 1974), and defend food sources and mates against competing groups (Wrangham 1980). These groups, however, rarely engage in organized collective action. As a result, the primate form of group living has only limited need for leaders, that is, individuals instrumental in initiating and coordinating group-level action with the approval and support of other group members. Instead, individuals vary in dominance based on motivation and pure physical prowess, and dominant males gain fitness at the expense of subordinate members of the group. This is especially true for our closest relatives, the genus *Pan*. As King et al. (2009) stress, other species do often have foraging leaders, but their power is based on hierarchical dominance rather than consensus. Despite the fact that such leaders of the hunt appropriate most of the spoils, followers must stick with the group to avoid predation while grabbing what little of the catch they can (King et al. 2008; Krauss et al. 2009).

In most primate species, both sexes form dominance hierarchies, in which more dominant individuals gain privileged access to food and mates, and as a result tend to have higher fitness (Vigilant et al. 2001; Maestripieri 2007; Majolo et al. 2012). In many primate species, dominant females depend on alliances to maintain their position, whereas the same is true for males in far fewer primate species (van Schaik 1996), most notably chimpanzees. Thus dominants rarely perform any group-level beneficial acts. One exception is male displays toward predators, a behavior seen in a variety of primate species, and generally linked to the protection of likely offspring. Another is triadic power interventions (e.g., Boehm 1994 and deWaal 1996) that end conflicts in apes and certain monkey species.

2.4.1   The Origins of Primate Socio-political Structure

Given the variety of contemporary primate socio-political structures, what can we say about the social structure of the most recent common ancestor of contemporary primates, the species from which the hominin species leading ultimately to *Homo sapiens* branched off? Our answer is based on the fact that traits shared by several closely related species were very likely shared by their most recent common ancestor. The challenge is that primates exhibit a wide variety of socio-political structures. However, if we limit our sample to species living in woodlands and open savannah that engage in collective defense and confrontational scavenging from large carnivores, which was the probable condition faced by the primates' most recent common ancestor, all extant species live in large, multi-male/multi-female groups.[1] Thus at least from *Homo habilis* on, hominins likely lived in large multi-male/multi-female groups (Foley 1996; Dunbar 2005).

Recently, sophisticated phylogenetic approaches have added precision to these inferences by reconstructing the origin of various kinds of social organization in deep time (Silk 2011). Shultz et al. (2011) completed a study based on the genetic distances and phenotypic social-structural similarities of 217 extant primate species, the most recent common ancestor of which is far more ancient than the ancestral *Pan*. Shultz et al. show that social organization tends to be similar among closely related species, which implies that social structure is determined largely by genes rather than environment in nonhuman primates. This finding runs counter to the alternative assumption that primate social structure is a response to the distribution of food resources or risks and is not affected by phylogenetic affiliation.

Shultz et al. (2011) conclude that the earliest primates lived some 72 Mya as solitary foraging individuals who came together only for mating. Multi-male/multi-female aggregations appeared some 52 Mya. We can infer from the social structure of contemporary nonhuman primate species living in multi-male/multi-female groups that mating was promiscuous and males formed a hierarchical power structure with a single alpha male at the apex. Indeed, most nonhuman primates that live in multi-male groups today exhibit this living pattern (Chapais 2008). While this social structure is highly stable and has persisted into the present, when suitably stressed it

---

[1]The grass- and savannah-living Patas monkey (Hall 1965) is the single exception to the rule that savannah-living primates exhibit a multi-male/multi-female social structure. They avoid predators by staying in trees as much as possible, cryptic behavior, wide group spread, and rapid flight.

broke down into two social forms in which a social group included only one male. The first, which may have appeared about 16 Mya, was the single-male harem while the second, appearing about the same time, was single pair-living.

The implication is that the earliest hominids lived in multi-male/multi-female promiscuous social bands, so *Pan* are archetypical species when it comes to reconstructing the origins of the human political system. Dominant male chimpanzees provide little leadership, and they provide virtually no parenting. In many primate species, dominant males have sufficiently high paternity certainty to induce them to provide protection to infants (Paul et al. 2000), but in chimpanzees paternity is much less concentrated in top-ranked males (Vigilant et al. 2001; Boesch et al. 2006), most likely because chimpanzee females prefer multiple matings and cannot be controlled by dominant males. Thus males tend to ignore rearing the young. The only clear service dominant males provide to the group is keeping the peace by intervening in disputes and leading predator mobbing (de Waal 1997; von Rohr et al. 2012). In short, the political structure of chimpanzee society, like that of primates generally, is largely a system for funneling fitness-enhancing resources to the apex of a social dominance hierarchy based on physical prowess and coalition-building talent. This holds basically for the bonobo as well, where monopolization of matings by particular males is even lower.

2.4.2   *Primate Coalitional Politics*

Chimpanzee males rely significantly on coalitions and alliances. There are two major types of coalition: rank-changing and leveling (Pandit and van Schaik 2003; van Schaik et al. 2006). Rank-changing occurs when a male relies on supporters to acquire and maintain hegemony (Goodall 1964; Nishida and Hosaka 1996; de Waal 1998), and hence may not have the highest individual fighting ability (de Waal 1998; Boesch et al. 1998). Leveling occurs when multiple lower-ranking males form coalitions to prevent the top male or males from appropriating too large a share of the resources. These coalitions do not change the dominance ranks of the participants. Females similarly form such leveling coalitions to counter the arbitrary power of dominant males, especially in captivity (Goodall 1986).

This pattern of political power based on the hierarchical dominance of the physically powerful along with a system of sophisticated political alliances to preserve or to limit the power of the alpha male (Boehm and Flack 2010) is carried over, yet fundamentally transformed, in human society (Knauft 1991; Boehm 2000).

The best predictor for male-male coalitions among primates is simply the fact that multiple males find themselves together and no single male can fully monopolize all matings (Bissonnette et al. 2014). Thus, there are broad similarities in social dominance and coalition-formation across all multi-male/multi-female primate species. This fact runs counter to traditional political theory. Aristotle's *zoon politikon* notwithstanding, political theorists have widely assumed that political structure involves purely cultural evolution, whereas the primate data show roots to political behavior going back millions of years. The primate evidence is important because it lays the basis for an evolutionary analysis of human political systems (de Waal 1998). Such an analysis may elucidate the role of basic human political predispositions in reinforcing and undermining distinct sorts of human socio-political structures.

2.5   The Evolutionary History of Primate Societies

It would be useful to be able to read ancient social structure from the historical record. But we cannot. The fossil record provides the most concrete answers to our evolutionary history, but is highly incomplete. There are, for instance, skeletal records of only about 500 individuals from our hominin past. Moreover, behavior does not fossilize and social structure leaves no direct marks in the earth. This is why we must resort to the relationship between phylogenetic proximity and social organization in living primate species (Shultz et al. 2011).

The hominin lineage branched off from the primate main stem some 6.5 million years ago or earlier (Wood 2010; Langergraber 2012). The watershed event in the hominin line was the emergence of bipedalism. Bipedalism is well-developed in *Australopithecus afarensis*, which appeared three million years after the origin of the hominin lineage. *Homo ergaster* (2.0 to 1.3 Mya) or *Homo erectus* (1.9 to 0.143 Mya) was the first currently documented specialized biped, having a relatively short arm/leg ratio that rendered brachiation infeasible.

Bipedalism in hominins was critically dependent upon the prior adaptation of the primate upper torso to life in the trees. The Miocene Hominoid apes were not true quadrupeds, but rather had specialized shoulder and arm muscles for swinging and climbing, as well as a specialized hand structure for grasping branches and manipulating leaves, insects, and fruit. When the hominin line was freed from the exigencies of arboreal life, the locomotor function of the upper limbs was reduced, so they could be reorganized for manipulative and projectile control purposes. Both a more efficient form of bipedalism and the further transformation of the arm, hand, and upper torso became possible.

Non-hominin primate species are capable of walking on hind legs, but only with difficulty and for short periods of time. Chimpanzees, for instance, cannot straighten their legs, and require constant muscular exertion to support the body. Moreover, the center of gravity of the chimpanzee body must shift with each step, leading to a pronounced lumbering motion with significant side-to-side momentum shifts (O'Neil 2012). The hominin pelvis was shortened from top to bottom and, by the time *Homo ergaster* emerged, had been rendered bowl-shaped to facilitate terrestrial locomotion without sideward movement, the hominin leg bones became sturdy, the leg muscles were strengthened to permit running, and the development of arches in the feet facilitated a low-impact transfer of weight from leg to leg (Bramble and Lieberman 2004). The specialized form of bipedality that arose around 2 Mya thus facilitates running efficiently for great distances, although not approaching the speed of many large four-footed mammals.

Today we celebrate specialized bipedality as the basis for human upper-body physical and psychomotor capacities for crafting tools and handicrafts. But another major contribution of these capacities, as we explain below, was for fashioning and using lethal weapons.

2.6 Fire and Social Sharing

The hominin control of fire cannot be accurately dated. We have firm evidence from about 400,000 years ago in Europe (Roebroeks and Villa 2011), and about 800,000 years ago in Israel (Alperson-Afil 2008), but it is likely that this key event had originated in Africa much earlier (Gowlett and Wrangham 2013). The control of fire had strong effects on hominin cultural and phylogenetic evolution. First, the transition to specialized bipedality is much easier to understand if the hominins that experienced this transition had control of fire (Wrangham and Carmody 2010). Prior to the control of fire, humans almost certainly took to the trees at night like most other primates, as a defense against predators. Because predators have an instinctive fear of fire, the control of fire permitted hominins, who were already bipedal, to abandon climbing almost completely.

Second, the practice of cooking food was a related cultural innovation with broad gene-culture coevolutionary implications. Cooking favors a central location to which the catch is transported, and hence requires abandoning the competitive, socially uncoordinated "tolerated theft" distribution of calories typical of food-sharing in nonhuman primate species, in favor of a distribution based on widely agreed-upon fairness norms (Isaac 1977; Blurton-Jones 1987). This major socio-psychological transition was probably made possible by the adoption of some form of cooperative breeding and hunting among hominins that had begun by the time *Homo erectus* emerged (Burkart and van Schaik 2010). In sum, while the early advent of cooking is not yet firmly established, it is likely that the control of fire and the practice of cooking were an important precondition of the emergence of a human moral order.

Hominins with access to cooked food did not require the large colon characteristic of other primates, which allowed them to reduce the amount of time spent chewing food from the four to seven hours a day characteristic of the great apes, to about one hour per day. With a smaller gut, less need for chewing, and more rapid digestion, hominins were liberated to develop their aerobic capacity and perfect their running ability (Wrangham and Carmody 2010).

2.7 From Gatherer to Scavenger

Beginning around 2.5 million years ago there was a major forking in the evolutionary path of our possible ancestors. The Australopithecines branched in at least two—perhaps more but the fossil record in this area is quite incomplete—very different evolutionary directions. One led to the robust Australopithecines and a genetic dead-end by about 1.4 million years ago, and the other very likely led to the first humans.

These diverging evolutionary paths appear to have been the response to novel environmental challenges. Coinciding with this hominin divergence was a shift in the global climate to frequently fluctuating conditions. Early hominins succeeded by learning to exploit the increased climatic instability (Potts 1996, 1998; Richerson et al. 2001; O'Connell et al. 2002).[2]

---

[2]DeMenocal (2011) notes that Darwin (1859) long ago speculated on the role of climate change in human evolution, as did Dart (1925), and that modern findings support the

The resulting adaptations enhanced hominin cognitive and socio-structural versatility. "Early bipedality, stone transport,...encephalization, and enhanced cognitive and social functioning," Potts (1998) argues, "all may reflect adaptations to environmental novelty and highly varying selective contexts."

A diet based significantly on the flesh and bone marrow of large animals provided a niche for emerging hominins quite distinct from that of other primates and thus selected for the traits that most distinguish humans from apes. This much was clear to Darwin in *The Descent of Man* (1871). However, until recently, most paleoanthropologists assumed that prey was acquired through hunting from the Australopithecine outset (Dart 1925; Lee and DeVore 1968; but see Binford 1985). In fact, it now appears that early hominins, in the transition from the Pliocene to the Pleistocene, were more likely scavenger-gatherers than hunter-gatherers, of which there is firm evidence dating from 3.4 Mya (McPherron 2010).

The first proponents of early hominins as scavengers believed that the scavenging was "passive," in that small groups of hominins took possession of carcasses only after other predators, upon being sated, abandoned their prey (Binford 1985; Blumenschine et al. 1994), but more recent evidence suggests the prevalence of "competitive" or "power" scavenging, in which organized groups of humans sporting primitive weapons chased the killers and appropriated carcasses in relatively intact shape (Dominguez-Rodrigoa and Barba 2006). The implicit argument is that the combination of coordinated collective action and the lethal weapons of the period were sufficient to drive off other predators, and hence presumably to kill certain live prey as well. While a large prey can be driven off a cliff or trapped in a box canyon, it requires powerful weapons to cripple or kill a large predator. Before the advent of poisoned stone-tipped spears and arrows, the active pursuit of large prey was likely impossible (Sahle et al. 2013). The earliest known use of wooden javelins (Keeley and Toth 1981; Thieme 1997) suggests medium-size prey.

---

importance of climate-based selection pressures (Vrba 1995; Potts 1998), and specifically, climate variability. Potts (1998) examined the environmental records of several hominin localities, finding that habitat-specific hypotheses are disconfirmed by the evidence. By contrast, the variability selection hypothesis, which states that large disparities in environmental conditions were responsible for important episodes of adaptive evolution, was widely supported.

Flaked stone toolmaking, butchering large animals, and expanded cranial capacity all appear around 3.4 Mya (McPherron 2010), but there is no evidence that Australopithecines hunted large game. *Australopithecus* and *Homo habilis* were in fact quite small, adult males weighing under 100 pounds and females about 75 pounds. Their tools were primitive, consisting of stone scrapers and rough hammerstones. They therefore lacked the sophisticated weapons for hunting large and swift-moving prey, and hence are unlikely to have hunted effectively, but they could well have scavenged. Modern chimpanzees and baboons are known to scavenge the kills of cheetahs and leopards (Medina 2007), so this behavior was likely in the repertoire of the earliest hominins. With highly cooperative and carefully coordinated maneuvers by use of weapons, they could have chased away even the most ferocious predators.

Hunting and scavenging small animals is not cost-effective for large non-human primates, while scavenging large animals requires group participation and efficiently coordinated cooperation, both in organizing an attack on predators feeding on a large prey, and protecting against predators while processing and consuming the carcass (Isaac 1978). Moreover, use of stones as weapons that might be used to scare off other predators and scavengers (Isaac 1987) has been questioned (Whittaker and McCall 2001), but most likely there was an array of tools made of softer materials, very probably including wooden spears, suitable for making bluffing attacks.

Unlike wooden weapons, stones could have been carefully amassed at strategic sites within a large scavenging area, so that when a scouting party located an appropriate food object to scavenge, it could call others to haul the stones to the site of the carcass, as a strategic operation preceding its appropriation (Isaac 1977). These could have been the first lethal weapons, but carrying wooden spears or clubs would have served equally well to intimidate competing predators, and also would have been useful in killing small game.

2.8   Primitive Lethal Weapons

Stones are used today in certain contexts by hunter-gatherers as found-objects, and possibly as fashioned projectiles. Barbara Isaac (1987) studied stones used by recent foragers, also found in concentrations at Olduvai sites by Mary Leakey (1971), some of which were carefully finished spheroids. She observes that the size and shapes of the Olduvai stones render them appropriate, to use for throwing. Recent foragers do use found-object stones quite effectively as fighting weapons. Isaac (1987) has documented devastating attacks by hunter-gatherers against early encroaching Europeans, when intensive stoning actually proved more effective than musketry in rapidly inflicting serious casualties. This took place at contact in various parts of the world, so the traditions were likely pre-existing.

In Africa, behaviorally modern humans could have used long-range projectile weaponry (atlatl darts and arrows) in conflict for at least 50,000 years (Shea 2006; Ambrose 2008; Wadley et al. 2009; Wynn 2009; Wilkins et al. 2012; Roach et al. 2013). The recent hunting evidence includes a Levalloisian spear point embedded in a prey skeleton (Boëda et al. 1999). Group conflict likely accounts for the limited sampling we do have for humans of Pleistocene death-by-projectiles (Keeley 1996; Thorpe 2003), which includes at Grimaldi a child with a point embedded in its spine (27,000-36,000 BP), in the former Czechoslovakia weapons traumas and cranial fractures on adult males (24,000-35,000 BP), in Egypt an adult male with a point embedded in his arm (20,000 BP), and a Nubian cemetery where 40% of the interred exhibited weapon traumas (12,000-14,000 BP). Tacon and Chippendale (1994) have documented Australian rock art dating back to 10,000 BP that depicts armed combat, with increasing numbers of combatants by 4000 BP. In the Holocene armed combat is well-documented and widespread, as in the work of Lambert (1997) on the remains of California Indians which exhibit plentiful head injuries and parrying fractures.

If behaviorally modern human beings have used long-range projectile weapons against prey for at least 50,000 years, doubtless they sometimes turned such weapons against other humans over the same period. A special instance of weapon use is documented in art from Spain's Remigia cave. Human stick figures are shown standing with bows held about their heads while a male lays on the ground with the same number of arrows pincushioning him. There are ten men in the largest of the groups. This may express a group execution theme, or possibly a raid carrying out an act of revenge (see Otterbein 2004). This art appears to date to the early Neolithic.

Technological developments such as atlatls, bows and arrows, shields, and body armor are all relatively recent. It has been widely suggested that the advent of the spear-thrower (atlatl) arrived rather late, about 30,000 BP, and the bow and arrow later still (e.g., Klein 1999). But there are recent reports (Lombard and Phillipson 2010) suggesting that bows and arrows may have been in use as early as about 60,000 BP. Some contemporary groups use poisoned projectiles, and their use in prehistory is now susceptible to study (d'Errico et al. 2012), but further research is needed.

This picture of Pleistocene weapon use is supported by the fact that the fossils of large animals that have markings on bones indicating hominin flaying and scraping with flaked stone tools are often found with stones that originated several kilometers away. Contemporary chimpanzees carry stones to nut-bearing trees that they use to crack the nuts (Boesch and Boesch-Achermann 2000), so this behavior was likely available to Australopithecines. Chimpanzees, however, carry stones only several hundred meters at most, whereas *Homo habilis* scavengers carried stones as far as ten kilometers, probably because they had invented portable containers (Mc-Grew 1992).

Neither the Oldowan tools of the early period nor the later and more sophisticated Acheulean tools, which are found from the early Pleistocene up to about 200,000 years ago, show any sign of being useful as hunting weapons. However, besides stones, human power scavengers of 500,000 years ago probably had sharpened and fire-hardened spears to ward off competitive scavengers and threatening predators, at least after the domestication of fire (Thieme 1997). These weapons could also have been used against conspecifics. By contrast, nonhuman primates use tools, but they do not use weapons in conflictual encounters (Huffman and Kalunde 1993; McGrew 2004). In these species there is simply no record of a fashioned or found-object weapon being used to injure or kill a conspecific.

The cognitive potential to invent and use lethal weapons is likely present in the two *Pan* species. However, in nature bonobos and chimpanzees fashion tools for extraction of insect or plant foods, while in both species intimidation displays merely involve found objects being brandished or dragged. Chimpanzees use sticks fashioned from tree branches to ferret bushbabies from their tree hollow hiding places (Pruetz and Bertolani 2007; Gibbons 2007), so the use of sharpened sticks was thus likely within the cognitive capacity of *Homo habilis*. However, there is a considerable distance between using sharp sticks as impaling devices and as well-aimed projectiles (Nishida 1973).

The first dedicated and unambiguously lethal weapons to appear with excellent preservation in the archeological record are the multiple all-wooden spears documented by Thieme (1997) at Schöningen, with over a dozen butchered wild horses and some bison located nearby. These javelins are both streamlined aerodynamically and well-balanced for effective throwing so they were projectile weapons capable of bringing down medium-sized game at a distance. They also provide a defense against dangerous prey, and they offer hunters a means of threatening other predators away from their kills. These considerations suggest that a paleo-record of lithic weaponry alone is seriously incomplete. What the lithic record does suggest, in its Acheulian continuity, is that this tradition of making wooden spears might also have had great longevity (see Kelly 2005).   The emergence of lethal weapons was likely important in the evolution of hominin social organization (Roach et al. 2013). In hunter-gatherer conflicts hunting weapons quickly become lethal, and even an outnumbered victim can inflict casualties (Lee 1979; see also Churchill and Rhodes 2009). Bingham (1999), Gintis (2000), Bingham and Souza (2009), and Boyd et al. (2010) stress the importance of the superior physical and psychomotor capacities of humans in clubbing and throwing projectiles as compared with other primates, citing Goodall (1964) and Plooij (1978) on the relative advantage of humans. Darlington (1975), Fifer (1987), and Isaac (1987) document the importance of these traits in human evolution. Bingham (1999), Boehm (1997), and Okada and Bingham (2008) document that humans have developed the ability to carry out collective punishment against norm violators, thus radically lowering the cost of punishing transgressors. Calvin (1983) argues that humans are unique in possessing the neural machinery for rapid manual-brachial movements that both allows for precision stone-throwing and lays the basis for the development of language, which like accurate throwing depends on the brain's capacity to orchestrate a series of rapidly changing muscle movements. Indeed, Roach et al. (2013) showed that *Homo erectus* had evolved this capacity for accurate overhead throwing, and recent work suggests that the origins of human language are also much older than commonly assumed (Dediu and Levinson 2013), originating in all likelihood more than 700,000 years ago.[3]

---

[3]The fossil evidence indicates that hominins developed speech on the order of one Mya. The hyoid bone is a key element of speech production in humans. Martinez et al. (2008) show that hominin hyoid bones from 540,000 years ago are similar, and hence were inherited from their last common ancestor, *Homo rhodesiensis*, which was from 700,000 to 1,000,000 years ago. Martinez et al. (2004) use evidence from the acoustical properties of Middle Pleistocene fossil remains of the hominin inner ear to argue that hominins of this period had auditory capacities similar to those of living humans.

2.9 Warfare

Fighting between groups ranges from single revenge killings, to careful raids in which safety of the raiders is as important as inflicting damage on the enemy, to intensive warfare with genocidal attacks and face-to-face large-scale battle (Keeley 1996; Kelly 2000; Otterbein 2004). Such fighting involves assessments of the relative fighting power of adversaries and of risk (Wrangham and Glowacki 2012), and the array of weapons available to each side obviously enters into these assessments. The result is an ethnocentric species (LeVine and Campbell 1972) whose members are predisposed to assume the risks associated with aggression, especially against outsiders, but also strive to minimize those risks.

All contemporary foragers arm themselves with lethal hunting weapons, and at times these weapons are deployed by individuals against within-group adversaries and by the group in executing serious deviants (Knauft 1991; Boehm 1997). Both types of homicide, while rare, are well documented despite a universal ethos that strongly discourages killing a group member (Brown 1991). To keep their systems of social cooperation viable, foragers strive to peaceably adjudicate conflicts within their midst (Boehm 2000).

These moral inhibitions are relaxed when inter-group rivalry comes into play. The use of weapons between groups can entail massive casualties when desired cooperative relations among groups fail and conflict gains the upper hand (Wiessner 1977). However, even given a pattern of recurrent ethnocentric fighting between groups, hunter-gatherers may succeed in managing these conflicts (Boehm 2013). While the active management of hostilities is universal within bands, such between-group efforts remain both sporadic and unpredictable. Weapons render forager bands very dangerous to one another, and some groups live with such hostilities with little effort expended to curtail them.

The history of human warfare remains a hotly controversial topic among anthropologists. The basic facts themselves are vigorously contested (Turchin 2015). Some argue that prior to the appearance of settled agriculture, humans approximated the "noble savage" picture drawn long ago by Jean-Jacques Rousseau. This view was definitively put to rest by Lawrence Keeley's *War Before Civilization* (1996), but continually pops up in the anthropological literature (Fry 2013). The opposing view is the Hobbesian picture of the distant past known as the "war of all against all" (Hobbes 1968[1651]). The evidence against this view is the documentation of extensive trade networks in hunter-gatherer societies (Adams 1974). Lying behind this controversy is the notion that if war is ancient, then making war is part of human nature, whereas if war is modern, then it is a purely cultural and environmental phenomenon that can be successfully countered by appropriate cultural changes.

But this is surely an illegitimate dichotomy. The idea that behavior is either innate or culturally determined was given up by sociobiologists long ago. As we have argued, early humans developed powerful lethal weapons, developed the skeletal and muscular morphology to use them skillfully, and learned how to cooperate in collective endeavors through creative politics and leadership. Moreover, anger and aggression are strong human predispositions. These human capacities allow humans to make war when ecological and social conditions render war profitable. Ancient or modern, war is part of how humans are defined as a species. War can be contained and controlled, but it cannot be ignored, whatever cultural structures govern future human societies.

2.10   Dominance and Reverse Dominance Hierarchies

James Woodburn (1982) classified hunter-gatherer societies into *immediate-return* and *delayed-return* systems. In the former, group members obtain direct return from their labor in hunting and gathering, with food lasting at most a few days. The tools and weapons they use are highly portable. In delayed-return foraging societies, individuals hold rights over valuable assets, such as means of production (boats, nets, beehives, and the like), and processed and stored food and materials. These societies exhibit forms of social stratification akin to those in modern societies: social dominance hierarchies in the form of lineages and clans. However, the fossil record suggests that delayed-return human society is a quite recent innovation, appearing some 10,000 years ago, although in ecologically suitable locations, it may have existed earlier (most such locations are now below sea level). *Homo sapiens* thus evolved predominantly in the context of immediate-return systems.

The important factor in "delayed return" is not the cognitive capacity for delayed gratification or long-range planning, which certainly existed in immediate return societies, but rather the availability of cumulable material wealth. Material wealth allows those who seek social dominance to control allies and resources and thereby thwart the capacity of subordinates to disable and kill them. As long as the material gains from a position of social dominance exceed the cost of coalition-building and paying guard labor, social dominance of the sort common in other primate societies can be reestablished in human society. In fact, the appearance of farming and private property in land led to high levels of political inequality in only a few societies, and states with a monopoly in coercive power emerged only after a millennium of settled agriculture. Nor were early farming societies more economically stratified than hunter-gatherer societies (Borgerhoff Mulder et al. 2009). The accumulation of material wealth is thus merely a precondition for the reestablishment of social dominance hierarchies. To avoid confusion, we will call societies that lack forms of material wealth accumulation *simple*, rather than *immediate-return*, societies.

Simple societies, Woodburn (1982) suggests, are "profoundly egalitarian ...systematically eliminat[ing] distinctions ...of wealth, of power and of status." Fried (1967), Service (1975), Knauft (1991) and others likewise comment on the egalitarian character of simple hunter-gatherer societies. The simple vs. delayed-return dichotomy is in fact somewhat overdrawn, as there is in fact a continuous range of variation between the two archetypes. Many Pleistocene humans used some storage even if they were nomadic and they remained strongly egalitarian. The majority of the 58 "Late Pleistocene Appropriate" foraging societies coded by Boehm (2012) (see discussion below), including the !Kung considered by Knauft (1991), are of an intermediate type. What factors are responsible for such unusual egalitarianism? Here, we will argue it is due to the combination of interdependence and ability to punish transgressors.

Cut marks on bones suggest that a major investment in large game hunting increased decisively only 250,000 years ago (Stiner 2002) and delegating sharing to a single butcher began 200,000 years ago (Stiner et al. 2009). In establishing timing of this transition to heavy reliance on medium-sized game in humans, Stiner (2002) uses multiple indices including the age structure of prey and cut marks to suggest that at this time ungulate hunting became prominent in human subsistence. However, cut marks on bones may not be a reliable indicator of how meat is shared (Lupo and O'Connell 2002). Indeed, if Wrangham and Carmody (2010) are correct in dating the control of fire by hominins and the cooking of meat, the problem of the fair distribution of meat among families, especially important in hard times when only medium- and small-size prey were available, may well have been solved much earlier. This was likely an early source of egalitarian sentiment, as well as providing the material substrate for the development of a social morality. Contemporary hunter-gatherer societies are often violent and competitive (Potts 1996), but they almost always distribute large game peacefully, if sometimes contentiously, based on a commonly accepted set of fairness principles (Kaplan and Hill 1985b; Kelly 1995; Boehm 2004).

The human ecological niche requires food sharing not only daily, but also on a longer-term basis due to the occasional injuries or illnesses to which even the best hunter or gatherer may be subjected (Sugiyama and Chacon 2000; Hill et al. 2011). Thus each individual forager, especially in the immediate-return form of foraging, is utterly dependent on the others in their camp, band, or even wider sharing unit. This strong interdependence dampens the tendency to free-ride on others' efforts, and favors strong individual tendencies toward egalitarianism, as well as sophisticated fairness norms concerning the division of the spoils (Whallon 1989; Kaplan and Hill 1985a).

Collective hunting in other species does not require a fairness ethic because participants in the kill simply eat what they can secure from the carcass, and because dominants are evolved to tolerate subordinates to a point that all the hunters are adequately nourished. However, the practice of bringing the kill to a central site for cooking, which became characteristic of hominin societies, is not compatible with uncoordinated sharing and eating. In the words of Winterhalder and Smith (1992),

> ...only with the evolution of reciprocity or exchange-based food transfers did it become economical for individual hunters to target large game. The effective value of a large mammal to a lone forager...probably was not great enough to justify the cost of attempting to pursue and capture it.... However, once effective systems of reciprocity or exchange augment the effective value of very large packages to the hunter, such prey items would be more likely to enter the optimal diet. (p. 60)

Fire and cooking thus coevolved with the emergence of a normative order and social organization based on ethical behavior.

The second element is that egalitarianism is imposed by the community, creating what Boehm (1999) calls a *reverse dominance hierarchy*. Hunter-gatherers share with other primates the striving for hierarchical power, but among mobile foragers, social dominance aspirations are successfully countered because individuals do not accept being controlled by an alpha male and are extremely sensitive to attempts of group members to accumulate power through coercion. When an individual appears to be stepping out of line by threatening or killing group members, he will be warned and punished. If this behavior continues and ostracism does not work, the group will delegate one member, usually a close relative of the offender, to kill him. Boehm's message in *Hierarchy in the Forest* (1999) is that "egalitarianism…involves a very special type of hierarchy, a curious type that is based on antihierarchical feelings."

We can regard this phenomenon as an extension of the leveling coalitions seen among primate males (Pandit and van Schaik 2003). Female chimpanzees in captivity act collectively to neutralize alpha male bullies (de Waal 1996), wild chimpanzees form large coalitions to banish, badly wound, or even kill high-ranking males. Bonobos in the wild have been observed to behave similarly. By comparison with humans, however, leveling coalitions among primates are limited to the genus *Pan* and generally quite small.

Because of the extremely long period during which humans evolved without the capacity to accumulate wealth, we have become constitutionally predisposed to exhibit these antihierarchical feelings. Of course, in modern democratic societies, there is still enough willingness to bend to authority in humans to ensure that a marked or tyrannical social dominance hierarchy remains a constant threat and often a reality.

Capable leadership in the absence of a strong social dominance hierarchy in band-level societies is doubtless of critical importance to their success, and leaders are granted by their superior position, and with the support of their followers, with fitness and material benefits. Leadership, however, is based not on physical prowess, but rather on the capacity to motivate, persuade, and help the band to reach a consensus. This account of the growth of intelligence is an elaboration on the Machiavellian Intelligence Hypothesis (Jolly 1972; Humphrey 1976; Byrne and Whiten 1988) that stresses the effect of encephalization on enhancing the mean fitness of group members, not simply advancing the interests of the leader. For recent evidence on leadership in hunter-gatherer societies, see von Rueden (2015) and von Rueden et al. (2014).

Reverse dominance hierarchy is documented in Boehm (2012). Boehm located 339 detailed ethnographic studies of hunter-gatherers, 150 of which are simple hunter-gatherer societies. He coded fifty of these societies from around the world. He calls these simple hunter-gatherer societies "Late

Pleistocene Appropriate" (LPA). Despite the fact that these societies have faced highly variable ecological conditions, Boehm finds that their social organization follows the pattern suggested by Woodburn (1982) and elaborated by Boehm (1997). The LPAs exhibit both reverse dominance hierarchy and subscribe to a common human social morality. This morality operates through internalized norms, so that individuals act prosocially because they value moral behavior for its own sake and would feel socially uncomfortable behaving otherwise.[4]

How do we explain this unique pattern of socio-political organization? Woodburn attributes this to humans' access to lethal weapons that neutralize a social dominance hierarchy based on coercion. "Hunting weapons are lethal," he writes, "not just for game animals but also for people. Effective protection against ambush is impossible...with such lethal weapons" (p. 436). Woodburn adds that "in normal circumstances the possession by all men, however physically weak, cowardly, unskilled or socially inept, of the means to kill secretly anyone perceived as a threat to their own well-being...acts directly as a powerful leveling mechanism. Inequalities of wealth, power and prestige...can be dangerous for holders where means of effective protection are lacking" (p. 436).

Boehm (2012) argues that his LPAs inherited from our ancient hunter-gatherer forbears the capacity to control free-riders through collective policing, using gossip and informal meetings as the method of collecting information concerning the behavior of group members. Moreover, according to our best evidence, the hunter-gatherer societies that defined human existence until some 10,000 years ago also were involved widespread communal and cooperative child rearing (Hrdy 1999, 2000, 2009) and hunting (Boehm 1999, 2012; Boyd and Silk 2002; Bowles and Gintis 2011), thus tightening the bonds of sociality in the human group and increasing the social costs of free-riding behavior.

Nonhuman primates never developed weapons capable of definitively controlling a dominant male. Even when sound asleep, a male chimpanzee reacts to being accosted by waking and engaging in a physical battle, ba-

---

[4]The notions of norms and norm internalization (Durkheim 1902; Parsons 1937) are common in the social sciences. According to the socio-psychological theory of norms, appropriate behavior in a social role is given by a social norm that specifies the duties, privileges, and expected behavior associated with the role. Adequate performance in a social role normally requires that the actor have a *personal commitment* to the role that cannot be captured by the self-regarding "public" payoffs associated with the role (Gintis 2003a; Gintis and Helbing 2015).

sically unharmed by surprise attack. In *Demonic Males* (1996) Richard Wrangham recounts several instances where even three or four male chimpanzees viciously and relentlessly attack a male for twenty minutes without succeeding in killing him (but see Watts et al. 2006). The limited effectiveness of chimpanzees in this regard can mainly be ascribed to their inability to effectively wield potentially dangerous natural objects, for instance stones and rocks. A chimpanzee may throw a large rock as part of a display, but only rarely will it achieve its target.

The human lifestyle, unlike that of chimpanzees, requires many collective decisions, such as when and where to move camp and which alliances to sustain or cut. This lifestyle thus requires a complex socio-political decision making structure and a sophisticated normative order. Many researchers incorrectly equate dominance, as found among chimpanzees, with leadership. In some species, such as gorillas, dominants can indeed initiate or influence group movements, because others rely on the dominant male as the main protector and value his proximity. In most human foragers, there are no such dominants.

Capable leadership in the absence of a social dominance hierarchy in egalitarian human societies is of critical importance to their success. However, despite their exceptionally generous treatment of band members, human leaders are granted by their superior position, and with the support of their followers, with certain material benefits and fitness (Price and Van Vugt 2014), such as superior mating opportunities. Leadership, as we have seen, is based not on physical prowess and coercion, but rather on the capacity to motivate and persuade. Eibl-Eibesfeldt (1989) and Wiessner (2006), among many others, have stressed the importance in hominin societies of leadership based on persuasion and coalition building. In discussing mobile foragers, Wiessner (2009) remarks, "Unlike nonhuman primates, for whom hierarchy is primarily established through physical dominance, humans achieve inequalities through such prosocial currencies as the ability to mediate or organize defense, ritual, and exchange" (pp. 197–198). Interestingly, our closest living relative, the chimpanzee, shows a tendency in the same direction, which is unusual among primates: successful top-ranked males are good social strategists (Goodall 1986; Nishida and Hosaka 1996).

It is important not to confuse reverse dominance hierarchy, which is based on a predisposition to reject being dominated, with a specific predisposition for egalitarian outcomes. Rather, persuasion and influence become a new basis for social dominance (Clutton-Brock 2009), which tends to be no less powerful for its subtlety. Wiessner (2006) observes that successful small-scale societies "encourage the capable to excel and achieve higher status on the condition that they continue to provide benefits to the group. In no egalitarian institutions can the capable infringe on the autonomy of others, appropriate their labor, or tell them what to do" (p. 198).

2.11   Are There Egalitarian Nonhuman Primates?

If there were a multi-male/multi-female primate society lacking a social dominance hierarchy and lacking lethal weapons, yet exhibiting reverse dominance hierarchy, the propositions offered in this chapter would be compromised. Does such a society exist? Here, an important distinction can be drawn between egalitarianism flowing from weak social interaction and a low level of social contestation on the one hand, and egalitarianism stemming from a high level of interdependence and some form of subordinate leverage over dominants (Sterck et al. 1997).

While there are clear behavioral patterns in nonhuman primates that serve as the basis for human reverse dominance hierarchy, all multi-male/multi-female nonhuman primate societies are in fact based on strongly expressed social dominance hierarchies. There may be variation in the degree to which female or male dominance relations are decided and thus their dominance hierarchies are more or less steep, depending on the strength of contest competition for resources (Sterck et al. 1997). It is often argued that bonobos (*Pan paniscus*) are more egalitarian than chimpanzees and more like humans (de Waal 1997; Hare et al. 2007). However, except for a female dominance hierarchy in feeding access for infants, the pattern of dominance in bonobos strongly resembles that of chimpanzees (Furuichi 1987, 1989, 1997), although estimates of the steepness of dominance hierarchies among males and females are not consistent across studies (Stevens et al. 2007; Jaeggi et al. 2010).

Similarly, reports indicate rather thoroughgoing egalitarianism among woolly spider monkeys, or muriquis (Strier 1992), which also live in sizeable multi-male/multi-female groups, much like those of bonobos and chimpanzees. They are highly promiscuous and males hardly compete for matings (Milton 1984; Strier 1987). In all the primate examples of egalitarianism in sizeable groups, there is a clear reduction in the intensity of male contest competition as a result of female reproductive physiology that leads to unpredictable ovulation and thus low potential monopolization of matings, and thus paternity concentration, by top-ranking males (van Schaik et al. 2004). Thus these egalitarian social relations are the result of scramble-like competition.

In none of these societies do we find the interdependence that we observe in human societies. The closest analogs are the societies of cooperative breeders, as in callitrichids, but these are rarely multi-male/multi-female. Among non-primates, wild dogs and wolves, which are both cooperative breeders and hunters (Macdonald and Sillero-Zubiri 2004), came closest, but even there we mostly, though not always, have a single breeding pair rather than multiple cooperating pairs. We conclude that, on the basis of available evidence, there are no multi-male/multi-female egalitarian primate societies except for *Homo sapiens*.

2.12 Governance by Consent

Following the development of lethal weapons and the suppression of dominance hierarchies based on physical prowess, successful social bands came to value individuals who could command prestige by virtue of their persuasive capacities. While it was by no means necessary that this behavior emerge from the collapse of a social dominance hierarchy based on force, it did in fact emerge in the human line, and no other solution to the problem of leadership has been observed in the primate order. As suggested in the Choreographer interview at the head of this chapter, the triumph of the gracile human skeleton over the robust Australopithecines and other hominids is a strong indication that brain and not brawn was conducive to individual fitness and best enhanced the fitness of human groups as well.

The human egalitarian solution emerged in the context of bands insisting that their leaders behave with modesty, generosity, and fairness (Boehm 1993). A sagacious and effective leader will attempt to parley an important social position into material and fitness benefits, but not so much as to induce followers to replace him with a less demanding leader. Persuasion was the name of the game, and excessive exercise of power would reverse the leader's fortunes. Persuasion depends on clear logic, analytical abilities, a high degree of social cognition (knowing how to form coalitions and motivate others), and linguistic facility (Plourde 2010). Leaders with these traits could be effective, but one intemperate move could lead to a fall from power. Thus in concert with the evolution of an ever more complex feeding niche (Kaplan et al. 2000), the social structure of hunter-gatherer life in typical gene-culture coevolutionary fashion contributed to the progressive encephalization and the evolution of the physical and mental prerequisites of effective linguistic and facial communication. In short, two million years of evolution of hyper-cooperative multi-family groups that deployed lethal weapons to hold down hierarchy gave rise to the particular cognitive and socio-political qualities of *Homo sapiens*.

The increased encephalization in humans was an extension of a long primate evolutionary history of increased brain size, usually associated with increased cognitive demands required by larger group size (Humphrey 1976; Jolly 1972; Byrne and Whiten 1988; Dunbar et al. 2010).[5] The argument presented here, which invokes coordinated collective action in cooperative foraging, made possible by a combination of interdependence and lethal weapons, extends this analysis to explain human exceptionalism in the area of cognitive and linguistic development.

This development in promoting egalitarian multi-male/multi-female bands explains the huge cognitive and linguistic advantage of humans over other species. The early students of human evolution interpreted human hypercognition as a process of runaway sexual selection, in which intelligent individuals were more successful in attracting mates but did not otherwise contribute to the fitness of band members. This was the favored theory of Charles Darwin (1871) and Ronald Fisher (1930), and more recently of Geoffrey Miller (2001). However, runaway selection is rare, and if it exists, it is generally a short-term deviation from fitness-maximizing behavior (Gintis 2009a; Pomiankowski 1987). Explaining human intelligence as a product of runaway sexual selection is a first-class just-so story, of the type so eloquently critiqued by Stephen Jay Gould and Richard Lewontin (1979). Our reading of the evidence suggests that human hypercognition, despite the extreme energy costs of maintaining a large brain, was fitness-enhancing because of increased cognitive and linguistic ability, which entailed heightened egalitarian leadership qualities. These leadership qualities increased the fitness of band members, who responded by ceding enhanced fitness benefits to leaders (Price and Van Vugt 2014).

The mating success of high cognition males was thus grounded in their contribution to the mean fitness of band members, and hence in the long

---

[5]Group size is certainly not the whole story. Multi-male/multi-female monkey groups are often as large as or larger than ape groups, although the latter have much larger brains and are considerably more intelligent. The full story concerning cephalization in mammals in general, and primates in particular, remains to be told (Navarrete et al. 2011).

run, to the evolutionary success of ancestral humans. In a sense, hominins evolved to fill a *cognitive niche* that was relatively unexploited in the early Pleistocene (Tooby and DeVore 1987; Pinker 2010).

2.13   Cooperative Mothering: The Evolution of Prosociality

In cooperative breeding, the care and provisioning of offspring is shared among group members. The standard estimate is that some 3% of mammals have some form of allomaternal care, but in the order Primates, this frequency rises to 20% or more (Hrdy 2009, 2010). In many nonhuman primates and mammals in general, cooperative breeding is accompanied by generally heightened prosociality, as compared with related species with purely maternal care (Burkart et al. 2014). The most plausible explanation is that cooperative breeding leads to a social structure that rewards prosocial behavior, which in turn leads to changes in neural structure that predisposes individuals to behaving prosocially (Burkart et al. 2009; Burkart and van Schaik 2010). An alternative possibility is that there is some underlying factor in such species that promotes prosociality in general, of which collective breeding is one aspect.

Human prosociality was strongly heightened beyond that of other primates living in large groups, including cooperative breeders, by virtue of the niche hominins occupied, involving coordination in scavenging and hunting, and sophisticated norms for sharing meat. This combination might account for the degree of cooperative breeding in the hominin line. As hominin brain size increased, the duration of immaturity did as well (Barrickman et al. 2008), and immatures had to learn an increasingly large number of foraging and other skills (Kaplan et al. 2000; Schuppli et al. 2012). Hominins evolved a unique system of intergenerational transfers that enabled the evolution of ever more complex cognitive abilities to support ever more complex subsistence skills (Kaplan et al. 2007). Our uniquely prosocial shared intentionality (Tomasello et al. 2005) can be traced back to the psychological changes involved in the evolution of cooperative breeding, and additionally, hunting (Burkart et al. 2009).

2.14   Lethal Weapons and Egalitarianism

In the Holocene, some Big Man societies have been relatively egalitarian, such as those of highlands New Guinea, where the Big Man serves the group in out-feasting other groups and cannot transmit wealth or prestige to descendants. Other Big Man societies are fully hierarchical, with prestige and power being transmitted to future generations. The latter could have led to chiefdoms (Service 1975; Flannery and Marcus 2012).

The slow but inexorable rise of the state, both as an instrument for exploiting direct producers and for protecting them against the exploitation of external states and bands of private or state-sanctioned marauders, was a synthesis of these two types of Big Man socio-political systems (Andreski 1968; Gies 1984). The hegemonic aspirations of states peaked in the thirteenth century, only to be driven back by the series of European population-decimating plagues of the fourteenth century. The period of state consolidation resumed in the fifteenth century, based on a new military technology: the use of cannon. In this case, as in some other prominent cases, technology becomes the handmaiden to establishing a social dominance hierarchy based on force.

In *Politics*, Book VI, Part VII, Aristotle writes, "There are four kinds of military forces—the cavalry, the heavy infantry, the light armed troops, the navy. When the country is adapted for cavalry, then a strong oligarchy is likely to be established [because] only rich men can afford to keep horses. The second form of oligarchy prevails when the country is adapted to heavy infantry; for this service is better suited to the rich than to the poor. But the light-armed and the naval elements are wholly democratic... An oligarchy which raises such a force out of the lower classes raises a power against itself."

The use of cavalry became dominant in Western Europe during the Carolingian period. The history of warfare from the late Middle Ages to the First World War was the saga of the gradual increase in the strategic military value of infantry armed with longbow, crossbow, hand cannon, and pike, which marked the recurring victories of the English and Swiss over French and Spanish cavalry in the twelfth to fifteenth centuries (Turchin and Korotayev 2006). Cavalries responded by developing dismounted tactics when encountering infantry, using heavy hand weapons such as two-handed swords and poleaxes. These practices extended the viability of cavalry to the sixteenth century in the French and Spanish armies, but gradually through the Renaissance, and with the rise of Atlantic trade, the feudal knightly warlords gave way to the urban landed aristocracy and warfare turned to the interplay of mercenary armies consisting of easily trained foot soldiers wielding muskets and other weapons based on gunpowder. Cavalry remained important in this era, but even in the eighteenth and nineteenth century, cavalry was used mainly to execute the coup de grace on seriously weakened infantry.

The true hegemony of the foot soldier, and hence the origins of modern democracy, began with the perfection of the hand-held weapon, with its improved accuracy and greater firing rate than the primitive muskets of a previous era. Until that point, infantry was highly vulnerable to attack from heavy artillery. By the early twentieth century, the superiority of unskilled foot soldiers armed with rifles was assured. World War I opened in 1914 with substantial cavalry on all sides, but mounted troops were soundly defeated by men with rifles and machine guns, and thus were abandoned in later stages of the war. The strength of the political forces agitating for political democracy in twentieth century Europe was predicated on the strategic role of the foot soldier in waging war and defending the peace (Bowles and Gintis 1986), simply because conscripted armies of foot soldiers lacked the moral resolve to defend a society from whose governance they were systematically excluded.

2.15   The Long-Term Evolution of Human Sociality

It is tempting to focus on the past several thousand years of human cultural history in modeling human socio-political organization because the changes that occurred in this period so radically and rapidly transformed the character of human society (Richerson and Boyd 2004; Pagel 2012). However, the basic genetic predispositions of humans underlying socio-political structure were forged over a much longer period of time, whence the million plus year perspective offered in this chapter.

The framework developed here is applicable to many spheres of human social life, although we have applied it only to the evolution of socio-political structure. The central tool is *gene-culture coevolution*, which bids us pay close attention to the long-term dynamic interplay between our phylogenetic constitution and our cultural heritage. The second important conceptual tool is the socio-psychological theory of norms, which we discuss in Chapter 6. Many social scientists reject this theory because it posits a causal social reality above the level of individual actors. This position is sometimes termed *methodological individualism*. Methodological individualism is not a philosophical, moral, or political principle, but an assertion about reality. As such, it is simply incorrect, because social norms are an emergent property of human society, irreducible to lower-level statements

(Durkheim 1902; Gintis 2009a). All attempts at explaining human culture without this higher-level construct fail.

In this context, we have suggested the following scenario for the long history of human socio-political dynamics. Our primate ancestors evolved a complex socio-political order based on a social dominance hierarchy in multi-male/multi-female groups. Enabled by bipedalism, environmental changes made a diet of meat from large animals fitness-enhancing in the hominin line. This, together with cultural innovation in the domestication of fire, the practices of cooking, and of collective child-rearing created a niche for hominins in which there was a high return to coordinated, cooperative, and competitive scavenging, as well as technology-based extractive foraging. This development was accompanied by the likely use of clubs, spears, and long-range projectiles as lethal weapons, and also led to the spread of specialized bipedalism and the reorganization of the upper torso, shoulders, arms, and hands to maximize the effectiveness of these weapons. There was also a growth of new neural circuitry allowing the rapid sequencing of bodily movements required for accurate weapon deployment.

The hominin niche increasingly required sophisticated coordination of collective meat procurement, the occasional but critical reliance on resources produced by others, a complementary willingness to provide others with resources, and procedures for the fair sharing of meat and collective duties. The availability of lethal weapons in early hominin society could have helped to stabilize this system because it undermined the tendencies of dominants to exploit others in society. Thus two successful socio-political structures arose to enhance the flexibility and efficiency of social cooperation in humans and likely their hominin ancestors. The first was the reverse dominance hierarchy, which required a brain large enough to enable a band's rank-and-files to create effective coalitions that could definitively put an end to alpha male hegemony and replace this with a lasting egalitarian order. Leaders were kept weak, and their reproductive success depended on an ability to persuade and motivate, coupled with the rank-and-file ability to reach a consensus with such leadership. The second was cooperative child rearing and hunting, which provided a strong psychological predisposition towards prosociality and favored internalized norms of fairness. This system persisted until cultural changes in the later Holocene fostered material wealth accumulation, through which it became once again possible to sustain a social dominance hierarchy based on coercion.

This scenario has important implications for political theory and social policy, for it suggests that humans are predisposed to seek individual dominance when this is not excessively costly, but also to form coalitions to depose pretenders to power. Moreover, humans are much more capable of forming large, powerful, and sustainable coalitions than other primates, due to our enhanced cooperative psychological propensities. Such coalitions also served to reinforce the moral order, as well as to promote cooperation in hunting, warding off predators, and raiding other human bands. This implies that many forms of socio-political organization are compatible with the particular human amalgam of hierarchical and anti-hierarchical predispositions that can result in either independent egalitarian bands or well-amalgamated large societies.

In particular, this implies that there is no inevitable triumph of liberal democratic over despotic political hierarchies. The open society will always be threatened by the forces of despotism, and a technology could easily arise that irremediably places democracy on the defensive. Perhaps the most important threat to freedom and democracy would be the development of robots that could replace foot soldiers in war and crowd management. The problem with current robot models is insufficient energy storage—nothing like mitochondria and ATP exist in the non-organic world (Gintis 2015). Nevertheless, the future of politics in our species, in the absence of concerted emancipatory collective action, could well be something akin to George Orwell's *1984*, or Aldous Huxley's *Brave New World*. However, humans appear constitutionally indisposed to accept a social dominance hierarchy based on coercion unless the coercive mechanism and its associated social processes can be culturally legitimated. It is somewhat encouraging that such legitimation is difficult except in a few well-known ways, based on patriarchy, popular religion, or principles of liberal democracy.

3

Distributed Effectivity: Political Theory and Rational Choice

> **C:** Why do people vote?
> **HG:** To help candidates get elected.
> **C:** But one vote never makes a difference.
> **HG:** Humans help even when each individually makes no difference. That is the nature of distributed effectivity.
> **C:** But are people not misguided in voting against their economic interests?
> **HG:** Voting is a morality game.
> **C:** But isn't that irrational?
> **HG:** I do not vote my economic interests. I doubt that you do. No, it is not irrational to vote your heart or your brain rather than your bank account.
>
> Choreographer interview

Behavioral disciplines are successful to the extent that they model individual behavior as rational choice. The rational actor model posits that an individual has a *preference ordering* over the outcomes that his actions bring about and *beliefs* concerning the relationship between actions and outcomes. His behavior can be modeled as maximizing an objective function given by this preference orderings, subject to the informational and material constraints he faces. As we show in Chapter 6, this approach is widely applicable and does not imply that rational actors are selfish, that they are omniscient, that their choices necessarily improve their welfare, or that they consciously maximize anything. Even bacteria are rational actors.

The standard rational choice models, however, do not explain large-scale voting or collective action. This is because in such cases individual actions do not affect outcomes, so the whole rational actor framework is inoperative. This inconvenient fact undermines, for instance, the classical defense of democracy in political theory.

It is generally held that rational choice theory provides a powerful defense of political democracy. It asserts that democracy gives people roughly the same power in public life that markets give them in private life: the power to implement preferred social outcomes. However, individual voters have virtually zero probability of affecting the outcome of an election. Thus standard rational choice models of voter behavior dramatically underpredict voter turnout in all but the smallest elections (Geys 2006). Fiorina (1990) has called this fact "the paradox that ate rational choice theory." It is the central target of Green and Shapiro's (1994) powerful critique of rational choice methodology in political science. The standard rational choice defense of democracy is thus incoherent.

Yet voters widely behave in strategically rational ways (Cox 1997). They vote for the candidates they want to win, they are attentive to and evaluate candidates' positions, they discuss who deserves their vote, and they often shun candidates whom they prefer but they think cannot win. This chapter proposes a form of *social rationality* that extends the classical rational actor model to include the behavior of individuals in voting and participating in collective actions. Socially rational voter behavior is a form of what I term *distributed effectivity*, according to which individuals behave as though they were members of a very small electorate where single votes really make a difference. Distributed effectivity explains the central statistical regularities concerning voter turnout and the historical regularities concerning large-scale collective action. Distributed effectivity can be interpreted broadly as *Kantian equilibrium* behavior (Roemer 2015), as described below in Section 3.3.

We can summarize distributed effectivity as follows, assuming a majority-rule election with two alternatives. A rational choice model of voting suggests that an individual will vote for his preferred alternative if

$$bp \geq c, \tag{3.1}$$

where b is the net payoff to winning, c is the cost of voting rather than abstaining, and p is the probability that the individual is a pivotal voter; that is, with probability p his preferred alternative wins if he votes but loses if he abstains. The rationality assumption places no constraints on b or c. The benefit b can include payoffs to the voter himself, to others about whose welfare he cares, or for purely moral concerns, such as equity and fairness. The cost c can include material costs, which increase c, as well as feelings of social obligation and social signaling, which lower c.

Note that equation (3.1) considers only the value of winning and losing an election. This abstracts from the desire to register support for a candidate independent from its contribution to the outcome. This signaling motive for voting could be included in our analysis with some added notational complexity.

In a large election, where the probability of a single voter being pivotal is infinitesimal, classical rationality implies that even if an agent believes the other voters uniformly embrace distributed effectivity, he will still not vote if it involves any personal cost. Distributed effectivity, which implies voting even when there are substantial costs of participation, thus entails a moral, materially costly but personally rewarding, commitment to collective action.

There is a tradition in political theory that identifies rationality with self-regarding instrumental behavior (Conn et al. 1973; Coleman 1990). Rational choice in this chapter, by contrast, is based on the standard treatment in analytical decision theory (von Neumann and Morgenstern 1944; Savage 1954), which stresses *choice consistency* and *Bayesian updating*. This notion of rationality can model both self-regarding and moral preferences (Elster 1985; Gintis et al. 2005; Bowles and Gintis 2011; Roemer 2015). We explore this issue in greater detail in Chapter 6.

3.1   Public and Private Spheres

The *private sphere* is the locus of everyday transactions in the lives of individuals operating in civil society. An agent's *private persona* is the set of preferences and beliefs that govern his behavior in the private sphere. The *public sphere* is the locus of activities that create, maintain, transform, interpret, enforce, and execute the rules of the game that define society itself. An individual's *public persona* is the set of preferences, beliefs, and social relations that govern his behavior in the public sphere. *Political theory* is the study of the public sphere.

The private and public spheres are closely interrelated in individual decision making. A public sphere transaction may have private sphere costs and benefits that a participant in the public sphere may take into account in deciding how to act. For instance, an individual may not vote if queues at the polling station are very long, or may decide to skip a collective action in which the probability of physical harm is very high.

By contrast with the private sphere, critically important public sphere choices are fundamentally *nonconsequential*: an agent's public sphere decisions have no individual payoffs and no discernible effect on social outcomes. Consider, for instance, voting. Estimates of the probability that a single voter's decision will determine the outcome of a large election are between one in ten million and one in one hundred million (Chamberlain and Rothschild 1981; Fischer 1999; Gelman et al. 1998; Good and Mayer 1975). In a compendium of close election results in Canada, Great Britain, Australia, and the United States, no election in which more than 40,000 votes were cast has ever been decided by a single vote. In the Massachusetts gubernatorial election of 1839, Marcus Morton won by two votes out of 102,066 votes cast. In the Winchester, UK, general election of 1997, Mike Oaten won by two votes out of 62,054 votes cast. The result was annulled and in a later by-election, Oaten won by 21,000 votes. In smaller elections, a victory by a very small margin is routinely followed by a recount where the margin is rarely less that twenty-five (Wikipedia, List of Close Election Results, November 2014).

There is thus virtually no loss in accuracy in modeling voting behavior in large elections as *purely nonconsequential* in the sense that a single individual's decision to vote or abstain, or for whom to vote, has no effect on the outcome of the election (Downs 1957a; Riker and Ordeshook 1968).

By a *canonical participant* in a decision process I mean an individual whose choice is nonconsequential: his behavior affects the outcomes infinitesimally or not at all. According to the data presented above, voters in a large election are canonical participants. Individuals who participate in large collective actions are similarly canonical participants. Of course, there are some public sphere activities that are non-canonical, such as running for office, organizing a voter registration drive, or contributing considerable amounts of money to a particular party or candidate. But voters in the public sphere are canonical. Ignoring the infinitesimal probabilities that canonical participants affect outcomes is a useful and harmless simplification, akin to ignoring the force of gravity in analyzing the electronic circuitry of a computer or ignoring the light from distant stars in calculating the effectiveness of a solar panel.

Canonical public sphere activities are at the center of the structure and dynamics of modern societies. If citizens did not vote, or voted in an uninformed or random manner, liberal democratic societies could not function. Moreover, modern liberal democracy was achieved through collective actions in thwarting the autocratic ambitions of despotic regimes over centuries. These collective actions have been successful because of the cumulative impact of canonical participants who incurred significant personal costs, often death, in opposing arbitrary authority (Tilly 1981; Bowles and Gintis 1986).

Canonical participants consider their behavior as rational goal-oriented behavior. When questioning someone in a queue at the polling booth as to why he is standing there, or when questioning someone in a group protesting political corruption why he is chanting and holding a sign, he will think the question absurd. He will reply that he is there, of course, to support various candidates for office, or to help topple a hated regime. After pointing out that his personal contribution will make no difference to the outcome, he might rightly respond that this reasoning is faulty because if everyone followed it, no one would vote and no one would fight to topple a hated regime. After persisting in asking why he *personally* votes, noting that the other participants do *not* follow this reasoning, and his abstention will *not* affect the decision of others, he may well judge this thinking process bizarre and illogical, precisely because accepting the same sort of reasoning would lead virtually all citizens to abstain from voting.

The classical axioms of rational choice theory cannot explain the behavior of a canonical participant in the public sphere because these axioms cover only situations in which meaningful choices are *consequential* in the sense of leading to distinct entries in the agent's preference function (von Neumann and Morgenstern 1944; Savage 1954). The behavior of canonical participants in the public sphere is thus not classically rational. We shall, however, argue that canonical participants are *socially rational* in an analytically clear sense.

3.2   Private and Public *Persona*

Rational actors exhibit three types of motives in their daily lives: *self-regarding*, *other-regarding*, and *universalist*. Self-regarding motives include seeking personal wealth, consumption, leisure, social reputation, status, esteem, and other markers of personal advantage. Other-regarding motives include valuing reciprocity and fairness, and contributing to the well-being of others. Universalist motives are those that are followed for their own sake rather than for their effects. Chief among universalist goals are *character virtues*, including honesty, loyalty, courage, trustworthiness, and considerateness. Of course, in the private sphere such universalist goals have consequences for those with whom one interacts, and for society as a whole. But one undertakes universalist actions *for their own sake*, beyond any consideration of their effects.

Agents will generally trade off among these various motives. For instance, being honest may be personally costly or reputationally rewarding, and may either hurt or benefit others whose well-being one values. Universalist motives thus do not reduce to self- or other-regarding motives, but they do trade off against these other motives. Rational choice is important because it allows us to model these trade-offs elegantly and insightfully.

The individual immersed in consequentialist everyday life expresses his *private persona*, while his behavior in the public sphere reveals his *public persona*. Individuals acting in the public sphere are then a different sort of animal, one which Aristotle called *zoon politikon* in his *Nicomachean Ethics* (2002[350BC]). The concept of a nonconsequentialist *public persona* suggests a two by three categorization of human motivations, as presented in Figure 3.1. In this figure, the three columns represent three modes of social interaction. The *self-regarding* mode represents the individual whose social behavior is purely instrumental to meeting his personal material comfort, while the *other-regarding* represents the individual who is embedded in a network of significant social interactions with valued others, and the *universalist* represents the individual who values moral behavior for its own sake. The two rows represent the agent's *private persona* of social relations in civil society, and the agent's *public persona* of political relationships in the public sphere.

	Self-regarding	Other-regarding	Universalist
Private *persona*	*Homo economicus*	*Homo socialis*	*Homo vertus*
Public *persona*		*Homo parochialis*	*Homo universalis*

Figure 3.1. A typology of human motivations

Homo economicus is the venerable rational selfish maximizer of traditional economic theory, *Homo socialis* is the other-regarding agent who cares about fairness, reciprocity, and the well-being of others, and *Homo vertus* is the Aristotelian bearer of non-instrumental character virtues and the Kantian follower of the categorical imperative. The new types of public *persona* are *Homo parochialis*, who votes and engages in collective action reflecting the narrow interests of the demographic, ethnic, and/or social status groups with which he identifies. There is strong evidence, in fact, that voters are generally swayed not so much by their personal experience, say with unemployment, but rather by the experiences of members, of the social networks within which they are embedded and with which they identify (Markus 1988; Abrams et al. 2011). Thus *Homo parochialis* is common in political life.

Finally, *Homo universalis* acts politically to achieve what he considers the best state for the larger society, perhaps reflecting John Rawls' (1971) *veil of ignorance*, John Harsanyi's (1977) *criterion of universality*, or John Roemer's (2015) Kantian equilibrium.

Non-canonical agents acting in the political sphere, for instance politicians, are properly located in the private *persona* row. The individual whose private *persona* is other-regarding is generally considered altruistic, whereas the individual whose public *persona* is other-regarding (*Homo parochialis*) is often considered selfish, acting in a partisan manner on behalf of the specific interests of the social networks to which he belongs. In terms of our typology, however, *Homo parochialis* is in fact altruistic, sacrificing on behalf of the interests of members of his social entourage.

3.3   Social Rationality

Several economists, decision theorists, and philosophers have explored a more socially relevant form of rationality than those embodied in the classical axioms of von Neumann and Morgenstern (1944) and Savage (1954). They term these forms variously "we-reasoning," "team reasoning," and "collective intentionality" (Bacharach 1987, 1992, 2006; Gilbert 1987, 1989; Searle 1995; Tuomela 1995; Bratman 1993; Hurley 2002; Sugden 2003; Bacharach et al. 2006; Colman et al. 2008). I will present several analytically clear examples of such choice behaviors that should appear in any plausible account of social rationality. There are many more subtle but equally or more weighty that are worthy of exploration.

The basic concept of social rationality I will use is that of a *Kantian equilibrium* (Fedderson 2004; Roemer 2015), which is an alternative to the standard Nash equilibrium of game theory. Fedderson (2004) defines a Kantian equilibrium in a symmetric n-player game as a strategy that all players would prefer "if everyone who shares their preferences were to act according to the same rule." Consider, for instance, the two-player symmetric pure coordination game depicted in Figure 3.2. This game has two pure-strategy Nash equilibria: {Up,Up} and {Down,Down}.

Alice

		Up	Down
Bob	Up	2,2	0,0
	Down	0,0	1,1

Figure 3.2. An elementary coordination game

There is also a mixed-strategy equilibrium with payoffs (2/3,2/3) in which both players choose Up with probability 1/3. There is no principle of classical rationality that would lead the players to coordinate on the higher-payoff, or any other, Nash equilibrium. Yet {Up,Up} is the obvious socially rational choice. This is also obviously the Kantian equilibrium of the game, and it has the added attraction that it is a Nash equilibrium, so even a self-regarding agent would happily play his part in this Kantian equilibrium.

Alice

		Up	Down
Bob	Up	0,0	1,0
	Down	0,0	1,2

Figure 3.3. An elementary game where {Down,Down} is socially rational

Figure 3.3 is another example where there is a clear socially rational choice, the {Down,Down} Nash equilibrium, which is just one of an infinite number of Nash equilibria of the game. There is no reason using classical rationality to choose this over any other move for Bob. This game is not obviously symmetric, but we can symmetrize it by viewing a strategy for each player as a choice of move depending upon whether they are assigned the Bob or the Alice position in the game. In this case, the only Kantian equilibrium is the socially rational Nash equilibrium. This equilibrium is also Nash, so neither player can gain by deviating from it.

Consider the slightly more sophisticated game depicted in Figure 3.4. In this case Down is the socially rational choice for Bob, and Down, while no longer a dominant strategy, is the preferred choice for Alice provided she knows that Bob is socially rational.

| | | Alice | |
		Up	Down
Bob	Up	0, 0	1, −3
	Down	0, 0	1, 2

Figure 3.4. A game where the socially rational strategies are chosen when there is mutual knowledge of Kantian commitment

A Kantian equilibrium can also require a costly moral commitment, as in the prisoner's dilemma shown in Figure 3.5. In this case there is a unique Nash equilibrium in which both players Defect and get (1,1), whereas if they cooperated and each played Coop, they would get (2,2). In this case, the Kantian equilibrium {Coop,Coop} is efficient and may be considered socially rational, but it is not a Nash equilibrium. Therefore playing this equilibrium is personally costly to both players, so each has a self-regarding interest in abandoning his part in the Kantian equilibrium.

| | | Alice | |
		Coop	Defect
Bob	Coop	2,2	0,3
	Defect	3,0	1,1

Figure 3.5. The Prisoner's Dilemma

In fact, experimental subjects often do play the Kantian equilibrium in this case. For instance, Kiyonari et al. (2000) ran an experiment with real monetary payoffs using 149 Japanese university students. The experimenters ran three distinct treatments, with about equal numbers of the subjects in each treatment. The first treatment was the standard "simultaneous" prisoner's dilemma of Figure 3.5, the second was a "second player" situation in which the subject was told that the first player in the prisoner's dilemma had already chosen Coop, and the third was a "first player" treatment in which the subject was told that his decision to play Coop or Defect would be made known to the second player before the latter made his own choice. The experimenters found that 38% of subjects cooperated in the simultaneous treatment, 62% cooperated in the second player treatment, and 59% cooperated in the first player treatment. The decision to cooperate in each treatment cost the subject about $5.00. This shows unambiguously that a majority of subjects chose the Kantian strategy when they knew their partner would do the same, and almost as many not only preferred the Kantian equilibrium but were willing to bet that their partners would be Kantians as well (59%). Under standard conditions, without this assurance, only 38% played the Kantian equilibrium strategy.

Brian Skyrms (2004) and Michael Tomasello (2014) have suggested a notion closely related to socially rational cooperation, which Tomasello calls *collaboration*. The idea is inspired by Rousseau's famous stag hunt game related in his *Discourse on Inequality* (1984). Consider n hunters who, if they all cooperate, can hunt stag with expected payoff 2 each, but if at least one of them chooses to go off alone and hunt rabbit, the lone hunters each have expected payoff 1, and those who hunt stag have payoff 0. Clearly this is a coordination game in which rationality and mutual belief in socially rational cooperation implies that all hunters go after the stag.

Skyrms (2004) and Tomasello (2014) argue that this coordination game characterizes the situation facing early hominins, and that this accounts for our success as a species. This reasoning leads these authors to assert that self-regarding collaboration in a coordination game rather than altruistic cooperation in a public goods game (Sober and Wilson 1998; Wilson and Wilson 2007; Bowles and Gintis 2011) accounts for our success as a species. This argument posits an implausible model of social coordination in which a single defector completely undermines the cooperative effort. In most cooperative endeavors involving several participants, and certainly in the cooperative hunting practiced by our hunter-gatherer forbears (see Chapter 2 and Whiten and Erdal 2012), any single hunter could defect with a high probability of not being observed free-riding, yet the hunt would still be sufficiently productive to render defection profitable. But in this case, cooperation cannot be sustained at all, because each hunter, if self-regarding, will defect.

The following is a more plausible model of collaborative rationality that combines the insights from the stag hunt game and the public goods game. I will present a two-player version of the game, but it will be clear how this might be extended to an *n*-player version. In Figure 3.6, a player can defect (D) hunting rabbit with payoff 2, or can collaborate with high (CH) or low (CL) energy.

	CH	CL	D
CH	4,4	1,5	−1, 2
CL	5,1	1,1	0,2
D	2, −1	2,0	2,2

Figure 3.6. Collaboration with Kantian commitment

If both players collaborate and play CH, they each earn 4, but each collaborator has an incentive to free-ride, playing CL and earning 5, while his CH partner then earns only 1. Thus the subgame involving only CH and CL is a prisoner's dilemma in which self-regarding agents will defect. Thus collaboration with self-regarding agents is less productive than uncooperative rabbit-hunting. If both players place a high moral commitment to collaboration, the CH/CH strategy profile will become a Nash equilibrium. This represents a form of non-self-regarding social rationality that helps explain our success as a species.

3.4   The Social Rationality of Voter Turnout

Socially rationality in the public sphere generally complements rather than contradicts classical rationality. The assertion that electoral behavior is rational is most clearly supported by the phenomenon of *strategic voting* (Cox 1994, 1997; Franklin et al. 1994). Strategic voting includes ignoring candidates that have no hope of winning, or voting for an unwanted candidate in order to avoid electing an even less preferred candidate (Cox 1994; Niemi et al. 1992). It also includes *Duverger's Law* (Duverger 1972), which asserts that plurality rule elections tend to favor a two-party system, whereas a double ballot majority system and proportional representation tend to multipartism. Voting also has a strong social element, including a rather ubiquitous *social network effect*: individuals who are more solidly embedded in strong social networks tend to vote at a higher rate (Edlin et al. 2007; Evren 2012), and they are likely to vote the interests of the social networks to which they belong (Weeden and Kurzban 2014).

To explore the nature of the social rationality of canonical participant behavior, I will follow Ledyard (1981) and Palfrey and Rosenthal (1985), who developed the currently most widely recognized, and I think most plausible, model of rational pivotal voter behavior. Suppose an election is determined by majority rule, with a tie vote decided by a fair coin toss. There are m agents who choose simultaneously to vote for Alternative 1, vote for Alternative 2, or abstain. Alternative 1 is preferred by $m_1 > 0$ voters and Alternative 2 is preferred by $m_2 = m - m_1 \geq 0$ voters. We assume $m_1 \geq m_2$. If b_i is the payoff to agent i who supports Alternative j if his alternative wins, with payoff zero if his alternative loses, and if c_i is the cost of voting as opposed to abstaining for agent i, then i will vote rather than abstain if

$$b_i \left(p_2^j + \frac{p_1^j}{2} \right) \geq c_i \tag{3.2}$$

where p_1^j is the probability that a supporter of Alternative j's vote leads to a tie, and p_2^j is the probability that this supporter's vote breaks a tie. Defining the pivotal probability

$$p^j = p_2^j + p_1^j/2 \tag{3.3}$$

and the *net cost of voting* $\gamma_i = c_i/b_i$ for agent i, (3.4) simplifies to

$$p^j \geq \gamma_i. \tag{3.4}$$

Note that (3.4) is a slight generalization of (3.1). It is clear that agent i will vote precisely when γ_i is below a threshold $\hat{\gamma}_i = p^j$. Recall that b_i need not be determined by pure self-interest, but may be affected by altruistic or spiteful attitudes towards others, as well as by purely ethical considerations concerning justice and equity. Similarly, c_i may include moral as well as self-interest motives, such as the citizen's duty to vote, signaling one's status as a good citizen, and garnering the good will of social network members.

Suppose $F_j(\cdot)$ is the cumulative distribution of γ_i for supporters of Alternative j. Then the probability q_j, $j = 1, 2$, that a supporter of Alternative j will vote is given by

$$q_j = F_j(p^j) \qquad \text{for } j = 1, 2, \tag{3.5}$$

where p^j is defined in equation (3.3). We assume for simplicity that the net cost of voting γ_i is uniformly distributed on the intervals $[0, b^1/c^1]$ and $[0, b^2/c^2]$, respectively.

Suppose agent i supports Alternative 1. The probability $p_1(k_1, k_2)$ that k_j votes are cast for Alternative j, $j = 1, 2$, not including the vote of agent i (should he vote), is given by

$$p_1(k_1, k_2) = \binom{m_1 - 1}{k_1}\binom{m_2}{k_2} q_1^{k_1} q_2^{k_2} (1-q_1)^{m_1-k_1-1}(1-q_2)^{m_2-k_2}. \quad (3.6)$$

The probability that Alternative 1 wins when agent i votes is then

$$\sum_{k_i=1}^{m_1-1} \sum_{k_2=0}^{k_1-1} p_1(k_1, k_2) + \sum_{k_1=0}^{m_2} p_1(k_1, k_1)$$

$$+ \frac{1}{2} \sum_{k_1=0}^{m_2-1} p_1(k_1, k_1 + 1). \quad (3.7)$$

The first (double) summation in (3.7) is the probability i's alternative wins but i is not pivotal, the second summation is the probability that i breaks a tie, and the final summation is the probability i creates a tie and the coin flip favors Alternative 1. The probability that Alternative 1 wins when this agent i abstains is similarly

$$\sum_{k_i=1}^{m_1-1} \sum_{k_2=0}^{k_1-1} p_1(k_1, k_2) + \frac{1}{2} \sum_{k_1=0}^{m_2} p_1(k_1, k_1). \quad (3.8)$$

Subtracting (3.8) from (3.7) we see that the probability of an Alternative 1 supporter changing the outcome of the election by voting rather than abstaining is given by

$$\pi_1(m_1, m_2) = \frac{1}{2}\left(\sum_{k_1=0}^{m_2} p_1(k_1, k_1) + \sum_{k_1=0}^{m_2-1} p_1(k_1, k_1 + 1) \right). \quad (3.9)$$

Therefore the probability q_1 that an Alternative 1 supporter votes is given by

$$q_1 = F_1(\pi_1(m_1, m_2)). \quad (3.10)$$

Repeating this argument for the case where agent i prefers Alternative 2, we have

$$\pi_2(m_1, m_2) = \frac{1}{2}\left(\sum_{k_1=0}^{m_2} p_i(k_1, k_1) + \sum_{k_1=0}^{m_2} p_i(k_1+1, k_1)\right). \quad (3.11)$$

Therefore the probability q_2 that an Alternative 2 supporter votes is given by

$$q_2 = F_2(\pi_2(m_1, m_2)). \quad (3.12)$$

Equations (3.9) and (3.11) jointly determine a Nash equilibrium of the voting game in which the probabilities $p_1(m_1, m_2)$ and $p_2(m_1, m_2)$ are mutually determined by the stipulation that Alternative 1 supporter i votes precisely when the expected gain $b_i p_1(m_1, m_2)$ exceeds his private cost of voting c_i.

This Nash equilibrium predicts the major empirical regularities of voter turnout, both in small-scale real-world and in laboratory (Levine and Palfrey 2007) elections. Rather than deriving the comparative statics of this voter turnout model analytically, which may or may not be possible, I have generated numerical solutions using the Mathematica software program (Wolfram Research 2014). These solutions assume the cumulative distribution $F_j(\cdot)$ of γ_i for $j = 1, 2$ is the uniform distribution on the unit interval.

The first regularity is the *electoral size effect*: voter turnout declines with increasing size of the electorate (Lijphart 1997). For instance, national elections have larger turnout rates than local elections. Figure 3.7 illustrates this phenomenon in our model with electorate sizes of 18 to 36, where the difference between the majority and minority voters is two. Similar results hold, however, for higher ratios of majority and minority voters. Note that these are clearly small election results because voter turnout declines rapidly towards zero for even moderately large electorate sizes.

The second and third electoral regularities are the *voting cost effect* and the *importance of election effect*: when the cost of voting increases, fewer people vote, and when the stakes are higher, more people vote. The reason given by our model is that an increase in the net cost of voting, *ceteris paribus*, entails a higher cutoff γ_j^*, and a higher benefit to a voter from winning lowers the net cost of voting cutoff γ_j^*. We can represent both of these effects, assuming they are experienced equally by all agents, by replacing

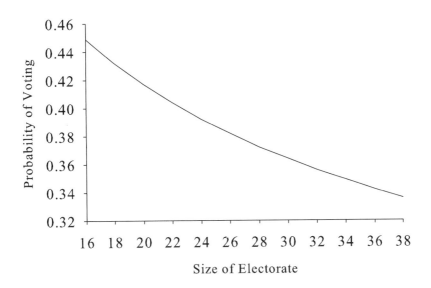

Figure 3.7. The Electoral Size Effect: Voter turnout rates decline as electorate size increases.

γ_i by $\alpha\gamma_i$ in equation (3.4), where $\alpha < 1$ corresponds to a decreased cost-benefit ratio. The results for an electorate of size 20 where the majority consists of between 11 and 20 voters is given in Figure 3.8. Note that there is positive turnout even when Alternative 2 has no supporters. This is because all agents want to avoid having the outcome determined by a coin flip.

The fourth electoral regularity is the *competition effect*: turnout is higher when the election is expected to be close (Shachar and Nalebuff 1999). The reason in our model is that the probability that there will be a pivotal voter is higher when the election is expected to be close. It should be noted that this appears to be true, as illustrated in Figure 3.9, but it is not obvious why this is the case.

3.5   The Logic of Distributed Effectivity

Comparative static voter turnout phenomena, verified analytically for a very small electoral size in the previous section, but reflecting voter behavior in large elections, lend support to the notion that *people behave in large elec-*

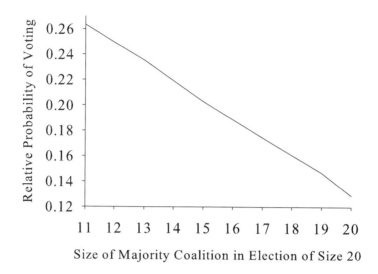

Figure 3.8. The Voting Cost and Importance of Election Effects: In an electorate of size 20, a higher cost to benefit ratio leads to lower turnout for all sizes of the majority faction. The vertical axis shows the percentage increase in turnout when the net cost of voting is lowered by one third. The horizontal axis shows the size of the majority coalition.

tions rationally and strategically as though they were actually involved in very small elections. People appear to follow a logic that may be described as *distributed effectivity*: in canonical public life, maximize utility assuming your probability of your having a pivotal effect on the outcome is high (Levine and Palfrey 2007).

I have stressed that distributed effectivity is rational and is so considered by those who embrace it. Acting on the false belief that large elections are small elections is, however, clearly substantively irrational. Moreover, it is not true that voters generally believe they are likely to be pivotal. The following is a more plausible analytical representation of distributed effectivity.

Suppose there are two candidates, $j = 1, 2$, and a relatively small number of *voter types* $\{s_{j1}, \ldots, s_{jm_j}\}$ for $j = 1, 2$ such that all voters of type s_{jk} prefer candidate j, have similar political philosophies and similar voting

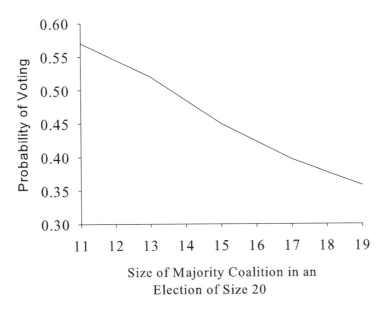

Figure 3.9. The Competition Effect: The closer in size the majority and minority coalitions, the higher the probability of voting. The election size is twenty.

costs. We can illustrate how distributed effectivity with a large electorate can lead agents to behave as though the electorate is small with a simple example. Suppose there are v members of each voter type, and voters of type s_{jk} share a common payoff from winning b_{jk} and cost of voting c_{jk}. Suppose the net cost of voting $\gamma_{jk} = c_{jk}/b_{jk}$ is uniformly distributed on the unit interval. Then each of the members of s_{jk} votes for candidate j provided the probability that the v votes of type s_{jk} voters is pivotal is greater than γ_{jk}. Then if we simply assume that each voter of each type contributes $1/v$ votes to the tally when voting for his preferred alternative, we obtain the same model as described in Section 3.4, but with an electorate of size m/v rather than m. This is of course simply the Kantian equilibrium behavior.

For a more detailed model of distributed effectivity, assume that voters of a given type s_{jk} accept a common net cost of voting cutoff γ_{jk} such that all members i of s_{jk} for whom $\gamma_i = c_i/b_i \leq \gamma_{jk}$ vote provided that γ_{jk} is less than the probability that type s_{ji} voters are pivotal for the election.

Suppose the frequency of type s_{jk} in the supporters of candidate j is α_{jk}, so the number of members of s_{jk} is $\alpha_{jk}m_j$. Consider a single voter type s_{1k}. Let q_{1k} be the fraction of type s_{1k} agents who vote, provided any type s_{1k} agents vote. Thus $v = \alpha_{1k}m_1q_{1k}$ type s_{1k} agents vote if any such agent votes.

The probability that candidate 1 wins if s_{1k} members vote is given by

$$\sum\{p_1(k_1,k_2)|k_1 = 1,\ldots,m_1 - v, k_2 = 0,\ldots,k_1 - 1\}$$
$$+ \frac{1}{2}\sum\{p_1(k_1,k_1 + v)|k_1 = 0,\ldots,m_2 - v\}$$
$$+ \sum\{p_1(k_1,k_2)|k_1 = 0,\ldots,m_1 - v, k_2 = 0,\ldots,k_1 + v - 1\}.$$

$$(3.13)$$

The first term represents the probability of winning without the s_{1k} votes, the second term is the probability of winning if there is a tie including the s_{1k} votes, and the third term is the probability that candidate 1 wins including the s_{1k} votes. If the s_{1k} types abstain, the probability of candidate 1 winning is given by

$$\sum\{p_1(k_1,k_2)|k_1 = 1,\ldots,m_1 - v, k_2 = 0,\ldots,k_1 - 1\}$$
$$+ \frac{1}{2}\sum\{p_1(k_1,k_1)|k_1 = 1,\ldots,m_2 - v\}.$$

$$(3.14)$$

The difference between (3.13) and (3.14) is the probability that voting will turn a defeat into a victory for candidate 1. In this difference, the probabilities of a tie are infinitesimal and can be ignored. The difference thus becomes

$$\pi_1 = \sum\{p_1(k_1,k_2)|k_1 = 0,\ldots,m_1-v, k_2 = 0,\ldots,k_1+v-1\}, \quad (3.15)$$

where we define $p_1(k_1,k_2) = 0$ for $k_2 > m_2$. The fraction of type s_{jk} members who vote is given by $q_{jk} = F_j(\gamma_{jk})$, and we have for $j = 1, 2$

$$q_j = \sum_{k=0}^{m_j}\alpha_{jk}q_{jk} = \sum_{k=0}^{m_j}\alpha_{jk}F_j(\gamma_{jk}). \quad (3.16)$$

Now (3.16) and (3.6) complete the specification of (3.15). Note that π_1 in (3.15) is equal to one for v sufficiently large. This analysis shows that distributed effectivity can explain high voter turnout even in large elections, where classical rationality implies extremely low turnout.

The importance of distributed effectivity is difficult to overstate. The character of our species as *Homo ludens* emerged from an extended evolutionary dynamic during which, until very recently, humans lived in small hunter-gatherer bands in which all political activity was doubtless consequential (see Chapter 2). In such settings the logic of distributed effectivity might well differ from classical rationality with self-interest supplemented by other-regarding and universalist preferences (see Section 3.2) in relatively minor ways. Even the rise of settled trade and agriculture some 10,000 years ago, followed by the appearance of states and empires, might well have proceeded with little need for so strong a notion of social rationality as that embodied in distributed effectivity. But the collective actions that overthrew despotic authorities and augured the emergence of democratic political orders dedicated to the rule of law and the protection of individual freedoms lie completely outside the range of classical rational choice theory. Distributed effectivity, nurtured in the formative years of our species' history, made the modern world possible.

We can confirm socially rational reasoning not only through the regularities of social behavior, but by the testimony of social actors themselves, for instance the conversation described above with a voter at the polling booth. By contrast, rule-consequentialism is a complex philosophical theory that is foreign to the minds of most canonical participants.

3.6   Situating Distributed Effectivity

The most obvious alternative to distributed effectivity is that canonical participants *believe* their actions are consequential even when they are not (Quattrone and Tversky 1988), so they act *as though* their actions determine outcomes with substantial probability. This is the most common, though rarely explicitly stated, assumption in the political science literature. For instance, Duncan Black's famous median voter theorem (Black 1948) implicitly assumes that a self-interested citizen will vote and this vote will register his personal preferences. Similarly, Anthony Downs, a pioneer in the application of the rational actor model to political behavior (Downs 1957a) describes his model as follows:

> Every agent in the model—whether an individual, a party or a private coalition, behaves rationally at all times; that is, it pro-

ceeds toward its goals with a minimal use of scarce resources and undertakes only those actions for which marginal return exceeds marginal cost. (Downs 1957b, p. 137)

And yet, almost immediately after stating this assumption, he writes:

[We assume that] voters actually vote according to (a) changes in their utility incomes from government activity and (b) the alternatives offered by the opposition (Downs 1957b, p. 138).

These two assumptions are compatible with classical rationality only if agents believe that their votes are consequential.

But in fact canonical participants generally do *not* believe that their behavior is consequential. For instance Enos and Fowler (2010) report a study in which the median respondent to the question as to the chance their vote will change the outcome of a presidential election gave the answer 1 in 1000, which, although small, is in fact too large by a factor of at least 10,000. The authors write:

However... over 40% of regular voters know that the chances of a pivotal vote are less than one in a million.... [Moreover], the less likely you are to think your vote will actually matter, the more likely you are to vote.

An alternative is that people consider voting a *social obligation* (Ali and Lin 2013; Fedderson and Sandroni 2006; Li and Majumdar 2010; Riker and Ordeshook 1968). Abstaining in this view is an unethical act of free-riding on the altruism of others. Indeed, in an American survey, when asked if the good citizen must always vote, the level of agreement is just slightly lower than obeying the law and paying taxes (Dalton 2008). In an Annenberg study of the 2000 election, 71% of Americans agreed that they felt guilty when they failed to vote. Even among those who reported that they had not voted, nearly half said they felt guilty. Blais (2000) reports that more than 90% of respondents in two Canadian provinces agree that "it is the duty of every citizen to vote." Clarke et al. (2004) present similar findings for British voters.

However, the duty to vote theory cannot explain the observed regularities of voting. Duty cannot explain strategic voting, or the size of election, competition, or importance of election effects described in Section 3.4. Duty can plausibly explain the social network effect and the voter cost effect, but the other indications of voter rationality make no sense when agents know that their actions are nonconsequential. Moreover, to the extent that voters are motivated by duty concerns, which some clearly are, they can be included in the cost of voting variable c_i.

Another theory is that canonical participants are *altruistic*, voting out of concern for the well-being of others who will be affected by the outcome of the electoral process (Edlin et al. 2007; Evren 2012; Faravelli and Walsh 2011; Fowler 2006; Fowler and Kam 2007). Even if voting is only infinites-imally consequential, when the election impacts millions of individuals, the extremely low probability of being a pivotal voter multiplied by the number of people thereby affected may become a large number. Formally, altruism can be expressed as a very large benefit b_i to agent i from winning, which is equivalent to a very small net cost of voting γ_i. Were the altruism assump-tion plausible, then our comparative static results could be reproduced even with large electorate size. However, it is implausible that large numbers of canonical participants act from a charity motive. Many canonical partici-pants have interests that are far narrower than the citizenry as a whole, and often act to promote the interests of one small group of citizens at the ex-pense of society as a whole. Indeed, it is common to hear a small group of voters deemed "selfish" because they promote their own parochial interest above the good of society. Perhaps more telling, the altruism model cannot explain the strategic rationality of voters and the comparative static results reported in Section 3.4. Moreover, the altruism aspect of voting, to the extent that it exists, is incorporated directly into the distributed effectivity model.

A related alternative is that voters *seek approval from their social network members* (Aytimur et al. 2014; Ben-Bassat and Dahan 2012; Fosco et al. 2011; Gerber et al. 2008; Gerber and Rogers 2009; Harbaugh 1996; Knack 1992). Like the altruism theory, this is likely to be minimally true, but people do not generally much care whether or not their colleagues, relatives, or neighbors vote. Moreover, even should voting send a desirable signal to others, there would be no reason to vote strategically. Nor would the observed comparative static results follow unless rather *ad hoc* assumptions concerning social approval are deployed. Finally, the approval effect can be incorporated in the cost of voting variable c_i.

A final alternative to distributed effectivity is *expressive theory*, according to which canonical participants abandon instrumental rationality in favor of expressive actions from which they derive direct utility (Brennan and Lomasky 1993; Hamlin and Jennings 2011; Rotemberg 2009; Schuessler

2000; Sears et al. 1980). Expressive models explain many of the key social aspects of canonical behavior in the public sphere, including the social network effect and the responsiveness of agents to exhortation by activists. But they do not explain why people consider participating a prosocial act and feel guilty having failed to participate. They also fail to explain why people are rewarded with social approval when they participate. Finally, they explain none of the rational behavior described in Section 3.4. Most important, they do not explain strategic voting without invoking *ad hoc* preferences.

4

Power and Trust in Competitive Markets

> In a perfectly competitive market it really doesn't matter who hires whom; so let labor hire capital.
>
> Paul Samuelson

> An economic transaction is a solved political problem. Economics has gained the title Queen of the Social Sciences by choosing solved political problems as its domain.
>
> Abba Lerner

> **C:** Who has power in a market economy?
> **HG:** The agent on the money side of a transaction.
> **C:** Why is that?
> **HG:** Because money is the third-party enforceable side of the transaction.
>
> Choreographer interview

Léon Walras was the nineteenth century Swiss creator of the model of general market exchange that culminated in the famous neoclassical general equilibrium model (Arrow and Debreu 1954). Walras defined the pure science to which he aspired as the study of *relationships among things, not people* and sought, with considerable success, to eliminate human relationships from his purview. His device for accomplishing this was the notion that interactions among economic agents might be represented as if they were relationships among *inputs* and *outputs*. Walras (1874, p. 225) writes:

> Assuming equilibrium, we may even go so far as to abstract
> from entrepreneurs and simply consider the productive services
> as being, in a certain sense, exchanged directly for one another.

Modern neoclassical textbook models of market exchange have wholly embraced Walras. James Buchanan, for instance, describes the anonymity of the market and the uncontested nature of claims in standard theory by reference to "a roadside stand outside Blacksburg," writing (1975, p. 17):

> I do not know the fruit salesman personally, and I have no
> particular interest in his well-being. He reciprocates this at-

titude... Yet the two of us are able to... transact exchanges efficiently because both parties agree on the property rights relevant to them.

A parallel logic suggests that while the modern enterprise appears to be a system of hierarchical authority that might be analyzed in political or sociological terms as a system of power, it is in fact nothing of the kind. Armen Alchian and Harold Demsetz capture this perspective in observing that (1972, p. 177)

> [the firm] has no power of *fiat*, no authority, no disciplinary action any different in the slightest degree from ordinary market contracting between any two people... [The firm] can fire or sue, just as I can fire my grocer by stopping purchases from him, or sue him for delivering faulty products.

Indeed, there is nothing in a model of general market exchange suggesting that owners and managers of the firm have any power over the firm's employees. As Paul Samuelson has noted (1957, p. 894),

> in a perfectly competitive market it really doesn't matter who hires whom; so let labor hire capital.

The result, expressed long ago by Joseph Schumpeter, is a decentralization of power to consumers (1934[1911], p. 21):

> The people who direct business firms only execute what is prescribed for them by wants.

These views taken together imply the apolitical conception of the economy described in a head quote to this chapter by Abba Lerner (1972, p. 259):

> An economic transaction is a solved political problem. Economics has gained the title of queen of the social sciences by choosing *solved political problems* as its domain.

4.1   The Short-Side Power Principle

Walras' reduction of relationships among people to relationships among things is indeed a powerful modeling device. We follow Walras in this respect in Chapter 6, although we radically violate it in Chapter 11, where we address market dynamics. However, the general equilibrium model cannot account for several key characteristics of the market economy. Most important, owners of capital and their representatives almost always control the firm. This is certainly not predicted by person-to-things reduction analysis. Moreover, trust and integrity are often central to efficient real-world transactions, but do not appear in the general equilibrium model.

Buchanan's insights concerning his grocer in Blacksburg may be correct, but the same cannot be said for Samuelson's assertion that labor could hire capital as easily as the other way around, or for Alchian and Demsetz' argument that the employer's relationship to his employee is no different from my relationship to my grocer, and certainly not for Lerner's assertion that there are no relations of power in the competitive economy. These arguments are correct for a Walrasian economy, but not for a real economy, where critically important exchanges are not guaranteed by a third party at no cost to the exchanging parties.

A second widely observed regularity is that while in general economic equilibrium supply equals demand for all goods and services, in real market economies some markets perpetually suffer either excess supply or excess demand. One is the labor market, where a positive and significant level of unemployment (excess supply) is normally observed. Another is the market for capital, in which there is generally excess demand. That is, at the on-going rate of interest, more people would like to borrow money than lenders are willing to lend. Indeed, some prospective borrowers cannot secure a loan at *any* interest rate. In effect, prospective borrowers line up and lenders inspect them carefully, choosing some and rejecting others.

Finally, consumer goods markets generally leave firms *quantity constrained*: firms would like to sell more at the going price than consumers are willing to buy at that price. Indeed, to the non-economist, success in business practically equates with success in selling more stuff than your competitors. In the general equilibrium model, by contrast, because markets clear, firms can sell as much as they please.

We can summarize these three cases by noting that in a real market economy employers have power over employees, lenders have power over borrowers, and consumers have power over their suppliers. This power in all three cases consists in the capacity to inflict losses on one's contracting partner by discontinuing the relationship (firing the worker, withdrawing credit from the borrower, switching to another supplier for the consumer).

Moreover, employees, borrowers, and consumer good firms are on the *long side* of a nonclearing market; i.e., on the side that would like to have more transactions at the prevailing price, but cannot find agents on the other side of the market with whom to transact. By contrast, employers, lenders, and consumers are on the *short side* of the market; i.e., on the side of the market for which the quantity desired is the lesser at the prevailing price, or equivalently, who can generally purchase more at the prevailing price if they so choose.

In each of these three examples, *money* is on one side of the transaction and a *promise* is on the other side. The employer pays a wage and the employee, rather than providing a contractually specified service, offers only a promise—the promise to work with due care and intensity on behalf of the employer. The lender transfers money to the borrower, and the borrower promises to take appropriate measures to ensure the money will be repaid. The consumer pays for a product and the producer promises that the product will be of good quality and serviceability. In each case, money is on the short side of a nonclearing market and a promise is on the other.

The key fact about such promises is that *they are not enforceable at reasonable cost by the exchanging parties.* Employers do not take their employees to court if their job performance is shoddy. Lenders cannot recoup their losses by suing a borrower who has defaulted if the borrower is bankrupt. And consumers rarely hire a lawyer when they are dissatisfied with a product they have bought at the supermarket or automobile dealership.

Let us call goods and services such as these, whose characteristics cannot be guaranteed by low-cost third-party enforcement (e.g., by the legal system), *variable quality goods*. Labor, capital, and consumer goods are all generally variable quality goods. We then have the *short-side power principle*:

> The market for a variable quality good does not clear. Power lies with the agents on the short side of the market, who trade money in exchange for the variable quality good.

In essence, the short-side agent *pays* for the power to enforce what he considers the proper contractual terms of the relationship.

The short-side power principle holds because where promises cannot be enforced by third parties, their being kept can only be secured by the integrity of the promiser, which we discuss in Section 4.3, or by the buyer, which requires that the buyer have the capacity to threaten the seller with harm if dissatisfied with his performance. We call this *endogenous enforcement*.

The most common form of endogenous enforcement is *contingent renewal*: the buyer promises to repeat the exchange periodically and indefinitely, provided he is satisfied with the quality of the product delivered. Thus the employer hires an employee for an indefinite number of work days, using the threat of dismissal to motivate the employee to work hard and with due care. Of course, this threat has force only if the employee has something to lose by being dismissed. This means the employer must pay the employee more than the employee could expect to get were he dismissed and were forced to seek alternative employment.

If all employers follow this strategy, motivating employees to supply high-quality work by offering them a wage that is sufficiently high that they cannot expect the equivalent elsewhere without incurring considerable search costs and suffering a period of unemployment, then it is clear that *full employment is impossible* (Bowles and Gintis 1993). Rather, a positive fraction of the labor force remains unemployed, and the threat of dismissal is therefore the threat to add the employee to the ranks of the unemployed for an extended period of time.

This explains the short-side power principle in the case of the labor market. Of course, the real world is a considerable elaboration of this principle. For instance, if the work is highly skilled, those frozen out of a position will generally be employed in a less-skilled job rather than being unemployed. Moreover, the employer can use the promise of advancement for satisfactory performance, or even seniority pay increases, to motivate performance while reinforcing the threat of dismissal.

In the case of the capital market, a similar contingent renewal strategy is often deployed by the lender, who offers the borrower a relatively small short-term loan with the right to keep tabs on the borrower's behavior, and renew the loan as long as he is satisfied with the borrower's behavior, especially in the area of risk exposure. Small business loans are often of this form. However, in the case of the capital market, the lender can also demand a certain level of *collateral* that is forfeited in case the loan is not repaid. The ubiquity of collateral requirements explains the otherwise rather perverse tendency of lenders to supply funds almost exclusively to the already wealthy. A third endogenous enforcement mechanism, applying mostly where loans are made to consumers, is a system of public information of credit history. Borrowers in this case are not sued for late payment or default, but their credit rating suffers, shutting off future borrowing opportunities.

The short-side power principle in the capital market holds because an increase in the interest rate, which is supposed to rise to a level that equates supply and demand, can *lower* rather than *raise* the profit of the lender. This is because an increase in the interest rate will drive away potential borrowers who have low-risk, low-return investments in mind, leaving only the high-risk, high-return investors. For instance, a high-risk, high-return borrower would be willing to pay a 100% interest rate to borrow $100 if his proposed investment pays $500 with probability 1/2 and zero otherwise. If he wins, he repays the lender $200 and keeps $300 for himself. If his investment were low-risk, low-return, say a 90% chance of returning $150, he loses money at a 100% interest rate. Knowing this situation, the lender will simply not make such high interest rate loans (Stiglitz and Weiss 1981).

The short-side power principle in the case of a consumer good, where it is infeasible for buyer and seller to write a third-party enforceable contract for the delivery of a product of a particular quality or serviceability, the buyer can implicitly promise to buy from the same supplier repeatedly so long as he remains satisfied with the quality of his previous purchases. The ubiquity of this implicit contract is reflected in the very term *customer*, implying a customary, habitually repeated interaction between buyer and seller. Of course, the customer's implicit threat to switch to a new supplier if dissatisfied is effective only if it is costly to the supplier. In the standard neoclassical general equilibrium model, this is not the case because the marginal cost of any single unit of the product is equal to its price, so the firm loses nothing when one of its buyers transacts elsewhere. Therefore the customer must be willing to pay a price *higher than marginal cost* for each transaction in order to render his threat to switch effective.

As in the case of the labor market, if all transactions in a variable quality market involve contingent renewal agreements, then even in market equilibrium price will exceed marginal cost for all firms, and hence each firm can increase its profits if it manages to attract a larger share of the market. This explains the ubiquity of advertising and marketing in modern economies, phenomena that make no sense in neoclassical general equilibrium. For a graphic example of consumer power in a market economy, observe that American cars were better than their Cold War era Russian equivalents because in Russia customers waited in line to purchase Volgas while in the United States, salesmen lined up to sell Fords.

In competitive neoclassical markets, firms are *price-takers* who have no incentive to undercut the market price and would lose all their sales if they raised their selling price by even a small amount. But in contingent renewal consumer goods markets, because buyers normally have somewhat heterogeneous preferences for product quality, a firm will generally behave like a monopolist with a downward sloping demand curve. Moreover, generally there will be idiosyncratic differences in product characteristics across firms that lead consumers to prefer one brand to another. For instance, all red wines of a certain quality may have the same price, but individual consumers may strongly prefer a small subset of offerings at that price. This condition leads to a less elastic firm demand curve and an increased gap between price and the firm's marginal cost, even further increasing the effectiveness of the customer's threat to switch to other suppliers.

4.2   Power in Competitive Markets

What is the meaning of the term "power" in the short-side power principle? Because of its close connection to value-laden words such as coercion and freedom the term "power" itself has proven to be controversial (Bachrach and Baratz 1962; Nozick 1969; Lukes 1974; Taylor 1976). Many political theorists regard sanctions as the defining characteristic of power. Lasswell and Kaplan (1950) make the use of "severe sanctions…to sustain a policy against opposition" a defining characteristic of a power relationship, and Parsons (1963) regards "the presumption of enforcement by negative sanctions in the case of recalcitrance" a necessary condition for the exercise of power. There is a standard and value-neutral definition of power that recognizes the centrality of sanctions. This definition is succinctly expressed by Robert Dahl (1957) (pp. 202–203):

> A has power over B to the extent that he can get B to do something that B would not otherwise do.

Transactions in standard economic theory are "solved political problems," to borrow from Abba Lerner's head quote to this chapter, because competition renders all market participants price-takers who never engage in personal interaction.

When the fruit market is perfectly competitive, James Buchanan and his fruit salesman at the roadside stand are faceless and anonymous to one another. Neither cares what the other one does because the quality of the fruit is perfectly known, one customer more or less makes no difference to the seller, and there are plenty of other fruit stands available to the buyer should he be displeased with some aspect of the transaction he now faces. Neither fruit salesman nor his client has any power in this situation. If we think of political phenomena as those involving the exercise of power, this transaction is truly a solved political problem.

In contingent renewal markets, whether they involve employers and employees, lenders and borrowers, or consumers and their suppliers, the agent on the money side of the transaction has power over his transacting agent precisely in Dahl's sense. Let us call the agent on the money side the *principal* and the agent on the other side of the exchange, offering only a promise, the *agent*. We can depict a contingent renewal transaction as one in which the principal offers monetary incentives to induce the agent to supply high quality services, with the threat of non-renewal in case the principal is dissatisfied. In this case, the principal has power over the agent in precisely the manner described by Dahl. The principal's threat of non-renewal is effective because if the customary principal-agent interaction is ended, and because there is a pool of available agents who do not have exchange partners (the unemployed workers, the credit constrained borrowers, and the quantity constrained consumer goods firms), the principal has no trouble finding a new agent, while the agent falls into the pool of unsatisfied agents in that market.

We say that the principal, who is on the money side of the transaction, has short-side *power* precisely because the conditions of contingent renewal contracts conform to Dahl's definition. Short-side power is political in a way that standard competitive exchanges are not. Having money in a standard neoclassical competitive market gives individuals power over *things* (goods and services) but not over *people* (suppliers of those goods and services).

4.3   Trust and Integrity

The simplicity and elegance of the neoclassical general equilibrium model, as we have seen, flows from the assumption that the agreed-upon particulars of a market exchange can be ensured by a contract enforceable by third parties at no cost to the transacting individuals. With this assumption, and abstracting from the presence of public goods, externalities, and increasing returns to scale industries, we can prove the so-called *First Theorem of Welfare Economics*: perfectly competitive market exchange leads to a

Pareto-efficient allocation, in which no alternative allocation could improve the payoff to one agent without reducing the payoff to at least one other agent.

But when such costlessly enforced contracts are not available, what do we do? As discussed above, one common form of endogenous contract enforcement is the sort of principal-agent mechanism in which one party, the principal, has power over the other party, the agent, by extending the latter favorable terms with the threat of dismissal in case of the principal's dissatisfaction. But many business agreements are one-time or very infrequent transactions rather that frequently repeated interactions. Kenneth Arrow once suggested (Arrow 1969) that norms of social behavior, including ethical and moral codes, might be "reactions of society to compensate for market failures" or "agreements to improve the efficiency of the economic system... by providing commodities to which the price system is inapplicable." His example was trust. Indeed the enforcement costs of a society without trust would be monumental.

Trust and integrity are highly valued characteristics that strongly promote economic efficiency (Zak and Knack 2001). When contracts are necessarily incomplete, contracting agents do well when they are mutually trusting and trustworthy. The most advanced economies in the world have achieved this status because they have fostered a high level of trust (Kosfeld et al. 2005; Gintis and Khurana 2008). A rather brilliant behavioral economics experiment conducted by Ernst Fehr, Simon Gächter, and Georg Kirchsteiger (Fehr et al. 1997) illustrates this point.

The authors conducted a series of incomplete contract experiments in which trust and integrity could possibly substitute for costlessly enforced perfect contracts. Several researchers had previously shown that trusting one's exchange partner can induce that partner to *reciprocate*, even at personal cost, and even when there is no opportunity for one's integrity to be rewarded through future interactions (Fehr et al. 1993, 1998). Recipients of generosity frequently respond by acting generously themselves.

The authors hypothesized that in the context of contract enforcement, mutual generosity might increase the payoffs to both parties. For example, by offering a high wage, a firm might induce a worker to provide an otherwise unattainable level of work effort. The resulting increase in worker productivity could more than offset the higher labor cost, increasing the firm's profit. Other researchers had shown that many subjects are *altruistic punishers*, who are willing to pay to punish others who have been unfair to them, even when there is no chance that they could gain in any way from that in the future, for instance by enhancing their reputations as hard bargainers (Güth and Tietz 1990; Roth 1995; Camerer and Thaler 1995). In the context of this experiment, this suggests that a subject who offered a generous contract but whose partner did not reciprocate by supplying high-quality services might be willing to punish the trading partner, even if he could not thereby gain monetarily either from that or any future transactions. A partner who anticipates this reaction, even if perfectly selfish, may then reciprocate the first party's generosity.

The game devised for this experiment had two types of players, "employers" and "employees." The rules of the game are as follows. If an employer hires an employee who provides effort e, where $0.1 \leq e \leq 1$, and receives an integer wage w, where $1 \leq w \leq 100$, the employer's payoff is $\pi = 100e - w$. The payoff to the employee is then $u = w - c(e)$, where $c(e)$ is the cost of effort function shown in figure 4.1. All payoffs involve real money that the subjects are paid at the end of the experimental session. Note that higher effort is costly to the employee but increases the payoff to the employers.

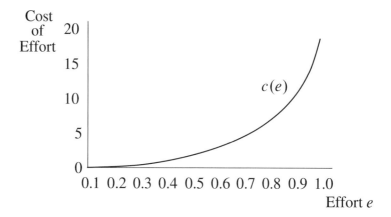

Figure 4.1. The cost-of-effort schedule in Fehr, Gächter, and Kirchsteiger (1997).

The sequence of actions is as follows. The employer first offers a "contract" specifying an integer wage w and a desired amount of effort e^*. A contract is made with the first employee who agrees to these terms. An employer can make a contract (w, e^*) with at most one employee. The experimenters created more employees than employers, so that some employees go without contracts in each period. The employee who agrees to the contract (w, e^*) receives the wage w and supplies an effort level e that *need* *not equal the contracted effort e**. In effect, the employee is not required to keep his promise, so the employee can choose an effort level, $e \in [0.1, 1]$. Although subjects may play this game several times with different partners, each employer-employee interaction is a unique nonrepeated event. Moreover, the identity of the interacting partners is never revealed.

If employees are purely self-regarding, as assumed in standard economic theory, they will choose the zero-cost effort level, $e = 0.1$, no matter what wage is offered them. If employers believe their employees are purely self-regarding, they will never pay more than the minimum necessary to get the employee to accept a contract, which is $w = 1$. Any potential employee will accept this offer and will set $e = 0.1$. Because $c(0.1) = 0$, the employee's payoff is $u = 1$. The employer's payoff is $\pi = 0.1 \times 100 - 1 = 9$.

The experiment had three different forms, which the authors call the *no reciprocity treatment*, the *weak reciprocity treatment*, and the *strong reciprocity treatment*. In the no reciprocity treatment, the effort level $e^* = e$ is set by the experimenters rather than the contracting parties. In this situation, if the given effort level is e, a profit-maximizing employer will offer wage w to the minimum value larger than to $c(e)$ because no employee will accept a lower wage, and since there are more potential employees than employers, rejecting the employer's offer leaves the individual simply unemployed for that period. Of course, employers may still be generous in this situation and simply transfer money to their employees out of good will.

In the weak reciprocity treatment, the employer was allowed to choose the whole contract (w, e^*) and an employee who accepted this contract was allowed to choose any permissible effort level e, as in our original description of the game. In the strong reciprocity treatment, the situation was the same as in the weak reciprocity treatment, except that after seeing the effort level provided by his employee, the employer was permitted to reward or fine the worker, at a cost to himself.

The experiment implemented four sessions of no reciprocity treatment, six sessions of weak reciprocity treatment, and two sessions of strong reciprocity treatment. Each no reciprocity and each weak reciprocity session had sixteen periods of contracting, and the strong reciprocity sessions had twelve periods of contracting. In each case, the anonymity of partners was maintained, and partners changed in each period.

In the no reciprocity treatment, as expected, employers almost always paid the minimum possible wage and employees had close to the minimum possible payoffs. The experimenters recorded the following results for the weak reciprocity treatment:

- Employers who requested higher effort levels also offered higher wages, and even higher worker payoff (wage minus cost of effort);
- Employees behave on average reciprocally, offering higher effort in return for a higher wage, even though this lowered their payoffs;
- Employers persistently tried to elicit higher effort by offering a wage above the minimum, but shirking rates were very high, so effort delivered was close to the minimum. Employers thus fared poorly in this situation.

The strong reciprocity treatment had the following results:

- Employers demand and enforce significantly higher effort levels than in weak reciprocity treatment. In other words, the fact that employers could punish shirking employees, and that the employees knew this, led to higher payoffs for both employers and employees, despite the fact that neither party could gain in future interactions for their reciprocal behavior.
- Employers rewarded workers who provided more than the contractual level of effort, and fined workers who provided less.
- Although employers demanded higher effort levels in the strong reciprocity treatment than in the weak, the rate of shirking was lower in the strong reciprocity treatment.

We conclude from this study that subjects who are assigned the role of employee conform to moral norms of reciprocity even when they are certain there are no material repercussions from behaving in a self-regarding manner. Moreover, subjects who are assigned the role of employer expect this behavior and are rewarded for acting accordingly. Finally, employers reward good behavior and punish bad behavior when they are allowed even though to do so is costly, and employees expect this from employers and adjust their own effort levels accordingly. In general, then, subjects behave with integrity not only because it is prudent or useful to do so, or because they expect to suffer some material loss if they do not, but also because they desire to do this *for its own sake*.

The next few sections of this chapter provide analytical models of contingent renewal markets of various types. The reader uninterested in mathematical detail may safely skip these sections, or simply read around the equations.

4.4   Reputational Equilibrium

Consider a consumer goods firm that can produce a quality good at any quality level $q \in [0, 1]$. If consumers anticipate quality q_a, their demand x is given by

$$x = 4 + 6q_a - p. \tag{4.1}$$

Suppose the firm knows this demand curve, and takes q_a as given but can set the quality q supplied. The firm has no fixed costs, and the cost of producing one unit of the good of quality q is $2 + 6q^2$.

At the start, each consumer chooses a supplier at random. In each period $t = 1, 2, \ldots$ the firm chooses a quality level q and a price p. Consumers see the price but do not know the quality until they buy the good. Consumers follow a strategy in which each buys the good in each period in which $q \geq q_a$, but if $q < q_a$ in some period, they switch to a different supplier.

We define a *reputational equilibrium* as one in which quality q_a is supplied in each period. To find the conditions for a reputational equilibrium, we assume the firm uses time discount factor $\delta = 0.9$, and note that if it is profitable for the firm to lie when it claims its product has quality $q > 0$, it might as well set its actual quality to 0, because the firm minimizes costs this way. If the firm lies and chooses its price to sell an amount x, its profits are then

$$\pi_f = (4 + 6q_a - x - 2)x = (2 + 6q_a - x)x. \tag{4.2}$$

Profits are maximized when

$$\frac{d\pi_f}{dx} = 2 + 6q_a - 2x = 0, \tag{4.3}$$

so maximum profit occurs when $x = 1 + 3q_a$, so $p = 3(1 + q_a)$ and $\pi_f = (1 + 3q_a)^2$.

Now suppose the firm tells the truth. Then, if π_t is per-period profits, we have

$$\pi_t = (2 + 6q_a - 6q_a^2 - x)x,$$
$$\frac{d\pi_t}{dx} = 2 + 6q_a - 6q_a^2 - 2x = 0,$$

so the profit-maximizing quantity is $x = 1 + 3q_a - 3q_a^2$, so $p = 3(1 + q_a - q_a^2)$ and $\pi_t = (1 + 3q_a - 3q_a^2)^2$. But total profits Π from truth telling are π_t forever, discounted at rate $\delta = 0.9$, or

$$\Pi = \frac{\pi_t}{1 - \delta} = 10(1 + 3q_a - 3q_a^2)^2. \tag{4.4}$$

Truth telling is profitable then when $\Pi \geq \pi_f$, or when

$$10(1 + 3q_a - 3q_a^2)^2 > (1 + 3q_a)^2. \tag{4.5}$$

Note that equation (4.5) is true for very small q_a (that is, q_a near 0) and false for very large q_a (that is, q_a near 1).

4.5   Contingent Renewal Labor Markets

In this section we develop a repeated game between employer and employee in which the employer pays the employee a wage higher than the expected value of his next best alternative, using the threat of termination to induce a high level of effort, in a situation where it is infeasible to write and enforce a contract for labor effort. When all employers behave in this manner, we have a nonclearing market in equilibrium.

Suppose an employer's income per period is $q(e)$, an increasing, concave function of the effort e of an employee. The employee's payoff per period $u = u(w, e)$ is an increasing function of the wage w and a decreasing function of effort e. Effort is known to the employee but is only imperfectly observable by the employer. In each period, the employer pays the employee w, the employee chooses effort e, and the employer observes a signal that registers the employee as "shirking" with probability $f(e)$, where $f'(e) < 0$. If the employee is caught shirking, he is dismissed and receives a fallback with present value z. Presumably z depends on the value of leisure, the extent of unemployment insurance, the cost of job search, the startup costs in another job, and the present value of the new job. The employer chooses w to maximize profits. The tradeoff the employer faces is that a higher wage costs more but it increases the cost of dismissal to the employee. The profit-maximizing wage equates the marginal cost to the marginal benefit.

The employee chooses $e = e(w)$ to maximize the discounted present value v of having the job, where the flow of utility per period is $u(w, e)$. Given discount rate ρ and fallback z, the employee's payoff from the repeated game is

$$v = \frac{u(w, e) + [1 - f(e)]v + f(e)z}{1 + \rho}, \tag{4.6}$$

where the first term in the numerator is the current period utility, which we assume for convenience to accrue at the end of the period, and the others measure the expected present value obtainable at the end of the period, the weights being the probability of retaining or losing the position. Simplifying, we get

$$v = \frac{u(w,e) - \rho z}{\rho + f(e)} + z. \tag{4.7}$$

The term ρz in the numerator is the forgone flow of utility from the fallback, so the numerator is the net flow of utility from the relationship, whereas $f(e)$ in the denominator is added to the discount rate ρ, reflecting the fact that future returns must be discounted by the probability of their accrual as well as by the rate of time preference.

The employee varies e to maximize v, giving the first-order condition

$$\frac{\partial u}{\partial e} - \frac{\partial f}{\partial e}(v - z) = 0, \tag{4.8}$$

which says that the employee increases effort to the point where the marginal disutility of effort is equal to the marginal reduction in the expected loss occasioned by dismissal. Solving (4.8) for e gives us the employee's best response $e(w)$ to the employer's wage offer w.

We assume that the employer can hire any real number n of workers, all of whom have the effort function $e(w)$, so the employer solves

$$\max_{w,n} \pi = q(ne(w)) - wn. \tag{4.9}$$

The first-order conditions on n and w give $q'e = w$ and $q'ne' = n$, which together imply

$$\frac{\partial e}{\partial w} = \frac{e}{w}. \tag{4.10}$$

This is the famous *Solow condition* (Solow 1979).

The best-response function and part of the employer's choice of an optimal enforcement strategy (w^*) are shown in figure 4.2, which plots effort against salary. The iso-v function v^* is one of a family of loci of effort levels and salaries that yield identical present values to the employee. Their slope, $-(\partial v/\partial w)/(\partial v/\partial e)$, is the marginal rate of substitution between wage and effort in the employee's objective function. Preferred iso-v loci lie to the right.

By the employee's first-order conditions (4.8), the iso-v loci are vertical where they intersect the best-response function (because $\partial v/\partial e = 0$). The negative slope of the iso-v functions below $e(w)$ results from the fact that

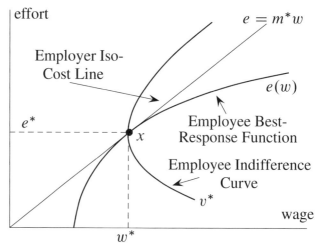

Figure 4.2. The employee's best-response function

in this region the contribution of an increase in effort, via $(\partial f/\partial e)(v - z)$, to the probability of keeping the job outweigh the effort-disutility effects. Above $e(w)$, the effort-disutility effects predominate. Because v rises along $e(w)$, the employee is unambiguously better off at a higher wage. One of the employer's iso-cost loci is labeled $e = m^*w$, where m^* is the profit-maximizing effort per dollar. The employer's first-order condition identifies the equilibrium wage w^* as the tangency between the employer's iso-cost function, $e = m^*w$ and the employee's effort function, with slope e', or point x in the figure.

It should be clear that the contingent renewal equilibrium at x is not first-best, because if the parties could write a contract for effort, any point in the lens-shaped region below the employee's indifference curve v^* and above the employer's iso-cost line $e = m^*w$ makes both parties strictly better off than at x. Note that if we populated the whole economy with firms like this, we would in general have $v > z$ in market equilibrium, because if $v = z$, (4.8) shows that $\partial u/\partial e = 0$, which is impossible so long as effort is a disutility. This is one instance of the general principle enunciated previously, that *contingent renewal markets do not clear in (Nash) equilibrium*, and the agent whose promise is contractible (usually the agent paying money) is on the long side of the market.

Perhaps an example would help visualize this situation. Suppose the utility function is given by

$$u(w, e) = w - \frac{1}{1 - e}$$

and the shirking signal is given by

$$f(e) = 1 - e.$$

You can check that $e(w)$ is then given by

$$e(w) = 1 - a - \sqrt{a^2 + \rho a},$$

where $a = 1/(w - \rho z)$. The reader can check that this function indeed has the proper shape: it is increasing and concave, is zero when $w = 2 + \rho(1 + z)$, and approaches unity with increasing w.

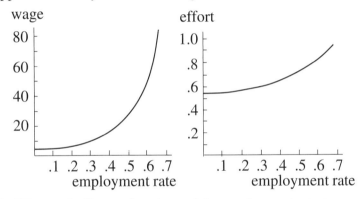

Figure 4.3. Wage and effort as functions of the employment rate in a contingent renewal labor market

The solution for the employer's optimum w, given by the Solow condition (4.10), is very complicated, so I will approximate the solution. Suppose $\rho = 0.05$ and the employment rate is $q \in [0, 1]$. An employee dismissed at the end of the current period therefore has a probability q of finding a job right away (we assume all firms are alike), and so regains the present value v. With probability $1 - q$, however, the ex-employee remains unemployed for one period and tries again afterward. Therefore we have

$$z = qv + (1 - q)z/(1 + \rho),$$

assuming the flow of utility from being unemployed (in particular, there is no unemployment insurance) is zero. Solving, we have

$$z = \frac{(1 + \rho)q}{q + \rho}v.$$

For a given unemployment rate q, we can now find the equilibrium values of w, e, v, and z, and hence the employer's unit labor cost e/w. Running this through Mathematica, the equilibrium values of w and e as the employment rate q goes from zero to 0.67 are depicted in figure 4.3.

Note that although effort increases only moderately as the unemployment rate drops from 100% to 33%, the wage rate increases exponentially as the unemployment rate approaches 33%. I could not find a solution for $q > 0.67$. The actual unemployment rate can be fixed by specifying the firm's production function and imposing a zero profit condition. However this is accomplished, there will be positive unemployment in equilibrium.

4.6   I'd Rather Fight than Switch

Consider a firm that produces a *quality good*, which is a good whose quality is costly to produce, can be verified only through consumer use, and cannot be specified contractually. In a single-period model, the firm would have no incentive to produce high quality. We develop a repeated game between firm and consumer, in which the consumer pays a price *greater* than the firm's marginal cost, using the threat of switching to another supplier to induce a high level of quality on the part of the firm. The result is a nonclearing product market, with firms enjoying price greater than marginal cost. Thus, they are quantity constrained in equilibrium.

Every Monday, families in Pleasant Valley wash their clothes. To ensure brightness, they all use bleach. Low-quality bleach can, with small but positive probability, ruin clothes, destroy the washing machine's bleach delivery gizmo, or irritate the skin. High-quality bleach is therefore deeply pleasing to Pleasant Valley families. However, high-quality bleach is also costly to produce. Why should firms supply high quality?

Because people have different clothes, washing machines, and susceptibility to skin irritation, buyers cannot depend on a supplier's reputation to ascertain quality. Moreover, a firm could fiendishly build up its reputation for delivering high-quality bleach and then, when it has a large customer base, supply low-quality bleach for one period, and then close up shop (this is called "milking your reputation"). Aggrieved families could of course sue the company if they have been hurt by low-quality bleach but such suits are hard to win and very costly to pursue. So no one does this.

If the quality q of bleach supplied by any particular company can be ascertained only after having purchased the product, and if there is no way to be compensated for being harmed by low-quality bleach, how can high quality be assured?

Suppose the cost to a firm of producing a gallon of the bleach of quality q is $b(q)$, where $b(0) > 0$ and $b'(q) > 0$ for $q \geq 0$. Each consumer is a customer of a particular supplier, and purchases exactly one gallon of bleach each Friday at price p from this supplier. If dissatisfied, the customer switches to another supplier at zero cost. Suppose the probability of being dissatisfied, and hence of switching, is given by the decreasing function $f(q)$. We assume an infinite time horizon with a fixed discount rate ρ.

Considering both costs $b(q)$ and revenue q as accruing at the end of the period, it is easy to show that the value $v(q)$ to a firm from having a customer is

$$v(q) = \frac{p - b(q)}{f(q) + \rho}. \tag{4.11}$$

Suppose the price p is set by market competition, so it is exogenous to the firm. It is then easy to show that the firm chooses quality q so that

$$p = b(q) + b'(q)g(q), \tag{4.12}$$

where $g(q) = -[f(q) + \rho]/f'(q)$, provided $q > 0$. It is also easy to show that quality is an increasing function of price, as we should expect.

Note that in this case firms are quantity constrained, because price is greater than marginal cost in a market equilibrium, and that consumers are on the long side of the market. Once again money confers short-side power.

This model raises an interesting question. What determines firm size? In the standard perfect competition model, firm size is determined by the condition that average costs be at a minimum. This is of course just silly, because a firm can always produce at any multiple of the "optimal firm size" simply by working the production process, whatever it might be, in parallel. The monopolistic competition model, in which a firm has a downward-sloping demand curve, is better, but it does not apply to a case like ours, where firms are price-takers, as in the perfect competition model, and firm size is determined by the dynamic process of movement of customers among firms. Here is one plausible model of such a process.

Suppose there are n firms in the bleach industry, all selling at the same price p. Suppose firm j has market share m_j^t in period t. Suppose for $j = 1, \ldots, n$, a fraction f_j of firm j's customers leave the firm in each period, and a fraction a_j of customers who have left firms are attracted to firm j. We say the bleach industry is *in equilibrium* if the market share of each firm is constant over time. We have the following.

THEOREM 4.1 *There is a unique asymptotically stable equilibrium in the bleach industry.*

PROOF: We normalize the number of customers in Pleasant Valley to one. Then, the number of customers leaving firm j is $m_j^t f_j$, so the total number of customers looking for new suppliers is $\sum_j m_j^t f_j$. A particular firm j attracts a fraction a_j of these. This assumes a firm can woo back a fraction a_j of its recently departed customers; the argument is the same if we assume the opposite. Thus, the net customer loss of firm j in period t is

$$f_j m_j^t - a_j \sum_{k=1}^{n} f_k m_k^t. \tag{4.13}$$

In equilibrium this quantity must be zero, and $m_k^t = m_k$ for all t and for $k = 1, \ldots, n$. This gives the equilibrium condition

$$m_j = \mu_j \sum_{k=1}^{n} f_k m_k, \tag{4.14}$$

where we have defined $\mu_k = a_k / f_k$. Note also that if we add up the n equations in (4.13), we get zero, so $\sum_k m_k^t = 1$ for all t, implying $\sum_k m_k = 1$. Summing (4.14), we arrive at the equilibrium condition

$$m_j = \frac{\mu_j}{\sum_k \mu_k}. \tag{4.15}$$

Thus, there exists a unique industry equilibrium. To prove asymptotic stability, we define the $n \times n$ matrix $B = (b_{ij})$, where $b_{ij} = a_i f_j$ for $i \neq j$, and $b_{ii} = a_i f_i + (1 - f_i)$, $i, j = 1, \ldots, n$. Then, writing the column vector $m^t = (m_1^t, \ldots, m_n^t)$, we have $m^{t+1} = B m^t$ and hence $m^t = B^t m^0$, where B^t is the tth power of B. The matrix B is a positive matrix, and it is easy to check that it has eigenvalue 1 with corresponding positive eigenvector $m = (m_1, \ldots, m_n)$. By Perron's theorem (see, for instance, Horn and Johnson 1985, section 8.2), 1 is the unique maximal eigenvalue of B. Also $(1, 1, \ldots, 1)$ is a right eigenvector of B corresponding to the eigenvalue 1. It follows that B^t tends to the matrix whose columns are each m (see Horn and Johnson 1985, theorem 8.2.8), which proves the theorem.

4.7   Regulating Market Power

It is easy to show that under many plausible conditions, the market for a variable quality good will not exist. For instance, without some regulation by the state (for instance enforcing truthful odometer readings and requiring sellers to provide minimal guarantees), the market for used cars may be extremely thin or nonexistent (Akerlof 1970). We can generalize to say that regulation can sometimes eliminate *pooling equilibria* (situations in which buyers cannot distinguish between high and low quality until the goods has been purchased) in variable quality markets, while sustaining *separating equilibria*, in which they can (Gintis 2009b). For example, requiring truth in advertising can often be justified by its effects on the efficiency of transactions. But there is no general theory of when intervention in variable quality markets will enhance economic efficiency.

While the power of employers over employees arises in our model as a rational strategy in the interests of profit-maximizing owners, the uses of power by managers may include assaults on the dignity of workers bearing no relationship whatsoever to the relatively benign objective of regulating labor effort. Sexual harassment comes to mind. Because owners necessarily exercise less than perfect control over the various levels of management in the firm, these and other uses of power that are arbitrary from the standpoint of the regulation of effort have substantial latitude. Assaulting the dignity of workers is not likely to be a profit-maximizing strategy, among other reasons, because it lowers the value of employment and hence the cost of job loss, but the power created by the short-side location of the employer permits the entrepreneur to trade off profits against illegitimate power over the worker (e.g., sexual harassment or ethnic discrimination). Moreover, in a corporate setting, the owners' inability to perfectly solve their own principal agent problem vis-a-vis management provides ample opportunity for managers to cater to their own personal objectives in exercising control over the worker.

5

Rational Choice Revealed and Defended

> The heart has its reasons of which reason knows nothing.
>
> Blaise Pascal

> Such is the way of all superstition...wherein men...mark the events where they are fulfilled, but where they fail, though this happens much oftener, neglect and pass them by.
>
> Francis Bacon

> **C:** Pierre Bourdieu rejects rational choice because action is patterned and interest-oriented at a tacit, prereflected level of awareness through time. What do you think?
> **HG:** Both are true.
>
> Choreographer interview

The previous chapters have illustrated the power of the rational actor model. The currently popular notion that humans are irrational is hopefully not for long in vogue. Choice behavior can generally be best modeled using the rational actor model, according to which individuals have a time-, state-, and social context-dependent *preference function* over outcomes, and *beliefs* concerning the probability that particular actions lead to particular outcomes. Individuals of course value outcomes besides the material goods and services depicted in economic theory. Moreover, actions may be valued for their own sake. For example, there are *character virtues*, including honesty, loyalty, and trustworthiness, that have intrinsic moral value, in addition to their effect on others or on their own reputation. Moreover, social actors generally value not only *self-regarding* payoffs such as personal income and leisure, but also *other-regarding* payoffs, such as the welfare of others, environmental integrity, fairness, reciprocity, and conformance with social norms. We developed these points in Chapter 3.

The rational choice model *expresses* but does not *explain* individual preferences. Understanding the content of preferences requires rather deep forays into the psychology of goal-directed and intentional behavior (Haidt 2012), evolutionary theory (Tooby and Cosmides 1992), and problem-solving heuristics (Gigerenzer and Todd 1999). Moreover, the social actor's preference function will generally depend on his current motivational state, his previous experience and future plans, and the social situation that he faces.

The first principle of rational choice is that in any given situation, which may be time-, state-, and social-context dependent, the decision-maker, say Alice, has a *preference relation* $\succ$ over choices such that Alice prefers x to y if and only if $x \succ y$. The conditions for the existence of such a relation, developed in Section 5.1 below, are quite minimal, the main condition being that Alice's choices must be *transitive* in the sense that if the choice set from which Alice must choose is X with $x, y, z \in X$, then if Alice prefers x to y, and also prefers y to z, then Alice must prefer x to z as well. An additional requirement is that if Alice prefers x to y when the choice set is X, she must continue to prefer x to y in any choice set that includes both x and y. This condition can fail if the choice set itself represents a substantive social context that affects the value Alice places upon x and y. For instance, Alice may prefer fish (x) to steak (y) in a restaurant that also serves lobster (z) because the fish is likely to be very fresh in this case, whereas in a restaurant that does not serve lobster, the fish is likely to be less fresh, so Alice prefers steak (y) to fish (x). For another commonplace example, Alice may prefer a $100 sweater to a $200 sweater in a store in which the latter is the highest price sweater in the store, but might reverse her preference were the most expensive sweater in the store priced at $500. In cases such as these, a more sophisticated representation of choice sets and outcomes both satisfies the rationality assumptions and more insightfully models Alice's social choice situation.

Every argument that I have seen for rejecting the rational actor model I have found to be specious, often disingenuous and reflecting badly on the training of its author. The standard conditions for rationality, for instance, do not imply that rational Alice chooses what is in her best interest or even what gives her pleasure. There are simply *no utilitarian or instrumental implications* of these axioms. If a rational actor values giving to charity, for instance, this does not imply that he gives to charity in order to increase his happiness. A martyr is still a martyr even though the act of martyrdom may be extremely unpleasant. Nor does the analysis assume that Alice is in any sense selfish, calculating, or amoral. Finally, the rationality assumption does not suggest that Alice is "trying" to maximize utility or anything else. The maximization formulation of rational choice behavior, which we develop below, is simply an analytical convenience, akin to the least action principle in classical mechanics, or predicting the behavior of an expert billiards player by solving a set of differential equations. No one believes that light "tries to" minimize travel time, or that billiards players are brilliant differential equation solvers.

The second principle of rational choice applies when Alice's behavior involves *probabilistic* outcomes. Suppose there are a set of alternative possible *states of nature* Ω with elements $\omega_1, \ldots, \omega_n$ that can possibly materialize, and a set of outcomes X. A *lottery* is a mapping that specifies a particular outcome $x \in X$ for each state $\omega \in \Omega$. Let the set of such lotteries be $\mathcal{L}$, so any lottery $\pi \in \mathcal{L}$ gives Alice outcome $x_i = \pi(\omega_i)$ in case ω_i, where $i = 1, \ldots, n$. By our first rationality assumption, Alice has a consistent preference function over the lotteries in $\mathcal{L}$. Adding to this a few rather innocuous assumptions concerning Alice's preferences (see Section 5.1), it follows that Alice has a consistent preference function $u(\pi)$ over the lotteries in $\mathcal{L}$ and also Alice attaches a specific probability $p(\omega)$ to each event in Ω. This probability distribution is called Alice's *subjective prior*, or simply her *beliefs*, concerning the events in Ω. Moreover, given the preference function $u(\pi)$ and the subjective prior $p(\omega)$, Alice prefers lottery π to lottery ρ, that is $\pi \succ \rho$, precisely when the *expected utility* of π exceeds that of lottery ρ (see equation 5.1).

The rational actor model does not hold universally (see Section 5.6). There are only two substantive assumptions in the above derivation of the expected utility theorem. The first is that Alice does not suffer from *wishful thinking*. That is, the probability that Alice implicitly attaches to a particular outcome by her preference function over lotteries does not depend on how much she stands to gain or lose should that outcome occur. This assumption is certainly not always justified. For instance, believing that she might win the state lottery may give Alice more pleasure while waiting for it to happen than the cost of buying the lottery ticket. Moreover, there may be situations in which Alice will *underinvest* in a desirable outcome unless she inflates the probability that the investment will pay off (Benabou and Tirole 2002). In addition, Alice may be substantively irrational, having excessive confidence that the world conforms to her ideological preconceptions.

The second substantive assumption is that the state of nature that materializes is not affected by Alice's choice of a lottery. When this fails the subjective prior must be interpreted as a *conditional probability*, in terms of which the expected utility theorem remains valid (Stalnaker 1968). This form of the expected utility theorem is developed in Section 5.1.

Of course, an individual may be rational in this decision-theoretic sense, having consistent preferences and not engaging in wishful thinking, and still fail to conform to higher canons of rationality. Alice may, for instance, make foolish choices that thwart her larger objectives and threaten her well-being. She may be poorly equipped to solve challenging optimization problems. Moreover, being rational in the decision-theoretic sense does not imply that Alice's beliefs are in any way reasonable, or that she evaluates new evidence in an insightful manner.

The standard axioms underlying the rational actor model are developed in von Neumann and Morgenstern (1944) and Savage (1954). The plausibility and generality of these axioms are discussed in Section 5.1. Section 6.10 explores the implications of replacing Savage's assumption that beliefs are purely personal "subjective probabilities" with the notion that the individual is embedded in a *network of social actors* over which information and experience concerning the relationship between actions and outcomes is spread. The rational actor thus draws on a network of beliefs and experiences distributed among the social actors to which he is informationally and socially connected. By the sociological principle of *homophily*, social actors are likely to structure their network of personal associates according to principles of social similarity, and to alter personal tastes in the direction of increasing compatibility with networked associates (McPherson et al. 2001; Durrett and Levin 2005; Fischer et al. 2013).

It is important to understand that the rational actor model says *nothing* about how individuals form their subjective priors, or in other words, their *beliefs*. This model does say that whatever their beliefs, new evidence should induce them to transform their beliefs to be more in line with this evidence. Clearly there are many beliefs that are so strong that such updating does not occur. If one believes that something is true with probability one, then no evidence can lead to the Bayesian updating of that belief, although it could lead the individual to revise his whole belief system (Stalnaker 1996). More commonly, strong believers simply discount the uncomfortable evidence. This is no problem for the rational actor model, which simply depicts behavior rather than showing that rational choice leads to objective truth.

5.1 The Axioms of Rational Choice

The word *rational* has many meanings in different fields. Critics of the rational actor model almost invariably attach meanings to the term that lie quite outside the bounds of rationality as used in decision theory, and incorrectly reject the theory by referring to these extraneous meanings. We here present a set of axioms, inspired by Savage (1954), that are sufficient to derive the major tools of rational decision theory, the so-called *expected utility theorem*.[1]

A *preference function* $\succeq$ on a choice set Y is a binary relation, where $\{x \succeq y | Y\}$ is interpreted as the decision-maker weakly preferring x to y when the choice set is Y and $x, y \in Y$. By "weakly" we mean that the decision-maker may be indifferent between the two. We assume this binary relation has the following three properties, which must hold for any choice set Y, for all $x, y, z \in Y$, and for any set $Z \subset Y$:

1. *Completeness*: $\{x \succeq y | Y\}$ or $\{y \succeq x | Y\}$;
2. *Transitivity*: $\{x \succeq y | Y\}$ and $\{y \succeq z | Y\}$ imply $\{x \succeq z | Y\}$;
3. *Independence of irrelevant alternatives*: For $x, y \in Z$, $\{x \succeq y | Z\}$ if and only if $\{x \succeq y | Y\}$.

Because of the third property, we need not specify the choice set and can simply write $x \succeq y$. We also make the rationality assumption that the actor chooses his most preferred alternative. Formally, this means that given any choice set A, the individual chooses an element $x \in A$ such that for all $y \in A$, $x \succeq y$. When $x \succeq y$, we say "x is weakly preferred to y" because the actor can actually be indifferent between x and y.

One can imagine cases where completeness might fail. For instance an individual may find all alternatives so distasteful that he prefers to choose none of them. However, if "prefer not to choose" is an option, it can be added to the choice set with an appropriate outcome. For instance, in the movie *Sophie's Choice*, a woman is asked to choose one of her two children to save from Nazi extermination. The cost of the option "prefer not to choose" in this case was having both children exterminated.

Note that the decision-maker may have absolutely no grounds to choose x over y, given the information he possesses. In this case we have *both*

---

[1]I regret using the term "utility" which suggests incorrectly that the theorem is related to philosophical utilitarianism or that it presupposes that all human motivation is aimed at maximizing pleasure or happiness. The weight of tradition bids us retain the venerable name of the theorem, despite its connotational baggage.

$x \succeq y$ and $y \succeq x$. In this case we say that the individual is *indifferent* between x and y and we write $x \sim y$. This notion of indifference leads to a well-known philosophical problem. If preferences are transitive, then it is easy to see that indifference is also transitive. However it is easy to see that because humans have positive sensory thresholds, indifference cannot be transitive over many iterations. For instance, I may prefer more milk to less in my tea up to a certain point, but I am indifferent to amounts of milk that differ by one molecule. Yet starting with one teaspoon of milk and adding one molecule of milk at a time, eventually I will experience an amount of milk that I prefer to one teaspoon.

The transitivity axiom is implicit in the very notion of rational choice. Nevertheless, it is often asserted that intransitive choice behavior is observed (Grether and Plott 1979; Ariely 2010). In fact, most such observations satisfy transitivity when the state dependence (see Gintis 2007c and Section 5.2 below), time dependence (Ahlbrecht and Weber 1995; Ok and Masatlioglu 2003), and/or social context dependence (Brewer and Kramer 1986; Andreoni 1995; Cookson 2000; Carpenter et al. 2005b) of preferences are taken into account.

Independence of Irrelevant Alternatives fails when the relative value of two alternatives depends on other elements of the choice set Y, but as suggested above, the axiom can usually be restored by suitably redefining the choice situation (Gintis 2009a).

The most general situation in which the Independence of Irrelevant Alternatives fails is when the choice set supplies independent information concerning the *social frame* in which the decision-maker is embedded. This aspect of choice is analyzed in Section 5.4, where we deal with the fact that preferences are generally state-dependent; when the individual's social or personal situation changes, his preferences will change as well. Unless this factor is taken into account, rational choices may superficially appear inconsistent.

When the preference relation $\succeq$ is complete, transitive, and independent from irrelevant alternatives, we term it *consistent*. It should be clear from the above that preference consistency is an extremely weak condition that is violated only when the decision-maker is quite lacking in reasonable principles of choice.

If $\succeq$ is a consistent preference relation, then there will always exist a utility function such that individuals behave as if maximizing their utility functions over the sets Y from which they are constrained to choose. Formally, we say that a utility function $u : Y \to \mathbf{R}$ *represents* a binary relation $\succeq$ if, for all $x, y \in Y$, $u(x) \geq u(y)$ if and only if $x \succeq y$. We have the following theorem, whose simple proof we leave to the reader.

THEOREM 5.1 *A binary relation $\succeq$ on the finite set Y can be represented by a utility function $u : Y \to \mathbf{R}$ if and only if $\succeq$ is consistent.*

As we have stressed before, the term "utility" here is meant to have no utilitarian connotations.

5.2   Choice Under Uncertainty

We now assume that an action determines a *statistical distribution* of possible outcomes rather than a single particular outcome. Let X be a finite set of outcomes and let $\mathcal{A}$ be a finite set of actions. We write the set of pairs (x, a) where x is an outcome and a is an action as $X \times \mathcal{A}$. Let $\succeq$ be a consistent preference relation on $X \times \mathcal{A}$; i.e., the actor values not only the outcome, but the action itself. By theorem 5.1 we can associate $\succeq$ with a utility function $u : X \times \mathcal{A} \to \mathbf{R}$.

Let Ω be a finite set of *states of nature*. For instance, Ω could consist of the days of the week, so a particular state $\omega \in \Omega$ can take on the values Monday through Sunday, or Ω could be the set of permutations (about 8×10^{67} in number) of the 52 cards in a deck of cards, so each $\omega \in \Omega$ would be a particular shuffle of the deck. We call any $A \subseteq \Omega$ an *event*. For instance, if Ω is the days of the week, the event "weekend" would equal the set {Saturday, Sunday}, and if Ω is the set of card deck permutations, the event "the top card is a queen" would be the set of permutations (about 6×10^{66} in number) in which the top card is a queen.

Following Savage (1954) we show that if the individual has a preference relation over lotteries (functions that associate states of nature $\omega \in \Omega$ with outcomes $x \in X$) that has some plausible properties, then not only can the individual's preferences be represented by a preference function, but also we can infer the probabilities the individual implicitly places on various events (his so-called *subjective priors*), and the expected utility principle holds for these probabilities.

Let $\mathcal{L}$ be a set of lotteries, where a *lottery* is now a function $\pi : \Omega \to X$ that associates with each state of nature $\omega \in \Omega$ an outcome $\pi(\omega) \in X$. We suppose that the individual chooses among lotteries without knowing the state of nature, after which the state $\omega \in \Omega$ that he obtains is revealed, so that if the individual chooses action $a \in A$ that entails lottery $\pi \in \mathcal{L}$, his outcome is $\pi(\omega)$, which has payoff $u(\pi(\omega), a)$.

Now suppose the individual has a preference relation $\succ$ over $\mathcal{L} \times \mathcal{A}$. That is, the individual values not only the lottery, but the action that leads to a particular lottery. We seek a set of plausible properties of $\succ$ that together allow us to deduce (a) a utility function $u : \mathcal{L} \times \mathcal{A} \to \mathbf{R}$ corresponding to the preference relation $\succ$ over $X \times \mathcal{A}$; and (b) there is a probability distribution $p : \Omega \to \mathbf{R}$ such that the expected utility principle holds with respect to the preference relation $\succ$ over $\mathcal{L}$ and the utility function $u(\cdot, \cdot)$; i.e., if we define

$$\mathbf{E}_\pi[u|a; p] = \sum_{\omega \in \Omega} p(\omega) u(\pi(\omega), a), \qquad (5.1)$$

then for any $\pi, \rho \in \mathcal{L}$ and any $a, b \in A$,

$$(\pi, a) \succ (\rho, b) \iff \mathbf{E}_\pi[u|a; p] > \mathbf{E}_\rho[u|b; p]. \qquad (5.2)$$

A set of axioms that ensure (5.2), which is called the *expected utility principle*, is formally presented in Gintis (2009a). Here I present these axioms more descriptively and omit a few uninteresting mathematical details. The first condition is the rather trivial assumption that

A1. If π and ρ are two lotteries, then whether $(\pi, a) \succ (\rho, b)$ is true or false depends only on states of nature where π and ρ have different outcomes.

This axiom allows us to define a *conditional preference* $\pi \succ_A \rho$, where $A \subseteq \Omega$, which we interpret as "π is strictly preferred to ρ, conditional on event A." We define the conditional preference by revising the lotteries so that they have the same outcomes when $\omega \notin A$. Because of axiom **A1**, it does not matter what we assign to the lottery outcomes when $\omega \notin A$. This procedure also allows us to define $\succeq_A$ and $\sim_A$ in a similar manner. We say $\pi \succeq_A \rho$ if it is false that $\rho \succ_A \pi$, and we say $\pi \sim_A \rho$ if $\pi \succeq_A \rho$ and $\rho \succeq_A \pi$.

The second condition is equally trivial, and says that a lottery that gives an outcome with probability one is valued the same as the outcome:

A2. If π pays x given event A and action a, and ρ pays y given event A and action b, and if $(x, a) \succ (y, b)$, then $\pi \succ_A \rho$, and conversely.

The third condition asserts that the decision-maker's subjective prior concerning likelihood that an event A occurs is *independent* from the payoff one receives when A occurs. More precisely, let A and B be two events, let (x, a) and (y, a) be two available choices, and suppose $(x, a) \succ (y, a)$. Let π be a lottery that pays x when action a is taken and $\omega \in A$, and pays some z when $\omega \notin A$. Let ρ be a lottery that pays y when action a is taken and $\omega \in B$, and pays z when $\omega \notin B$. We say event A is *more probable than* event B, given x, y, and a if $\pi \succ \rho$. Clearly this criterion does not depend on the choice of z, by **A1.** We assume a rather strong condition:

A3. If A is more probable than B for some x, y, and a, then A is more probable than B for any other choice of x, y, and a.

This axiom, which might be termed the *no wishful thinking condition*, is often violated when individuals assume that states of nature tend to conform to their ideological preconceptions, and where they reject new information to the contrary rather than update their subjective priors (Risen 2015). Such individuals may have consistent preferences, which is sufficient to model their behavior, but their wishful thinking often entails pathological behavior. For instance, a healthy individual may understand that a certain unapproved medical treatment is a scam, but change his mind when he acquires a disease that has no conventional treatment. Similarly, an individual may attribute his child's autism to an immunization injection and continue to believe this in the face of extensive evidence concerning the safety of the treatment.

The fourth condition is another trivial assumption:

A4. Suppose the decision-maker prefers outcome x to any outcome that results from lottery ρ. Then the decision-maker prefers a lottery π that pays x with probability one to ρ.

We then have the following expected utility theorem:

THEOREM 5.2 *Suppose A1–A4 hold. Then there is a probability function p on the state space Ω and a utility function $u : X \to \mathbf{R}$ such that for any $\pi, \rho \in \mathcal{L}$ and any $a, b \in \mathcal{A}$, $(\pi, a) \succ (\rho, b)$ if and only if $\mathbf{E}_\pi[u|a; p] > \mathbf{E}_\rho[u|b; p]$.*

We call the probability p the individual's *subjective prior* and say that A1–A4 imply *Bayesian rationality*, because they together imply Bayesian probability updating. Because only A3 is problematic, it is plausible to accepted Bayesian rationality except in cases where some form of wishful thinking occurs, although there are other, rather exceptional, circumstances in which the expected utility theorem fails (Machina 1987; Starmer 2000).

5.3 Bayesian Updating with Radical Uncertainty

The only problematic axiom among those needed to demonstrate the expected utility principle is the "wishful thinking" axiom A3. While there are doubtless many cases in which at least a substantial minority of social actors engage in wishful thinking, there is considerable evidence that Bayesian updating is a key neural mechanism permitting humans to acquire complex understandings of the world given severely underdetermining data (Steyvers et al. 2006).

For instance, the spectrum of light waves received in the eye depends both on the color spectrum of the object being observed and the way the object is illuminated. Therefore inferring the object's color is severely underdetermined, yet we manage to consider most objects to have constant color even as the background illumination changes. Brainard and Freeman (1997) show that a Bayesian model solves this problem fairly well, given reasonable subjective priors as to the object's color and the effects of the illuminating spectra on the object's surface.

Several students of developmental learning have stressed that children's learning is similar to scientific hypothesis testing (Carey 1985; Gopnik and Meltzoff 1997), but without offering specific suggestions as to the calculation mechanisms involved. Recent studies suggest that these mechanisms include causal Bayesian networks (Glymour 2001; Gopnik and Schultz 2007; Gopnik and Tenenbaum 2007). One schema, known as *constraint-based learning*, uses observed patterns of independence and dependence among a set of observational variables experienced under different conditions to work backward in determining the set of causal structures compatible with the set of observations (Pearl 2000; Spirtes et al. 2001). Eight-month-old babies can calculate elementary conditional independence relations well enough to make accurate predictions (Sobel and Kirkham 2007). Two-year-olds can combine conditional independence and hands-on information to isolate causes of an effect, and four-year-olds can design purposive interventions to gain relevant information (Glymour et al. 2001; Schultz and Gopnik 2004). "By age four," observe Gopnik and Tenenbaum (2007), "children appear able to combine prior knowledge about hypotheses and new evidence in a Bayesian fashion" (p. 284). Moreover, neuroscientists have begun studying how Bayesian updating is implemented in neural circuitry (Knill and Pouget 2004).

For instance, suppose an individual wishes to evaluate a hypothesis h about the natural world given observed data x and under the constraints of a background repertoire T. The value of h may be measured by the Bayesian formula

$$\mathrm{P}_T(h|x) = \frac{\mathrm{P}_T(x|h)\mathrm{P}_T(h)}{\sum_{h' \in T} \mathrm{P}_T(x|h')\mathrm{P}_T(h')}. \tag{5.3}$$

Here, $\mathrm{P}_T(x|h)$ is the likelihood of the observed data x, given h and the background theory T, and $\mathrm{P}_T(h)$ gives the likelihood of h in the agent's repertoire T. The constitution of T is an area of active research. In language acquisition, it will include predispositions to recognize certain forms as grammatical and not others. In other cases, T might include physical, biological, or even theological heuristics and beliefs.

5.4   State-Dependent Preferences

Preferences are obviously state-dependent. For instance, Bob's preference for aspirin may depend on whether or not he has a headache. Similarly, Bob may prefer salad to steak, but having eaten the salad, he may then prefer steak to salad. These state-dependent aspects of preferences render the empirical estimation of preferences somewhat delicate, but they present no theoretical or conceptual problems.

We often observe that an individual makes a variety of distinct choices under what appear to be identical circumstances. For instance, an individual may vary his breakfast choice among several alternatives each morning without any apparent pattern to his choices. Is this a violation of rational behavior? Indeed, it is not.

Following Luce and Suppes (1965) and McFadden (1973), I represent this situation by assuming the individual has a utility function over bundles $x \in X$ of the form

$$u(x) = v(x) + \epsilon(x) \tag{5.4}$$

where $v(x)$ is a stable underlying utility function and $\epsilon(x)$ is a random error term representing the individual's current idiosyncratic taste for bundle x. This utility function induces a probability distribution π on X such that the probability that the individual chooses x is given by

$$p_x = \pi\{x \in X | \forall y \in X, v(x) + \epsilon(x) > v(y) + \epsilon(y)\}.$$

We assume $\sum_x p_x = 1$, so the probability that the individual is indifferent between choosing two bundles is zero. Now let $B = \{x \in X \mid p_x > 0\}$, so B is the set of bundles chosen with positive probability, and suppose B has at least three elements. We can express the Independence of Irrelevant Alternatives in this context by the assumption (Luce 2005) that for all $x, y \in B$,

$$\frac{p_{yx}}{p_{xy}} = \frac{P[y \mid \{x, y\}]}{P[x \mid \{x, y\}]} = \frac{p_y}{p_x}.$$

This means that the relative probability of choosing x vs. y does not depend on whatever other bundles are in the choice set. Note that $p_{xy} \neq 0$ for $x, y \in B$. We then have

$$p_y = \frac{p_{yz}}{p_{zy}} p_z \tag{5.5}$$

$$p_x = \frac{p_{xz}}{p_{zx}} p_z, \tag{5.6}$$

where $x, y, z \in B$ are distinct, by the Independence of Irrelevant Alternatives. Dividing the first equation by the second in (5.5), and noting that $p_y/p_x = p_{yx}/p_{xy}$, we have

$$\frac{p_{yx}}{p_{xy}} = \frac{p_{yz}/p_{zy}}{p_{xz}/p_{zx}}. \tag{5.7}$$

We can write

$$1 = \sum_{y \in B} p_y = \sum_{y \in B} \frac{p_{yx}}{p_{xy}} p_x,$$

so

$$p_x = \frac{1}{\sum_{y \in B} p_{yx}/p_{xy}} = \frac{p_{xz}/p_{zx}}{\sum_{y \in B} p_{yx}/p_{zy}}, \tag{5.8}$$

where the second equality comes from (5.7).

Let us write

$$w(x, z) = \beta \ln \frac{p_{xz}}{p_{zx}},$$

so (5.8) becomes

$$p_{x,B} = \frac{e^{\beta w(x,z)}}{\sum_{y \in B} e^{\beta w(y,z)}}. \tag{5.9}$$

But by the Independence of Irrelevant Alternatives, this expression must be independent of our choice of z, so if we write $w(x) = \ln p_{xz}$ for an arbitrary $z \in B$, we have

$$p_x = \frac{e^{\beta w(x)}}{\sum_{y \in B} e^{\beta w(y)}}. \tag{5.10}$$

Note that there is one free variable, β, in (5.10). This represents the degree to which the individual is relatively indifferent among the alternatives. As $\beta \to \infty$, the individual chooses his most preferred alternative with increasing probability, and with probability one in the limit. As $\beta \to 0$, the individual becomes more indifferent to the alternative choices.

This model helps explain the compatibility of the *preference reversal* phenomenon (Lichtenstein and Slovic 1971; Grether and Plott 1979; Tversky et al. 1990; Kirby and Herrnstein 1995; Berg et al. 2005) with the rationality postulate. As explained in Gintis (2007c), in the cases discussed in the experimental literature, the experimenters offer only alternative lotteries with expected values that are very close to being equal to one another. Thus decision-makers are virtually indifferent among the choices based on the expected return criterion, so even a small influence of the social frame in which the experimenters embed the choice situation on the subjects' preference state may strongly affect their choices. For experimental support for this interpretation, see Sopher and Gigliotti (1993).

5.5 Networked Minds and Distributed Cognition

I have stressed that there is one assumption in the derivation of the rational actor model that conflicts with the repeatedly observed fact that human minds are not isolated instruments of ratiocination, but rather are networked and cognition is distributed over this network. We return to this point in Section 6.10. I propose here an analytical tool, based on a refinement of the rational actor model proposed by Gilboa and Schmeidler (2001), for representing distributed cognition. Following Gilboa and Schmeidler we assume there is a single decision-maker, say Alice, who faces a *problem p* such that each *action a* that Alice takes leads to some *result r*. Alice does not know the probability distribution of outcomes following action a, so she searches her memory for similar problems she has faced in the past, the action she has taken for each problem, and the result of her action. Thus her memory M consists of a set of *cases* of the form (q, a, r), where q is a problem, *a* is the action she took facing this problem, and *r* was the result of the action. Alice has a utility function $u(r)$ defined over results, and a *similarity function* $s(p, q)$ representing how "similar" her current problem *p* is to any past problem *q* that she has encountered. Gilboa and Schmeidler then present a set of plausible axioms that imply Alice will choose her action *a* to maximize the expression

$$\sum_{(q,a,r)\in M_a} s(p,q)u(r) \qquad (5.11)$$

where M_a is the subset of Alice's memory where she took action *a*.

Several empirical studies have shown that this case-based decision approach is superior to other more standard approaches to choice under radical uncertainty (Gayer et al. 2007; Golosnoy and Okhrin 2008; Ossadnik et al. 2012; Guilfoos and Pape 2016). To extend this to distributed cognition, we simply replace Alice's personal memory bank by a wider selection of cases distributed over her social network of minds. It would also be plausible to add a second similarity function indicating how similar the individual who actually took the action is Alice herself.

5.6   Limitations of the Rational Actor Model

One often hears that a theory fails if there is a single counterexample. Indeed, this notion was the touchstone of Karl Popper's famous interpretation of the scientific method (Popper 2002[1959]). Because biological systems are inherently complex, this criterion is too strong for the behavioral sciences (Godfrey-Smith 2006, 2009; Weisberg 2007; Wimsatt 2007). Despite its general usefulness, the rational actor model fails to explain choice behavior in several well-known situations. Two examples are the famous Allais and Ellsberg Paradoxes. These are of course not paradoxes, but rather violations of rational choice.

5.6.1   The Allais Paradox

Maurice Allais (1953) offered the following scenario as a violation of rational choice behavior. There are two choice situations in a game with prizes $x = \$2,500,000$, $y = \$500,000$, and $z = \$0$. The first is a choice between lotteries $\pi = y$ and $\pi' = 0.1x + 0.89y + 0.01z$. The second is a choice between $\rho = 0.11y + 0.89z$ and $\rho' = 0.1x + 0.9z$. Most people, when faced with these two choice situations, choose $\pi \succ \pi'$ and $\rho' \succ \rho$. Which would you choose?

This pair of choices is not consistent with the expected utility principle. To see this, let us write $u_h = u(2500000)$, $u_m = u(500000)$, and $u_l = u(0)$. Then if the expected utility principle holds, $\pi \succ \pi'$ implies $u_m > 0.1u_h + 0.89u_m + 0.01u_l$, so $0.11u_m > 0.10u_h + 0.01u_l$, which implies (adding $0.89u_l$ to both sides) $0.11u_m + 0.89u_l > 0.10u_h + 0.9u_l$, which says $\rho > \rho'$.

Why do people make this mistake? Perhaps because of *regret*, which does not mesh well with the expected utility principle (Loomes 1988; Sugden 1993). If you choose π' in the first case and you end up getting nothing, you will feel really foolish, whereas in the second case you are probably going to get nothing anyway (not your fault), so increasing the chances of getting nothing a tiny bit (0.01) gives you a good chance (0.10) of winning the really big prize. Or perhaps because of *loss aversion*, because in the first case, the anchor point (the most likely outcome) is $500,000, while in the second case the anchor is $0. Loss-averse individuals then shun π', which gives a positive probability of loss whereas in the second case, neither lottery involves a loss, from the standpoint of the most likely outcome.

The Allais paradox is an excellent illustration of problems that can arise when a lottery is consciously chosen by an act of will and one *knows* that one has made such a choice. The regret in the first case arises because if one chose the risky lottery and the payoff was zero, one knows for certain that one made a poor choice, at least ex post. In the second case, if one received a zero payoff, the odds are that it had nothing to do with one's choice. Hence, there is no regret in the second case. But in the real world, most of the lotteries we experience are chosen by default, not by acts of will. Thus, if the outcome of such a lottery is poor, we feel bad because of the poor outcome but not because we made a poor choice.

5.6.2   The Ellsberg Paradox

Another classic violation of the expected utility principle was suggested by Daniel Ellsberg (1961). Consider two urns. Urn A has 51 red balls and 49 white balls. Urn B also has 100 red and white balls, but the fraction of red balls is unknown. One ball is chosen from each urn but remains hidden from sight. Subjects are asked to choose in two situations. First, a subject can choose the ball from urn A or urn B, and if the ball is red, the subject wins \$10. In the second situation, the subject can choose the ball from urn A or urn B, and if the ball is white, the subject wins \$10. Many subjects choose the ball from urn A in both cases. This violates the expected utility principle no matter what probability the subject places on the probability p that the ball from urn B is white. For in the first situation, the payoff from choosing urn A is $0.51u(10)+0.49u(0)$ and the payoff from choosing urn B is $(1-p)u(10)+pu(0)$, so strictly preferring urn A means $p > 0.49$. In the second situation, the payoff from choosing urn A is $0.49u(10)+0.51u(0)$ and the payoff from choosing urn B is $pu(10)+(1-p)u(0)$, so strictly preferring urn A means $p < 0.49$. This shows that the expected utility principle does not hold.

Whereas the other proposed anomalies of classical decision theory can be interpreted as the failure of linearity in probabilities, regret, loss aversion, and epistemological ambiguities, the Ellsberg paradox appears to strike even more deeply because it implies that humans systematically violate the following principle of first-order stochastic dominance (FOSD).

> Let $p(x)$ and $q(x)$ be the probabilities of winning x or more in lotteries A and B, respectively. If $p(x) \geq q(x)$ for all x, then $A \succeq B$.

The usual explanation of this behavior is that the subject *knows* the probabilities associated with the first urn, while the probabilities associated with the second urn are *unknown*, and hence there appears to be an added degree of risk associated with choosing from the second urn rather than the first. If decision-makers are risk-averse and if they perceive that the second urn is considerably riskier than the first, they will prefer the first urn. Of course, with some relatively sophisticated probability theory, we are assured that there is in fact no such additional risk, so it is hardly a failure of rationality for subjects to come to the opposite conclusion. The Ellsberg paradox is thus a case of performance error on the part of subjects rather than a failure of rationality.

5.6.3 Failures of Judgment

Contemporary behavioral economics has developed a powerful critique of the standard assumption that people are instrumentally rational (Ariely 2010; Thaler and Sunstein 2008). In fact, human decision-makers are close to instrumentally rational when they are sufficiently informed and the cost of exploring alternative strategies is low (Gintis 2009a; Gigerenzer 2015).

Nevertheless, the behavioral economics critique of the assumption of instrumental rationality is important and well-taken.

But as we have seen, the rational actor model depicts *formal rationality*, not *instrumental rationality*. That is, it assumes that people have consistent preferences and update according to Bayes rule, but it does not assume that rational behavior is oriented towards any particular end state or goal, and certainly not that rational behavior furthers the fitness or welfare interests of the decision-maker. Let us review the major claims made by behavioral economists supporting the notion that choice behavior is fundamentally irrational.

- *Logical Fallibility*: Even the most intelligent decision-makers are prone to commit elementary errors in logical reasoning. For example, in one well-known experiment performed by Tversky and Kahneman (1983), a young woman, Linda, is described as politically active in college and highly intelligent. The subject is then asked the relative likelihood of several descriptions of Linda, including the following two: "Linda is a bank teller" and "Linda is a bank teller and is active in the feminist movement." Many subjects rate the latter statement more likely than the former, despite the fact that the most elementary reasoning shows that if p implies q, then p cannot be more likely than q. Because the latter statement implies the former, it cannot be more likely than the former.

- *Anchoring*: When facing extreme uncertainty in making an empirical judgment, people often condition their behavior on recent but irrelevant experience. For instance, suppose a subject is asked to write down a number equal to the last two digits of his social security number and then to consider whether he would pay this number of dollars for particular items of unknown value. If he is then asked to bid for these items, he is likely to bid more if the number he wrote down was higher.

- *Cognitive Bias*: If you ask someone to estimate the result of multiplying $1 \times 2 \times 3 \times 4 \times 5 \times 6 \times 7 \times 8$, he is likely to offer a lower estimate than if you had presented him with $8 \times 7 \times 6 \times 5 \times 4 \times 3 \times 2 \times 1$. Similarly, if you ask someone what fraction of English words end in "ng" and give the example "gong," you will probably get a lower estimate than if you gave the example "going."

- *Availability Heuristic*: People tend to predict the frequency of an event based on how often they have heard about it. For example, most people believe that homicides occur with more frequency than suicides, although the reverse is the case. Similarly, they believe that certain cancers cluster in certain communities because of environmental pollutants, where in fact such clusters may occur no more frequently than chance, but are more likely to be reported.

- *Status Quo Bias*: Decision-makers tend to follow a certain traditional pattern of behaviors even after there is strong credible evidence that a superior course of action is available. For instance, in an early well-known experiment, Samuelson and Zeckhauser (1988) presented subjects with a task in which several financial assets were listed and the subjects were asked to choose one that they prefer to invest in. A second set of subjects was given the same list of financial assets, but one was presented as the *status quo*. They found that the asset listed as the *status quo* was chosen at a much higher frequency than when it was presented just as one among several randomly presented alternatives.

- *Herd Mentality*: People are heavily influenced by the actions of others. For instance, Solomon Asch (1951) showed that peer pressure can induce subjects to offer clearly false evaluations, even when the subject and his peers do not know each other and will likely never meet outside the laboratory. Groups were formed consisting of eight college students, all but one of whom were confederates of the experimenter. Each student was shown a card with a black line on it, and a second card with three black lines, one of which was the same length as the line on the first card, and the other two were of very different lengths, one longer and the other shorter. Each student was asked to say out loud which line on the second card matched the line on the first card, the seven confederates going first and choosing an obviously incorrect line. More than a third of subjects agreed with the obviously wrong answer.

- *Framing Effects*: A framing effect is a form of cognitive bias that occurs when choice behavior depends on the wording of two logically equivalent statements. Take, for example, the classic example of the physician and his heart patient, analyzed by McKenzie et al. (2006) and Thaler and Sunstein (2008). The patient must decide whether to have heart surgery or not. His doctor tells him either (A) "Five years after surgery, 90% of patients are alive," or (B) "Five years after surgery, 10% of patients are dead." The two statements are of course logically equivalent, but subjects are far more likely to accept surgery with frame (A) than with frame (B).

- *Default Effects*: In choosing among various options, if one is offered as the default option, people tend to choose it with high frequency. A most dramatic example is organ donation (Johnson and Goldstein 2003). Countries in Europe that have a presumed consent default have organ donation rates that are about 60% higher than countries with explicit consent requirements. Another famous example involves registering new employees in a company 401(k) savings plan. When participating is the default, participation is considerably higher than when the default is non-participation (Bernheim et al. 2011).

The logical fallibility argument would of course be devastating to rational choice theory, which implicitly assumes that decision-makers are capable of making logical deductions. There are certainly complex logical arguments that the untrained subject is likely to get wrong. Indeed, even a mistake in mathematical computation counts as an error in logical reasoning. But what appear to be the elementary errors of the type revealed by the Linda the Bank Teller example are more likely to be errors of interpretation on the part of the experimenters. It is important to note that given the description of Linda, the probability that an individual is Linda if we know that the individual is a bank teller is much lower than the probability that an individual is Linda if we know that she is a feminist bank teller. This is because Linda is probably a feminist, and there are far fewer feminist bank tellers than there are bank tellers. Subjects in the experiment might reasonably assume that the experimenters were looking for a conditional probability response rather than a simple probability response because they supplied a mass of information that is relevant to conditional probability, but is quite irrelevant to simple probability.

Indeed, in normal human discourse, a listener assumes that any information provided by the speaker is relevant to the speaker's message (Grice 1975). Applied to this case, the norms of discourse reasonably lead the subject to believe that the experimenter wants Linda's politically active past to be taken adequately into account (Hilton 1995; Wetherick 1995). Moreover, the meaning of such terms as "more likely" or "higher probability" are vigorously disputed even in the theoretical literature, and hence are likely to have a different meaning for the average subject versus for the expert. For example, if I were given two piles of identity folders and asked to search through them to find the one belonging to Linda, and one of the piles was "all bank tellers" while the other was "all bank tellers who are active in the feminist movement," I would surely look through the latter (doubtless much smaller) pile first, even though I am well aware that there is a "higher probability" that Linda's folder is in the former pile rather than the latter one. In other words, conditional rather than straight probability is the appropriate concept in this case.

However important anchoring, cognitive bias, and the availability heuristic may be, they are clearly not in conflict with the rational actor model because they do not compromise any of the rational choice axioms. In particular, they do not imply preference inconsistency or the failure of Bayesian updating. The *status quo* bias may seem to contradict Bayesian updating, but it does not. For one thing, if one is satisfied with a particular choice, it may plausibly appear excessively costly to evaluate properly new information. Herbert Simon (1972) called this reasonable behavior "satisficing." It is clearly compatible with rational updating. For another, one may reasonably ignore new information on the grounds that it is unreliable. Models of Bayesian updating simply assume that the new information is rigorously factual, which is often not the case.

The framing effects literature is more challenging. Indeed, some argue that because it is impossible to avoid framing effects, there are no true underlying preferences, so the rational actor model fails. This conclusion is unwarranted. In this book we specified from the outset that preferences are generally state-, time-, and social frame-dependent. In particular, preferences are frame-dependent because individual choices, except perhaps for Robinson Crusoe before he meets Friday, occur within a social context, and that context is the social frame for choice behavior. Indeed, even the absence of a social frame is a social frame.

Consider, for instance, the physician and his heart patient scenario described above. Thaler and Sunstein (2008) interpret this as showing that many decision-makers are irrational. But it is more accurate to interpret these results as patients simply following the implicit suggestion of the physician, the expert on whom their well-being depends. Note first that neither (A) nor (B) gives the patient sufficient information to make an informed choice because the physician does not provide the equivalent survival and death rates *without* surgery. The only reasonable inference is that the patient believes the doctor is recommending surgery in case (A), and recommending against surgery in case (B).

The physician and his heart patient example is not an isolated case of the tendency for behavioral economists to ignore the intimately social nature of choice, and to interpret completely reasonable behavior as irrational. Gigerenzer (2015), who documents several additional examples of this tendency, concludes:

> Research...indicates that logical equivalence is a poor general norm for understanding human rationality...Speakers rely on framing in order to implicitly convey relevant information and make recommendations, and listeners pay attention to these. In these situations, framing effects clearly do not demonstrate that people are mindless, passive decision-makers.

Similarly, default effects do not illustrate the decision-maker's irrationality, but rather the tendency to treat the default as a recommendation by experts whose advice it is prudent to follow unless there is good information that the default is not the best choice (Johnson and Goldstein 2003). Indeed, Gigerenzer (2015) reports that a systematic review of hundreds of framing studies could not find a single one showing that framing effects incur real costs in terms.

6

An Analytical Core for Sociology

> Social life comes from a double source, the likeness of
> consciences and the division of social labor.
>
> Émile Durkheim

> **C:** What is the state of contemporary sociology?
> **HG:** Applied sociology is very strong.
> But there is no core sociological theory.
> Rather than building an analytical core,
> each theoretical tradition hawks its wares
> and denigrates the others.
>
> Choreographer interview

Modern societies are complex systems whose institutions are modified through legislation and popular collective action. This chapter offers an analytical framework for modeling the structure and dynamics of modern societies. Standard dynamical systems theory suggests that we first specify the conditions for social equilibrium, and then study the dynamical principles that govern disequilibrium behavior. The resulting *general social equilibrium model* is patterned after the highly successful Walrasian general equilibrium model (Arrow and Debreu 1954), and its dynamical principles can be modeled using evolutionary game theory (Weibull 1995; Helbing 1995; Gintis 2009b) and agent-based Markov models based on variants of the replicator dynamic (Helbing 1996, 2010; Gintis 2007a, 2013). We bring these two research tools together in Chapter 11.

Talcott Parsons initiated the formal modeling of modern societies in *The Structure of Social Action* (1937) and *Toward a General Theory of Action* (1951). As we explain in Chapter 7, this brilliant effort foundered for reasons unrelated to the scientific value of his project. Briefly, Parsons lacked analytical decision theory, stemming from Savage (1954), as well as game theory, which developed following Nash (1950). He also lacked an appreciation for general equilibrium theory, which came to fruition in the mid-1950s (Arrow and Debreu 1954). These powerful tools together allow us to formulate an analytical core for sociology. Second, Parsons followed Vilfredo Pareto (1896, 1906) in maintaining a strict separation between preferences over *economic values*, based on material self-interest on the one hand

and *social, political, and moral values*, involving concern for social life in the broadest sense, on the other. This separation precludes any general model of rational choice and social action (Lindenberg 1983, 2004; Fehr and Gintis 2007; Gintis 2009a).

Several scientific traditions contribute to a core analytical model for sociology. The first is the work of sociologists Max Weber, Émile Durkheim, George Herbert Mead, Ralph Linton, Talcott Parsons, and others, whose insights have so far largely escaped analytical expression and are little known, despite their extreme relevance, beyond the sociology discipline.

The second is a model of individual choice behavior, which is a broadened version of rational decision theory (Savage 1954; Fishburn 1970; Gintis 2009a). The two behavioral disciplines that include a core analytical model, biology and economics, are built around the notion of rational choice. This theory is useful in conjunction with game theory which, while widely applied in sociobiology in general (Alcock 1993; Krebs and Davies 1997; Dugatkin and Reeve 1998), is especially important for humans (Bowles and Gintis 2011; Wilson 2012; Tomasello 2014) because *Homo sapiens* is not only *Homo socialis*, but also *Homo ludens*—Man, the game player. As developed in Chapter 3, our species has the capacity to construct novel games with great flexibility and its members possess the cognitive and moral requirements for game-playing. A major innovation in this respect is our expansion of Thomas Schelling's notion of a *focal point equilibrium* (Schelling 1960) by proposing the *correlated equilibrium*, rather than the more standard Nash equilibrium, as the basis of an analytical model of social norms (Aumann 1987a; Gintis 2009a).

The third tradition is the *general equilibrium model* of Léon Walras (1874), Kenneth Arrow and Gérard Debreu (1954), and others, which is analytically rigorous and mathematically elegant. Despite its appearance of extreme abstraction, it is in fact capable of a surprisingly straightforward and plausible extension to a general social equilibrium model of considerable sophistication.

Modeling social dynamics is significantly more challenging than modeling social equilibrium because society is a complex dynamical system: it consists of many structurally similar, strongly interacting, and intricately networked units (social actors) operating in parallel with little centralized structural control (Miller and Page 2007). Such complex systems generically exhibit emergent properties at the macrosystem level that resist analytical derivation from the behavior of the individual parts (Morowitz 2002).

The fourth intellectual strand is *evolutionary game theory*, a field that did not exist until recently (Aumann 1987b; Weibull 1995; Gintis 2009b; Grund et al. 2013), as well as agent-based modeling of stochastic behavior (Helbing 1995, 2012; Gintis 2009b, 2013). Chapter 11 deploys these tools in the context of a simple model of general market exchange. We show that such an economy always has a stable equilibrium in which supply equals demand in all markets, but the degree of volatility of prices and quantities in such an economy can only be assessed through computer modeling (Mandel and Gintis 2014; Mandel and Gintis 2016).

The fifth foundational element is *behavioral game theory* (Camerer 2003; Gintis 2009a; Dhami 2016), based on laboratory (Fehr and Gintis 2007) and field (Carpenter et al. 2005a; Herbst and Mas 2015) experimentation into choice and social interaction. Behavioral game theory, which provides the empirical basis for the generalization of rational choice theory to include moral, social, and other-regarding values (Camerer and Fehr 2004; Fehr and Gintis 2007; Dhami 2016).

6.1 Game Theory

Game theory studies how rational players interact when the choices of each player affect the payoffs to other players. Game theory is a general lexicon that applies to all life forms. Strategic interaction neatly separates living from nonliving entities and in an important sense defines life itself. Indeed, strategic interaction is the sole concept commonly used in the analysis of living systems that has no counterpart in physics or chemistry. The concept of strategic interaction is central to understanding the behavior of living creatures, from the single-celled bacterium to the most complex and highly evolved creature.

Game theory forces us to supply the precise information we need to explain social interaction, including the characteristics of the players, the rules of the game, the information available to the players, and the payoffs associated with particular player choices. Game theory thus contributes to the analytical framework underlying all the behavioral disciplines.

The most famous equilibrium concept in game theory is the *Nash equilibrium*, which is a choice of a strategy by each player such that no player can gain by switching to a different strategy, holding fixed the strategies of the other players (Nash 1950). I will suggest later that a superior concept is that of the *correlated equilibrium*, described below.

One of the most important contributions of game theory is its role in the methodology of behavioral research in laboratory and field. With *behavioral game theory*, experimentalists specify the exact conditions to which subjects are exposed, and hence render the results of experiments relatively easy to replicate, revise, and extend (Plott 1979; Smith 1982; Sally 1995; Dhami 2016). Behavioral game theory allows us to use the results of laboratory experiments to generate plausible and testable models of real-life social behavior (Fehr and Gintis 2007).

It is impossible to draw systematic inferences from behavioral game theory without presupposing some version of the rational actor model. That is, we cannot model the behavioral regularities of subjects unless we assume their choices reflect underlying preferences and beliefs. For instance, in the *ultimatum game*, discussed in Chapter 2, responders reject very unequal offers, even though they lose money thereby. If we assume the rational actor model, then responder Alice must have some positive entry in her utility function to offset the monetary loss from rejecting a positive offer. To see that Alice values punishing an unfair offer by proposer Bob, we note that if the offer was generated by a computer rather than by Bob, and if Alice knows this, she will generally accept even very low offers (Blount 1995). The reason simply is that the unequal offer was not Bob's fault, so Alice has no basis for holding this against him.

Some have criticized this interpretation on the grounds that Alice may simply be unused to one-shot anonymous games, and hence might incorrectly consider her rejection as establishing a reputation as a hard bargainer that would be useful in future interactions (Binmore and Shaked 2010). However, even very sophisticated players turn down positive offers in the ultimatum game, and players clearly react strategically to subtle changes in experimental conditions. Moreover, when responders are asked why they rejected positive offers, they often reply that the proposer was unfair (Eckel and Gintis 2010).

The central solution concept in game theory is the *Nash equilibrium*, which is a situation in which no player has an incentive to change his behavior, given the behavior of the other players (Nash 1950). Several insightful theorists have modeled social norms as Nash equilibria of games played by rational agents, including David Lewis (1969), Michael Taylor (1976, 1982, 1987), Robert Sugden (1986, 1989), Cristina Bicchieri (1993, 2006), and Ken Binmore (1993, 1998, 2005). However, for sociological theory, the concept of a *correlated equilibrium*, developed by Robert Aumann (1987a), is more appropriate.

The model of social norms as correlated equilibria has an attractive property lacking in the notion of social norms as Nash equilibria: the conditions under which rational agents play Nash equilibria are generally complex and implausible, whereas rational agents in a very natural sense play correlated equilibria, provided they have common knowledge of the correlating device. For instance, Thomas Schelling's notion of a *focal point* equilibrium can be interpreted as a correlated equilibrium. Consider the situation of two friends who agree to have lunch in the city but fail to state exactly where and at what time to meet. There are an infinite number of Nash equilibria for this situation, one for each time and place in the city. The chances the two friends will agree on which Nash equilibrium to implement are extremely small.

The insight underlying the Nash equilibrium approach to social norms is that if agents play a game with several Nash equilibria, a social norm can serve to choose a single one among them. The Kantian equilibrium explored in Chapter 3 is of this form. While this insight applies to several important social situations, it is insufficiently broad for a core analytical model of social norms. The broader concept of *correlated equilibrium* (Aumann 1974, 1987a) in fact better captures the notion of a social role.

A correlated equilibrium consists of a *correlating device*, which I sometimes call the *Choreographer*, that sends a signal indicating a suggested action to each social actor, and perhaps implementing sanctions if the actor does not take a recommended action, such that the actor, for both material and moral reasons, does best by obeying the Choreographer's suggestion, provided the other relevant social actors do so as well. While the notion of a Choreographer accurately captures the effect of a correlating device's fostering of social cooperation, it is wrong to think of the Choreographer as a dictator who rules by force. Social norms generally will not be followed when they are not considered legitimate, whatever the social sanctions entailed by the discovery of violations. Moreover, social norms generally are instantiated and changed through collective action, so that the Choreographer itself is the product of a social will (Gintis 1975; Winter et al. 2012).

Whereas the epistemological requirements for rational agents playing Nash equilibria are very stringent and usually implausible, the requirements for a correlated equilibrium amount to the existence of *common priors*, which we interpret as induced by the cultural system of the society in question. In this view, human beings may be modeled as rational agents with special neural circuitry dedicated to reacting to, evaluating, and sustaining social norms by recognizing and responding to Choreographer signals and incentives.

When the Choreographer has at least as much information as the players, we need in addition only to posit that individuals obey the social norm when this maximizes their payoffs. When players have some information that is not available to the Choreographer, so that not all social roles can be properly carried out by self-regarding agents, role occupants must have a moral predisposition to follow the norm even when it is costly to do so. The latter case explains why social norms are associated with prosocial preferences. For instance, a system of traffic lights can serve as the Choreographer for controlling vehicular traffic, but when a police officer issues a traffic ticket to a driver, both must internalize the immorality of bribe-taking and bribe-offering for the institution to work properly.

Social norms as correlating devices are not explained by game theory and the rational actor model, but rather are irreducible expressions of social organization. Social norms provide a dimension of causal efficacy to social theory, whereas standard game theory alone recognizes no causal efficacy above the level of individual choice behavior. Because of the independent causal effectivity of social norms, the methodological individualism of standard economic theory is untenable. In particular, social norms are predicated upon certain mental predispositions, a *social epistemology*. This social epistemology fosters the interpersonal sharing of mental concepts, and justifies the assumption of common priors upon which the efficacy of the correlated equilibrium rests.

There is also a serious game-theoretic weakness of the standard rational choice model: there is no appreciation for the concept of *social rationality* in its defining principles. Several economists, decision theorists, and philosophers have explored a more socially relevant form of rationality than those embodied in the standard von Neumann-Morgenstern and Savage axioms. They term these forms variously "we-reasoning," "team reasoning," and "collective intentionality" (Bacharach 1987, 1992, 2006; Bacharach et al. 2006; Bratman 1993; Colman et al. 2008; Gilbert 1987, 1989; Hurley 2002; Searle 1995; Sugden 2003; Tuomela 1995). Several analytically clear examples of such choice behaviors that should appear in any plausible account of social rationality applicable to an analytical core for the behavioral disciplines were presented in Chapter 3.

6.2 Complexity

Complexity theory is the study of the emergent properties of nonlinear dynamical systems, of which complex social systems are prime examples. Complexity theory complements the analytical methods of game theory and the rational actor model, dealing with society in more macrolevel, interpretive terms, and developing schemas that shed light where analytical models cannot penetrate. Anthropological and historical studies fall into this category, as well as macroeconomic policy and comparative economic systems (Tesfatsion and Judd 2006). Agent-based modeling of complex dynamical systems is useful in dealing with emergent properties of complex systems.

  We develop below the concept of a *general social equilibrium*, which is a natural generalization of the concept of *general market equilibrium*. Because the market economy is a prime example of a complex dynamical system (see Chapter 11), it follows that society as a whole is a complex system. Society, like the market economy, follows an *evolutionary dynamic*. A complex society is never in equilibrium, but is constantly subjected to shocks, both exogenous and endogenous, that affect its short-term movements. There are frequent local nonlinear resonances that lead to significant deviations of social variables (for instance, in the economy, prices, quantities, wages, asset prices) from their equilibrium values even in the absence of strong or systematic perturbations to the system. We see such deviations quantitatively in many economic time series, which often have the "fat tails" characteristics of the power laws of complex systems, as opposed to the Gaussian distributions of neoclassical systems (Farmer and Lillo 2004).

  General social dynamics are quite poorly understood, but there have been notable contributions to complexity economics in recent years. These include Eric Beinhocker's study of complex macrodynamics (Beinhocker 2006), Brian Arthur's work on increasing returns (Arthur 1994), Peyton Young and Mary Burke's analysis of crop sharing (Young and Burke 2001), evolutionary models inspired by Richard Nelson and Sidney Winter (Nelson and Winter 1982) and Geoffrey Hodgson (1998), William Brock and Stephen Durlauf's study of social interaction (Brock and Durlauf 2001), Edward Glaeser, Bruce Sacerdote, and Jose Scheinkman's treatment of crime (Glaeser et al. 1996), Samuel Bowles' treatment of institutional evolution (Bowles 2004), Robert Axtell's study of firm size (Axtell 2001), Alan Kirman and his colleagues' models of financial markets (Kirman et al. 2005), and models of the evolution of other-regarding preferences (Gintis 2000; Bowles et al. 2003) and the agent-based simulation of general equilibrium and barter exchange (Gintis 2007a, 2013). Tesfatsion and Judd (2006) is a comprehensive overview of computational methods in complexity economics.

The following is a useful summary of social complexity theory.

a. *Dynamics*: The complex society is thermodynamically open, dynamic, nonlinear, and generally far from equilibrium, whereas an equilibrium system is thermodynamically closed, static, and smooth in the sense that it can be understood using manifold and classical dynamical systems theory.

b. *Agents*: In a complex society, agents have limited information and face high costs of information processing. However, under appropriate conditions, they evolve non-optimal but highly effective heuristics for operating in complex environments. There is no assurance that when faced with novel environments, individuals will shift efficiently to new heuristics. In equilibrium, by contrast, agents have perfect information and can costlessly optimize.

c. *Networks*: Agents in a complex society participate in sophisticated overlapping networks that allow them to compensate for having limited information and facing formidable information processing costs.

d. *Emergence*: In a complex society, macrosocial patterns are emergent properties of micro-level interactions and behaviors, in the same sense as the chemical properties of a complex molecule, such as carbon, is an emergent property of its nuclear and electronic structure, or that thermodynamics is an emergent property of many-particle systems. In such cases we cannot analytically derive the properties of the macro system from those of its component parts, although we can apply novel mathematical techniques to model the behavior of the emergent properties. In the case of a complex society, these higher-level modeling constructs are currently largely absent, although agent-based modeling may provide the data needed to develop the appropriate mathematical tools. By contrast, the neoclassical macroeconomic model, for instance, has no global properties that cannot be derived from its micro properties (such as the First and Second Welfare Theorems).

e. *Evolution*: In a complex society, the evolutionary process of differentiation, selection, and amplification provides the system with novelty and is responsible for the growth in order and complexity. In equilibrium, there is no mechanism for creating novelty or growth in complexity.

This description applies well to institutional and organizational development, cultural change, and even scientific discovery. The evidence for this view is that almost all attempts at technical or institutional innovation fail, and few individuals are responsible for more than one successful innovation. How, then, are great thinkers possible? When asked how he was able to make so many discoveries, Linus Pauling replied: "You must have lots of ideas and just throw away the bad ones." Great thinkers for the most part simply are attuned to generating mutant ideas, they evaluate more effectively the prospects for a new mutant idea, and discard more rapidly the defective mutations. A similar argument likely obtains for technical change, institutional innovation, and product innovation.

My own foray into modeling general equilibrium as a complex system, as we shall see in Chapter 11, suggests that the general market equilibrium model of Walras will emerge enriched rather than replaced as a result of such research. General equilibrium theory captures important long-term aspects of a market economy, and many of the basic insights of the Walrasian model will be retained. Even in the long run, there will be a strictly positive rate of unemployment, supply will exceed demand for quality goods and services, efficiency will be considerably less than 100%, and there will be other deviations from equilibrium due to incomplete information and "frictions" amplified by local nonlinear resonances. Moreover, the Walrasian assumption that agents are price-takers, that complete contracts can be written for all important exchanges and can be costlessly enforced by a third party, are all unrealistic. Hence, the Walrasian system is a very poor guide to micro-modeling real economic transactions. In particular, as we have seen in Chapter 4, the Walrasian assumptions concerning labor markets, capital markets, and consumer goods markets are misleading (Bowles and Gintis 1993; Gintis 2002).

6.3   Roles, Actors, and the Division of Social Labor

A society includes a network of *social roles* (Mead 1934; Linton 1936; Parsons and Shils 1951). The *content* of a social role is a set of rights, duties, expectations, material and symbolic rewards, and behavioral norms. In equilibrium, the content of social roles is public information shared by all members of society, and this content influences the mutual expectations of individuals involved in social interaction. In periods of social change, the content of particular roles are subject to conflicting forces and the process of re-establishing a common understanding of role-contents involves dialog, collective action, and the exercise of power.

Role-occupants are *actors* who fill many different and contrasting roles in the course of performing their daily activities. An individual may act as spouse preparing breakfast, as parent advising children on the day's activities, as sales manager in an enterprise, as school committee and church member, and as voter.

We model actors as rational decisions-makers who maximize their preference functions subject to the content of the social roles they occupy, and given a belief system that is context-dependent and governed by the expectations defined by the actor's social location. These decisions determine the social actors' role-specific behaviors. For instance, when one engages a taxi in a strange city, both the driver and the client may know exactly what is expected of each, so no time or energy is wasted bargaining or otherwise adjudicating mutually acceptable behavior.

Social roles generally promote various forms of cooperation and competition among social actors, and hence tend to be bundled into what we may call *social frames*. For instance, a checkout line at the supermarket, a restaurant, or a public restroom are social frames. Social norms and conventions generally regulate appropriate behavior in a social frame. An especially important social frame is an *organization*, such as a firm, a hospital, a social service agency, or an organized sport. Organizations are conveniently modeled as noncooperative games, where the formal rules, conventions, and payoffs are set largely by those who control the organization. These rules and conventions determine the role-structure of the organization, and players are individual actors who fill the organizationally defined roles. These players choose behaviors based on the game structure as well as their personal moral values and social commitments. Of course, the full game structure includes informal interactions among role-players, with associated rewards and sanctions (Aoki 2010).

Social actors filling particular social roles can be modeled by appropriately enriching the *general equilibrium model* of economic theory (Walras 1874; Arrow and Debreu 1954). In this general economic equilibrium model, actors are *owners of productive resources*, which they supply to firms, and they are *consumers of the goods and services* produced by firms. Productive resources include capital goods, raw materials, and various sorts of labor services. Firms combine productive resources to generate marketable commodities, choosing a pattern of inputs and outputs to maximize profits, given the prices they face. Social actors choose their pattern of consumption, as well as their supply of services to firms, to maximize their preference functions at given prices. An economic equilibrium occurs when prices are such that the plans of all agents are simultaneously satisfied.

The firm in a market economy is a game in which owners, managers, and employees are players. In this sense, the firm is just one among an array of organizational forms. We can then view general economic equilibrium as a special case of *general social equilibrium*. This sociological broadening of general economic equilibrium is quite natural, because it is reasonable to view a position in the firm as a social role whose content includes not only the salary and the employee's obligation to come to work, but also a set of rights and behavioral norms, as well as a pattern of symbolic rewards and sanctions determined by the culture of the firm and the larger society (Aoki 2010).

While interpreters generally stress the price system as the key element in adjudicating among the interests of economic actors, the theory becomes more powerful if the general content of social roles is viewed as adjusting when out of equilibrium (Granovetter 1985, 1995; DiMaggio 1994, 1998; Hechter and Kanazawa 1997; Hedström and Bearman 2009). The general economic equilibrium model recognizes only one social institution: profit-maximizing firms. Families in this model are treated as "black boxes," as is government, if it is treated at all. The general social equilibrium model must add at a minimum *families* and *communities*, as well as *public institutions* and *private associations*, such as governmental, religious, scientific, charitable, and cultural organizations. These organizations are constrained in their internal organization of social roles to maintain a positive balance sheet, but otherwise can determine their organization of social roles according to criteria other than profitability. A theory of the family, for instance, would suggest how the limits of family membership are determined, what social roles are occupied by family members, and how content of these roles is determined.

The general economic equilibrium model assumes that in equilibrium all agents have appropriate information concerning the nature of the goods and services they exchange and the prices at which they exchange. The same must be true of general social equilibrium. Out of equilibrium, however, the content of social roles, including their material, social, and moral attributes, are statistical distributions over which individuals have subjective and networked probability distributions. This corresponds to the fact that in the general economic equilibrium model, out of equilibrium there is no basis for forming price expectations except by networked experience, which may differ significantly across economic agents (Gintis 2007a). For instance, in deciding whether to take a job at a particular wage, the worker must consider the return to continuing job search, which will depend on the statistical distribution of demand for labor in the economy. The worker has only his networked experience to estimate this distribution, and such experience can vary widely among workers with similar credentials and demographics.

The general economic equilibrium model embodies many assumptions that render the model amenable to analytical modeling. Some of these can be relaxed if we desire finer detail, but one assumption is especially important in this regard. This is the assumption that the characteristics of the goods exchanged are *completely known* to all parties. This is only possible if all contracts between two parties are costlessly and perfectly enforced by a third party—presumably the judicial system. As we have seen in Chapter 4, this assumption is seriously violated in dealing with labor services, where the effort and care of the employee is not subject to third-party enforcement. It is similarly violated in dealing with capital transactions, where the promise of a borrower to repay cannot be enforced if the borrower is bankrupt. Most basic aspects of a market economy are precisely a response to the need for endogenous contract enforcement.

In general social equilibrium, each actor maximizes his preference function in the sense that no change of role will increase his expected payoff, taking into account possible search and relocation costs, and the pattern of supply and demand for social roles will be such that expected payoffs will not change over time. In addition, if there are institutions, such as firms, hospitals, families, communities, or governments, these institutions may have certain social conditions that must be satisfied in equilibrium, such as a balance between expenditures and receipts, or achievement of certain institutional goals. When an organization is modeled as a game, in equilibrium all members play their part in a Nash or correlated equilibrium of the game (see Section 6.6).

In proposing the actor/role model, sociologists have traditionally held that the major difference between social and economic roles is that social roles function properly only by virtue of the moral commitments of role-occupants, whereas economic roles function independently from role-occupants' social conscience and moral commitments. To achieve its purported independence from moral commitment, general economic equilibrium models make the implausible assumptions of *complete contracts*, meaning that any contract between individuals, however complex, covers all possible contingencies and can be enforced by a third party (the judicial system) at no cost to the contracting parties (see Chapter 4). After dropping this assumption from the general economic equilibrium model, moral commitments become as salient in economic life as they are in social life in general.

The major effect of conceiving of the general network of social roles as an expansion of the general economic equilibrium model is the clarification it lends to the distinction between equilibrium and dynamic models of society. The general economic equilibrium model is a static construct that gives no suggestion as to how equilibrium might be attained. This is a critical limitation, just as is the parallel limitation of the general social equilibrium model developed in this chapter. While Chapter 11 provides a plausible dynamic for the general economic equilibrium model and proves the stability of equilibrium for this dynamic, this proof does not extend to the general social equilibrium model.

6.4   The Socio-psychological Theory of Norms

Émile Durkheim (1902) was the first to recognize the social tension in modern society caused by an increasingly differentiated social role structure, the social division of labor, and the need for a common base of social beliefs and values, which he terms *collective consciousness*, to promote social harmony and efficient cooperation. Durkheim's theme was developed into a theory of *social norms* by Ralph Linton (1936) and George Herbert Mead (1934), and integrated into a general social theory by Talcott Parsons (1937). Social norms are often promulgated by a nexus of system-wide cultural institutions and social processes that in equilibrium produce a consistent set of expectations and normative predispositions across all social actors. The *socio-psychological theory of norms* models this social subsystem and accounts for their effectivity. Other social norms govern well-defined subsets of the population, such as religious groups, professional associations, and sports. Out of equilibrium, conflicting social norms often vie for dominance, and cultural dynamics are often the result of these conflicts (Coser 1956; Winter et al. 2012).

In the first instance, the complex of social rules has an *instrumental* character devoid of normative content, serving merely to associate rewards and penalties with behavior, and as an informational device that coordinates the behavior of rational agents (Lewis 1969; Gauthier 1986; Binmore 2005; Bicchieri 2006). A social rule with this character we term a *convention*. Conventions thus supply the general factual descriptions of the content of many standard social roles (employer, worker, mother, judge, traffic cop, taxi driver, and the like), allowing social actors to coordinate their behavior even when dealing with unfamiliar social partners in novel situations. Conventions thus create *common subjective priors* that facilitate general social cooperation.

However in many social roles high-level performance requires that the actor have a *personal commitment* to role performance that cannot be captured by the self-regarding "public" rewards and penalties associated with the role (Conte and Castelfranchi 1999; Gintis 2009a). For instance, a physician may be obligated to ignore personal gain when suggesting medical procedures, only the most egregious of violations of which will incur serious social sanctions. The need for a normative content to social roles follows from the fact that (a) a social actor may have private, publicly inaccessible payoffs that conflict with the public payoffs associated with a role, inducing him to act counter to appropriate role-performance given by the content of the social role (e.g., corruption, favoritism, aversion to specific tasks); (b) the signal used to determine the public payoffs may be inaccurate and unreliable (e.g., the performance of a teacher or physician); and (c) the public payoffs required to gain compliance by self-regarding actors may be higher than those required when there is at least partial reliance upon the moral commitment of role incumbents (e.g., it may be less costly to employ personally committed rather than purely materially motivated physicians and teachers). In such cases, self-regarding actors who treat social norms purely instrumentally will behave in a socially inefficient and morally reprehensible manner.

The normative aspect of social roles is motivating to social actors because to the extent that social roles are considered *legitimate*, role-occupants normally place an intrinsic positive ethical value on role-performance (Andreghetto et al. 2013). This may be termed the *normative bias* associated with role-occupancy (Bicchieri 2006; Gintis 2009a). Second, human ethical predispositions include *character virtues*, such as honesty, trustworthiness, promise-keeping, and obedience, that may increase the value of conforming to the duties associated with role-incumbency (Aristotle, 2002[350 BC], Ullmann-Margalit 1977). Third, humans are predisposed to care about the esteem of others even when there can be no future reputational repercussions (Smith 1759; Mead 1934; Masclet et al. 2003), and take pleasure in punishing others who have violated social norms even when they can gain no personal advantage thereby (Güth et al. 1982; Gintis 2000; Fehr and Fischbacher 2004). These normative traits by no means contradict rationality, because individuals trade off these values against material reward, and against each other, just as described in the economic theory of the rational actor (Gneezy and Rustichini 2000; Andreoni and Miller 2002).

6.5   Socialization and the Internalization of Norms

Society is held together by *moral values* that are transmitted from generation to generation by the process of *socialization*. A *social norm* is based on generally accepted moral values. Thus obedience to legitimate authority, being quiet in a library, or bribing a police officer may be social norms. Moral values are instantiated through the *internalization of norms* (Parsons 1967; Grusec and Kuczynski 1997; Nisbett and Cohen 1996; Rozin et al. 1999), a process in which the initiated instill values into the uninitiated, usually the younger generation, through an extended series of personal interactions, relying on a complex interplay of affect and authority. Through the internalization of norms, initiates are supplied with moral values that induce them to conform voluntarily and even oftimes enthusiastically to the duties and obligations of the role-positions they are expected to occupy. In addition, the adherence to social norms is socially reinforced by the approval and rewards offered by prosocial individuals, and the decentralized punishment of norm violation by concerned individuals (Gintis 2000; Fehr and Fischbacher 2004). Moreover, humans acquire social norms simply through the action of homophily, imitation of the behavior and acquiring the value of social peers (Kandel 1978; McPherson et al. 2001; Durrett and Levin 2005).

The internalization of norms of course presupposes a genetic predisposition to moral cognition that can be explained only by gene-culture coevolution (Boyd and Richerson 1985, 2004; Gintis 2003a, 2011; Haidt 2001).

It is tempting to treat some norms as inviolable *constraints* that lead the individual to sacrifice personal welfare on behalf of morality, but virtually all norms are violated by individuals under some conditions, indicating that there are tradeoffs that could not exist were norms merely constraints on action. In fact, internalized norms are accepted not as instruments towards achieving other ends, but rather as ends in themselves—*arguments in the preference function that the individual maximizes*. For instance, an individual who has internalized the value of speaking truthfully will do so even in cases where the net payoff to speaking truthfully would otherwise be negative. Such fundamental human emotions as shame, guilt, pride, and empathy are deployed by the well-socialized individual to reinforce these prosocial values when tempted by the immediate pleasures of such deadly sins as anger, avarice, gluttony, and lust.

The human openness to socialization is perhaps the most powerful form of epigenetic transmission found in nature. This preference flexibility accounts in considerable part for the stunning success of the species *Homo sapiens*, because when individuals internalize a norm, the frequency of the desired behavior will be higher than if people follow the norm only instrumentally—i.e., when they perceive it to be in their best interest to do so on self-regarding grounds. The increased incidence of prosocial behaviors are precisely what permits humans to cooperate effectively in groups (Gintis et al. 2005).

There are, of course, limits to socialization (Wrong 1961; Gintis 1975; Tooby and Cosmides 1992; Pinker 2002), and it is imperative to understand the dynamics of emergence and abandonment of particular values, which in fact depend on their contribution to fitness and well-being, as economic and biological theory would suggest (Gintis 2003a, 2003b). Moreover, there are often swift society-wide value changes that cannot be accounted for by socialization theory. For instance, movements for gender and racial equality have been highly successful in many countries, yet initially opposed all major socialization institutions, including schools, churches, the media, and the legal system.

6.6   A Model of Norm Internalization

For analytical specificity, we study the dynamics of a single altruistic norm that has a payoff disadvantage for those who adopt it, but is transmitted vertically by parents and obliquely through socialization institutions. We allow altruism to be either beneficial or harmful to the group, and we admit four types of cultural change.

- Individuals mate and have offspring. Families who use lower payoff strategies have fewer offspring (biologically adaptive dynamics).

- Families pass on their cultural traits, self-interested or altruistic, to their offspring (vertical transmission) through internalization.
- A fraction of self-interested offspring are induced to adopt altruistic norms by socialization institutions (oblique transmission).
- Some members of the resulting population change their cultural values to conform to the behavior of other individuals who have higher payoffs (replicator dynamics).

This model yields two general conclusions.

- In the absence of oblique transmission of the altruistic norm, altruism is driven out by self-interested behavior. When oblique transmission of altruism is present, a positive frequency of altruism can persist in cultural equilibrium.
- A high level of cooperation can be sustained in cultural equilibrium by the presence of a minority of agents who adopt the altruistic norm of what I call *strong reciprocity*: cooperating unconditionally and punishing defectors at a personal cost, the remaining agent being self-interested.

The first assertion states what might be called the Fundamental Theorem of Sociology: *extra-familial socialization institutions are necessary to support altruistic forms of prosociality*. The second assertion expresses the insight that cooperation is robustly stable when antisocial behavior is punished by the voluntary, and largely decentralized, initiative of group members (Helbing et al. 2010).

Because social norms generally have a strong moral component, construction dynamic models of the evolution of social norms is an inherently complex and ill-understood process. For instance, social norms concerning gender roles or inter-ethnic relationships can persist for many generations and then change extremely rapidly. Such changes are virtually unpredictable given the current state of social theory. Conventions, by contrast, may be more or less desirable on social efficiency grounds, but because they lack a moral component, they are more easily modeled and understood.

A *convention* is a correlated equilibrium of a coordination game. A *coordination game* is defined as follows. Suppose there is some social activity that requires the cooperation of one or more types of social actor. For instance, the activity may be building a wall. The types of social actor may be "bricklayer" and "assistant." Cooperation is successful when the bricklayer asks for a piece of building material and the assistant provides the proper material. The social convention may be that the bricklayer shows one finger when he wants a brick, two fingers when he wants some mortar, and three fingers when he wants a bucket of water. A second convention may be to show one finger for a bucket of water, two fingers for some mortar, and to say "ladrillo" for a brick. It does not much matter what the particular sign is for each of the three possibilities, just so both the bricklayer and his assistant agree, and the assistant has some incentive to obey the requests of the bricklayer.

There are several plausible models of the evolution and transformation of conventions (Kandori et al. 1993; Young 1993, 1998) based on the notion of a Markov process. Section 6.7 provides a simple but representative example of this approach to modeling the evolution of conventions.

6.7   The Evolution of Social Conventions

A *Markov process* $\mathcal{M}$ consists of a finite number of *states* $S = \{1, \ldots, n\}$, and an n-dimensional square matrix $P = \{p_{ij}\}$ such that p_{ij} represents the probability of making a transition from state i to state j. A *path* $\{i_1, i_2, \ldots\}$ determined by Markov process $\mathcal{M}$ consists of the choice of an initial state $i_1 \in S$, and if the process is in state i in period $t = 1, 2, \ldots$, then it is in state j in period $t + 1$ with probability p_{ij}. Despite the simplicity of this definition, finite Markov processes are remarkably flexible in modeling dynamical systems, although characterizing their long-run properties becomes challenging for systems with many states.

I will use the Markov process as a tool to model the evolution of money as a convention in trade among many individuals. Consider a rudimentary economy in which there are g goods, and each social actor produces one unit of one of these goods in each period. After production takes place, individuals encounter one another randomly and they trade equal amounts of their wares if each wants what the other is offering. However, it often happens that one of the pair does not consume what the other produces, so no direct trade is possible. However, suppose each social actor is willing to accept one of the g goods not for consumption, but rather to use as *money* in trading with other producers. The use of money increases the efficiency of the economy because the frequency of welfare-increasing trades is higher with the use of money. Moreover, it is clear that the highest efficiency would be attained if all social actors were willing to accept the *same* good as money. Under what conditions might this occur without a central government or other macrosocial institution bringing this about?

To pose the question more formally, what is the long-run distribution of the fraction of the population accepting each of the g goods as money? To answer this question, we must make some assumption concerning how individual traders decide to change the good they are willing to accept as money. We simply assume that one of the n traders in the economy in each period switches to the money type of a randomly chosen trading partner. We represent the state of the economy as $(w_1 \ldots w_g)$, where w_i is the number of agents who accept good i as money. The total number of states in the economy is thus the number of different ways to distribute n indistinguishable balls (the n agents) into g distinguishable boxes (the g goods), which is $C(n + g - 1, g - 1)$, where

$$C(n, g) = \frac{n!}{(n - g)!g!}$$

is the number of ways to choose g objects from a set of n objects. For instance, if there are 100 social actors ($n = 100$) and ten goods ($g = 10$), then the number of states S in the system is $S = C(109, 9) = 4{,}263{,}421{,}511{,}271$.

To verify this formula, write a particular state in the form

$$s = x \ldots x A x \ldots x A x \ldots x A x \ldots x$$

where the number of x's before the first A is the number of agents choosing type 1 as money, the number of x's between the $(i - 1)$th A and the ith A is the number of agents choosing type i as money, and the number of x's after the final A is the number agents choosing type g as money. The total number of x's is equal to n, and the total number of A's is $g - 1$, so the length of s is $n + g - 1$. Every placement of the $g - 1$ A's represents a particular state of the system, so there are $C(n + g - 1, g - 1)$ states of the system.

Suppose in each period two agents are randomly chosen and the first agent switches to using the second agent's money type as his own money. This gives a determinate probability p_{ij} of shifting from one state i of the system to any other state j. The matrix $P = \{p_{ij}\}$ is called a *transition probability matrix*, and the whole stochastic system is clearly a finite Markov process.

What is the long-run behavior of this Markov process? Note first that if we start in state i at time $t = 1$, the probability $p_{ij}^{(2)}$ of being in state j in period $t = 2$ is simply

$$p_{ij}^{(2)} = \sum_{k=1}^{S} p_{ik} p_{kj} = (P^2)_{ij}. \tag{6.1}$$

This is true because to be in state j at $t = 2$ the system must have been in some state k at $t = 1$ with probability p_{ik}, and the probability of moving from k to j is just p_{kj}. This means that the two period transition probability matrix for the Markov process is just P^2, the matrix product of P with itself. By similar reasoning, the probability of moving from state i to state j in exactly r periods is P^r. Therefore, the time path followed by the system starting in state $s^0 = i$ at time $t = 0$ is the sequence $s^0, s^1, \ldots$, where

$$P[s^t = j | s^0 = i] = (P^t)_{ij} = p_{ij}^{(t)}.$$

The matrix P in this example has $S^2 \approx 1.818 \times 10^{15}$ entries. The notion of calculation P^t for even small t is quite infeasible. There are ways to reduce the calculations by many orders of magnitude (Gintis 2009b, Ch. 13), but these methods are completely impractical with so large a Markov process.

Nevertheless, we can easily understand the dynamics of this Markov process. We first observe that if the Markov process is ever in the state

$$s_*^r = (0_1, \ldots, 0_{r-1}, n_r, 0_{r+1}, \ldots, 0_k),$$

where all n agents choose type r money, then s_*^r will be the state of the system in all future periods. We call such a state *absorbing*. There are clearly only g absorbing states for this Markov process.

We next observe that from any non-absorbing state s, there is a strictly positive probability that the system moves to an absorbing state before returning to state s. For instance, suppose $w_i = 1$ in state s. Then there is a positive probability that w_i increases by 1 in each of the next $n - 1$ periods, so the system is absorbed into state s_*^i without ever returning to state s. Now let $p_s > 0$ be the probability that the Markov process never returns to state s. The probability that the system returns to state s at least q times is thus at most $(1 - p_s)^q$. Since this expression goes to zero as $q \to \infty$, it follows that state s appears only a finite number of times with probability one. We call s a *transient* state.

We can often calculate the probability that a system starting out with w_r agents choosing type r as money, $r = 1, \ldots, g$, is absorbed by state r.

Let us think of the Markov process as that of g gamblers, each of whom starts out with an integral number of coins, there being n coins in total. The gamblers represent the types and their coins are the agents who choose that type for money, there being n agents in total. I have shown that in the long run, one of the gamblers with have all the coins, with probability one. Suppose the game is fair in the sense that in any period a gambler with a positive number of coins has an equal chance to increase or decrease his wealth by one coin. Then the expected wealth of a gambler in period $t + 1$ is just his wealth in period t. Similarly, the expected wealth $E[w^{t'}|w^t]$ in period $t' > t$ of a gambler whose wealth in period t is w^t is $E[w^{t'}|w^t] = w^t$. This means that if a gambler starts out with wealth $w > 0$ and he wins all the coins with probability q_w, then $w = q_w n$, so the probability of being the winner is just $q_w = w/n$.

We now can say that this Markov process, despite its enormous size, can be easily described as follows. Suppose the process starts with w_r agents holding good r. Then in a finite number of time periods, the process will be absorbed into one of the states $1, \ldots, g$, and the probability of being absorbed into state r is w_r/n. In all cases, a single good will eventually evolve as the universal medium of exchange.

Of course the assumption that all traders are willing to adopt any good as money may be unrealistic. For instance, the producers of a particular good i can benefit from having good i as money because it increases their demand. If exactly one of the producer types simply refused to accept any good but their own as money, while all other groups were unbiased in their choice of money, eventually good i will be the universal money good. However, if more than one type of producer adopts this intransigent strategy, an irreducible conflict must obtain.

6.8   The Omniscient Choreographer and Moral Preferences

As we saw in Section 6.1, social norms are more insightfully and effectively represented as correlated equilibria rather than the Nash equilibria of standard game theory. Many socially efficient social norms are purely *conventional*, in the sense that the Choreographer's signals will be obeyed by all prudent self-regarding rational agents.

For example, consider a town with a North-South/East-West array of streets. In the absence of a social norm, whenever two cars find themselves in a condition of possible collision, both stop and each waits for the other to go first. Obviously not a lot of driving will get done. There are a myriad of Nash equilibria of this game and no way for drivers to coordinate on any one, much less on a socially efficient one. However, consider a correlated equilibrium in which (a) all cars drive on the right, (b) at an intersection both cars stop and the car that arrived first proceeds forward, and (c) if both cars arrive at an intersection at the same time, the car that sees the other car on its left proceeds forward. This is one of several social norms that will lead to an efficient use of the system of streets, provided there is not too much traffic. The social norm serves as a Choreographer giving rise to a self-enforcing correlated equilibrium among rational self-regarding drivers.

Suppose, however, that there is so much traffic that cars spend much of their time stopping at crossings. We might then prefer the correlated equilibrium in which we amend the above social norm to say that cars traveling North-South always have the right of way and need not stop at intersections. However, if there is really heavy traffic, East-West drivers may never get a chance to move forward at all using this social norm.

Note that our correlated equilibrium in this case is simply a Nash equilibrium because there is no explicit Choreographer issuing signals and applying sanctions. To handle the heavy traffic problem, however, we may implement a true Choreographer in the form of a set of traffic signals at each intersection that indicate "Go" or "Stop" to drivers moving in one direction and another set of "Go" or "Stop" signals for drivers moving in the crossing direction. We can then correlate the signals so that when one set of drivers see "Go," the other set of drivers see "Stop." The social norm then says that "if you see Go, do not stop at the intersection, but if you see Stop, then stop and wait for the signal to change to Go." We add to the Choreographer property that the system of signals alternates sufficiently rapidly and there is a sufficiently effective surveillance and penalty system that no driver has an incentive to disobey the social norm even when pressed for time.

Conventional correlated equilibria, however, cannot always achieve socially efficient solutions. Consider, for example, that police in a certain town are supposed to apprehend criminals, where it costs police officer Bob a variable amount f to file a criminal report. For instance, if the identified perpetrator is in Bob's ethnic group, or if the perpetrator offers Bob a bribe to be released, f might be very high, whereas an offender from a different ethnic group, or one who does not offer a bribe, might entail a low value of f. How can this society erect incentives to induce the police to act in a non-corrupt manner?

Assuming Bob is self-regarding, he will report a crime only if $f \leq w$, where w is the reward for filing an accurate criminal report (accuracy can be guaranteed by fact-checking). A conventional correlated equilibrium that requires that all apprehended criminals be prosecuted cannot then be sustained because all officers for whom $f < w$ with positive probability will at least at times behave corruptly. Suppose however officers have a *normative predisposition* to behave honestly, in the form of a police culture favoring honesty that is internalized by all officers. If $f < w + \alpha$ with probability one for all officers, where α is the strength of police culture, the social norm can be sustained.

We can summarize the lesson learned from these two examples by saying that when the Choreographer is *omniscient*, socially efficient social norms can be implemented as conventional correlated equilibria. But when players can take actions that are not observed by the Choreographer, the socially efficient social norms must involve normative correlated equilibria. To generalize from the above example, suppose Bob's payoff consists of a *public component* that is known to the Choreographer and a *private component* that reflects the idiosyncrasies of the agent and is unknown to the Choreographer. Suppose the maximum size of the private component in any state for Bob is α, but Bob's inclination to follow the Choreographer has strength greater than α. Then Bob continues to follow the Choreographer's signals whatever the state of his private information. Formally, we say Bob has an α-*normative predisposition* towards conforming to the social norm if he strictly prefers to play his assigned strategy so long as all his pure strategies have payoffs no more than α greater than when following the Choreographer. We call an α-normative predisposition a *social preference* because it facilitates social coordination but violates self-regarding preferences for $\alpha > 0$. There are evolutionary reasons for believing that humans have evolved such social preferences for fairly high levels of α through gene-culture coevolution, as outlined in Chapter 1 (Bowles and Gintis 2011; Grund et al. 2013).

6.9 The Evolution of Norm Internalization

Why do we have the generalized capacity to internalize norms? From a biological standpoint, internalization may be an elaboration upon the imprinting and imitation mechanisms found in several species of birds and mammals. But its elaborately developed form in humans indicates it had great adaptive value during our evolutionary emergence as a species. Moreover, the everyday observation that people who exhibit a strongly internalized moral code lead happier and more fulfilled lives than those who subject all actions to a narrow calculation of personal costs and benefits of norm compliance suggests it might not be prudent to be self-interested.

In Chapter 10 we show that *if* internalization of *some* norms is personally fitness-enhancing (e.g., preparing for the future, having good personal hygiene, positive work habits, and/or control of emotions), *then* genes promoting the capacity to internalize can evolve. Given this genetic capacity, altruistic norms will be internalized as well, provided their fitness costs are not excessive. In effect, altruism "hitchhikes" on the personal fitness-enhancing capacity of norm internalization. This mechanism was asserted by Simon (1990), we might note, who instead of "internalization of norms," used the term "docility," in the sense of "capable of being easily led or influenced."

Why, however, should the internalization of *any* norms be individually fitness-enhancing? The answer is that we humans have primordial drives and needs some of which do not well serve our fitness interests in complex social settings. These primordial drives are more or less successfully overridden by our internalized norms. Sigmund Freud (1933) somewhat fancifully but accurately described this as the Superego (internalized values) ordering the Ego (rational decision-maker) to suppress the urges of the Id (primitive drives). These primitive drives know little of thinking ahead in a sophisticated manner, but rather satisfy immediate desires. Lying, cheating, killing, stealing, and satisfying short-term bodily needs (e.g., wrath, lust, greed, gluttony, sloth) are all actions that produce immediate pleasure and drive-reduction, at the expense of our overall well-being in the long run.

Internalization alters the agent's *goals*, whereas instrumental and conventional cultural forms merely aid the individual in attaining *pre-given* goals. Through internalization, the individual's immediate needs are satisfied by behaviors that are in the long run fitness-enhancing. These internalized values cannot be represented in the genes because cultural transmission and the nature of man as *Homo ludens* produces rapidly changing social environments, thus conferring high fitness value on *non-genetic mechanisms for altering the agent's goals* in a fitness-enhancing direction. Internalization is limited to our species, moreover, because no other species is defined by gene-culture coevolution and no other species maintains and transforms systematically the rules of the game that define social life.

This evolutionary argument is meant to apply to the long period in the Pleistocene during which the human character was formed, as outlined in Chapter 2. Social change since the agricultural revolution some 10,000 years ago has been far too swift to permit even the internalization of norms to produce a close fit between utility and fitness. Indeed, with the advent of modern societies, the internalization of norms has been systematically diverted from *fitness* (expected number of offspring) to *welfare* (net degree of contentment) maximization. This, of course, is precisely what we would expect when humans obtain control over the content of ethical norms. Indeed, this misfit between welfare and fitness is a necessary precondition for a high level of *per capita* income. This is true because, were we fitness maximizers, every technical advance would have been accompanied by an equivalent increase in the rate of population growth, thus nullifying its contribution to human welfare, as predicted long ago by Malthus (1798). The demographic transition, which has led to dramatically reduced human birth rates throughout most of the world, is a testimonial to the gap between welfare and fitness (Borgerhoff Mulder 1998). Perhaps the most important form of prosocial cultural transmission in the world today is the norm of having few, but intensively supported, offspring.

6.10   Modeling Networked Minds

There are many plausible ways to model the cognition of social actors as networked across a range of significant others (Coleman 1988; Rauch 1996; Bowles and Gintis 2004; Di Guilmi et al. 2011; Gintis 2013). We described one of these in our discussion of case-based decision theory in Section 5.5. The following model is offered as a more fully articulated version of the case-based model.

Suppose there are social actors $i = 1, \ldots, m$ and there is a network of information flows among the actors. Let $\mathcal{P}_i$ be the set of actors to whom actor i is directly linked. Suppose there are n traits, such as gender, ethnicity, occupation, religion, social position, physical attributes, family relationships, cultural beliefs, and demographic characteristics. Suppose each social actor has a *social trait vector* $a = (a_1, \ldots, a_n)$ where each a_j takes the value zero and one. We interpret $a_j = 0$ as meaning that the individual does not have trait j, and $a_j = 1$ means the individual has trait j. An actor i

with personal traits vector $a^i \in A$ has available a set of *trait filters*, where a trait filter $b_i \in A$ represents the set of traits that i considers relevant in polling others in a particular decision context. I interpret $b_{ij} = 1$ as meaning members of $\mathcal{P}_i$ satisfying the filter have trait j, and $b_{ij} = 0$ as meaning that members of $\mathcal{P}_i$ may or may not have trait j. For some decisions, i will consider only other actors with the same personal characteristics, so $b_i \leq a_i$, in the sense that $b_{ij} \leq a_{ij}$ for all traits j. However, in other cases i may defer to experts or highly experienced network members with personal traits that differ in important ways.

In facing a particular decision, actor i evaluates information from other social actors in his network $\mathcal{P}_i$, using a trait filter b_i that is dependent on the nature of the decision. The *strength* $\rho(b_i)$ of a trait filter b_i is the number of positive entries in b_i. The stronger the trait filter, the closer others must be in social space for their experience to count in the actor's decision. The strength of a trait filter is a partial order on A in the obvious sense. I write $b_i(\mathcal{P}_i)$ for the set of network links to i that conform to the filter b_i.

Let k_i be the number of actors in $\mathcal{P}_i$, and let $k_i(b_i)$ be the number of actors in $\mathcal{P}_i$ who conform to the filter b_i, which is decreasing in the strength of the filter b_i. Thus $k_i(b_i)/k_i$ is the fraction of social actors in i's network who have the traits b_i. Let $q_i(b_i)$ be the probability that a social actor with traits b_i provides correct information allowing i to choose an action that maximizes i's payoff. Because the use of a stronger filter cannot improve the decision-maker's information unless it also increases the probability of receiving correct information, we may safely assume that for a given decision problem, decision-maker i considers only filters that belong to a totally ordered sequence of increasingly strong filters $b^{i1}, b^{i2}, \ldots$ such that $q_i(b^{ij})$ is increasing in j. Let q_i^* be the probability that i chooses correctly without information.

We suppose individual i queries a particular member of his network with traits b_i, who tells him the correct action if he knows it, which occurs with probability $q_i(b_i)$. Otherwise the queried actor gives no information. We can then express the probability that the individual receives the correct information as $p_i(b_i) = \alpha(b_i)q_i(b_i) + (1 - \alpha(b_i))q_i^*$, where $\alpha(b_i) = k_i(b_i)/k_i$. The decision-maker can then choose the filter b_i to maximize the probability of obtaining useful information (Bowles and Gintis 2004).

6.11   Class Structure in General Social Equilibrium

An elaboration on the general social equilibrium model of the previous section illustrates how wealth inequality can translate into a stratified distribution of social classes. This model is a variant of Eswaran and Kotwal (1986) and Bowles (2004, Ch. 10), who apply a method initiated by Roemer (1982). Suppose all families face the household production function

$$q = f(k, l) \qquad (6.2)$$

where k is capital and l is labor. We assume $f(k, l)$ is increasing and concave in its arguments; i.e., there is decreasing marginal productivity of both labor and capital in household production. However, there is a startup capital cost $\kappa > 0$ for household production. A family can apply its own labor l_f, it can hire labor l_h, and it can sell labor l_w to other households and to firms in the market sector. If the household hires labor l_h, it must supervise this labor, incurring a supervisory cost in personal labor time $s(l_h)$. We assume $s(l_h)$ is increasing and convex in the amount of labor hired, with $s(0) = 0$. With supervision, hired workers are as productive as the household labor, so total effective labor in household production is simply $l = l_h + l_f$.

   We assume households are credit constrained, with the maximum amount a household with wealth k_f can borrow is $c(k_f)$, where $c(k_f)$ is increasing in k_f with $c(0) = 0$, meaning that a family with no wealth cannot borrow at all. Let w and r be the wage rate and the rate at which capital can be borrowed or loaned. If a household chooses to produce, the credit rationing constraint requires that

$$c(k_f) \geq w(l_f + l_h) + r(k - k_f) + \kappa, \qquad (6.3)$$

where k is the amount of capital the household uses in production. This inequality assumes that all production costs must be paid at the start of the period.

   We assume a simple household payoff $y + u(\rho)$, where y is income and ρ is the amount of leisure consumed, and where $u(\rho)$ is increasing and concave (decreasing marginal utility of leisure). We also assume $u'(0)$ is sufficiently negative that the household always chooses a positive amount of leisure. Then an individual who chooses to enter into household production has payoff

$$\pi_f = f(k, l_f + l_h) - (1 + r)[w(l_h - l_w) + v(k - k_f) + \kappa] + u(\rho), \quad (6.4)$$

where the $(1 + r)$ term represents the total amount of the loan that must be paid at the end of the period.

An individual who hires out as a worker rather than engaging in household production will have payoff

$$\pi_w = (1 + r)(wl_w + v\kappa) + u(\rho), \tag{6.5}$$

assuming wages are paid at the start of the production period.

An individual who undertakes household production, such that (6.4) holds, must choose k, ρ, l_w, l_h, l_f, and l to maximize (6.4) subject to the credit constraint (6.3), the inequality constraints k, l_h, $l_f \geq 0$, and a labor constraint given by

$$l_f = 1 - s(l_h) - l_w - \rho \geq 0, \tag{6.6}$$

where we have normalized the individual's labor endowment to unity. The Lagrangian for this optimization problem is given by

$$\mathcal{L} = f(k, l_f + l_h) - (1 + r)[wl_h + vk + \kappa] + \pi_w + \tag{6.7}$$

$$\lambda[c(\kappa) - w(l_f + l_h) + r(k - k_f) + \kappa] + \tag{6.8}$$

$$\mu[1 - s(l_h) - l_f - \rho]. \tag{6.9}$$

The first-order conditions for this problem are

$$\mathcal{L}_k = f_k - (1 + r + \lambda)v = 0 \tag{6.10}$$

$$\mathcal{L}_{l_h} = f_l(1 - s'(l_h)) - (1 + r + \lambda)v - \mu s'(l_h) \leq 0 \tag{6.11}$$

$$\mathcal{L}_\rho = -f_l + u'(\rho) - \mu = 0 \tag{6.12}$$

$$\mathcal{L}_{l_f} = -f_l + w(1 + r + \lambda) - \mu \leq 0, \tag{6.13}$$

where (6.11) is an equality if any labor is hired ($l_h > 0$) and (6.13) is an equality if the agent himself works in domestic production ($l_f > 0$). The value of λ determined by these equations is the shadow price of borrowed capital, and is strictly positive if the demand for capital in the household sector is positive, which will be the case when the market wage w is not so high that household production is never superior to working in the market sector. In this case $1 + r + \lambda$ is the real cost of borrowing (note that the capital itself is used up in production), and (6.10) says that if household production is undertaken, the marginal productivity of capital used by households will equal the marginal cost of capital.

If the household supplies its own labor, then $l_f > 0$, so the constraint (6.6) is not binding, and hence $\mu = 0$. In this case, (6.13) asserts that if the household also works in the market sector, the marginal product of labor will be equal to the cost of labor $w(1 + r + \lambda)$. Note that the cost of labor is the wage w, plus the interest that must be paid on this, rw, plus the constraint cost of the wage λw.

Wealth	Class	Borrows	Activities	μ	λ
$0 \leq k_f < k_1$	pure wage	No	$l_w > 0$	$\lambda = 0$	$\mu > 0$
$k_1 < k_f < k_2$	wage and domestic	Yes	$l_w, l_f, k > 0$	$\lambda > 0$	$\mu = 0$
$k_2 < k_f < k_3$	pure domestic	Yes	$l_f, k_f > 0$	$\lambda > 0$	$\mu = 0$
$k_3 < k_f < k_4$	small capitalist	Yes	$l_f, l_h, k > 0$	$\lambda > 0$	$\mu = 0$
$k_4 < k_f < k_5$	large capitalist	Yes	$l_h, k > 0$	$\lambda > 0$	$\mu > 0$
$k_5 < k_f$	financial	No	Pure Lender	$\lambda > 0$	$\mu > 0$

Table 6.1. Class Structure in a Market and Domestic Production System

In this model, then, there will be six classes of households, a household's status being a function of its wealth k_f. Indeed, there is a sequence of increasing wealth levels $0 < k_1 < k_2 < k_3 < k_4 < k_5$ such that households with wealth $k_f < k_1$ are *pure wage workers*, hiring no labor or capital and working only in the market sector ($l_w > 0$). If these households have any capital ($k_f > 0$), they lend it to others. Households with $k_1 < k_f < k_2$ are *mixed wage workers and domestic producers*, working in the market sector ($l_w > 0$) but also in domestic production ($l_f > 0$) using their own capital ($k_f > 0$). Households with wealth $k_2 < k_f < k_3$ are *pure domestic producers*, using only their own labor ($l_f > 0$) and no capital ($k_f < 0$). Households with $k_3 < k_f < k_4$ are *small capitalist producers*, using their own labor ($l_f > 0$) and supervising hired labor ($l_h > 0$), while borrowing ($k > 0$) to achieve a higher capital input to production than possible with their own wealth. Households with $k_4 < k_f < k_5$ are *large capitalist producers* who hire labor and capital ($l_h, k > 0$), supervise the hired labor, but otherwise do not engage in production ($l_f = 0$) and of course do not work for others ($l_w = 0$).

Finally, households for which $k_5 < k_f$ are *financial capitalists* who do no work themselves and do not engage in production, but rather lend all their capital and live off the proceeds. Table 6.1 illustrates this social equilibrium.

6.12   Resurrecting Sociological Theory

A scientific discipline attains maturity when it has developed a core analytical theory that is taught to all fledgling practitioners, is accepted by a large majority of seasoned practitioners, and is the basis for intradisciplinary communication. Theoretical contributions then consist of additions to and emendations of this core theory. Occasionally the core paradigm may come under attack and be replaced by a more powerful core theory that includes all of the insights of the older doctrine, and new insights as well (Kuhn 1962). Physics, chemistry, astronomy, and many of their subfields attained this status by the last quarter of the nineteenth century, biology developed a core theory with the synthesis of Mendelian and population genetics in the first half of the twentieth century, and economics followed in the last half of the twentieth century with the general equilibrium model (Arrow and Hahn 1971) and neoclassical microeconomic theory (Samuelson 1947; Mas-Colell et al. 1995).

Sociology, anthropology, and social psychology have never developed core analytical theories, and indeed it is not clear why they have not coalesced into a single discipline. Sociology and anthropology have the same object of study—human society. There is no plausible justification for considering the focus of sociology on highly institutional societies and of anthropology on small-scale societies a good reason for maintaining contrasting and barely overlapping theoretical and empirical literatures. Moreover, the practice in social psychology of treating individual social behavior as capable of explanation independent of general social theory is not defensible. All these fields have suffered by separating themselves from sociobiology, which is the study of social life in general (Maynard Smith 1982; Wilson 1975; Alcock 1993; Krebs and Davies 1997).

Sociology moved haltingly towards a general analytical core with the early work of Talcott Parsons, but Parsons himself strayed into relatively tangential territory in his later work, and no one came along to pick up where Parsons left off in creating an analytical basis for sociology. Moreover, there developed a strong antagonism between economists and sociologists, which prevented sociologists from developing an analytical core that is synergistic with economic theory, while economic theory accepted unrealistic assumptions that allowed economists to model social behavior without the need for sociological notions (Gintis 2009a). Both fields are worse for their studied mutual antipathies, but sociology has fared worse, because sociological theory since Parsons has become unacceptably fragmented (Turner 2006).

7

The Theory of Action Reclaimed

Quid leges sine moribus vanae proficient? [Of what use
are laws without morals?]

Horace

C: If you could converse with Talcott Parsons
      today, what would you say?
HG: I would explain succinctly his only serious mis-
      take and suggest a way to correct this mistake.
C: Would he agree?
HG: In my dreams.

Choreographer interview

The analytical core for sociological theory proposed in the previous chapter includes a rational actor model inspired by Talcott Parsons' *voluntaristic
theory of action* (1937). My elaboration of this model, however, followed a
different path from that taken by Parsons in his later work, which wandered
away from the microfoundations of human behavior into the dusty realm of
structural-functionalism. Big mistake.

This chapter explains where and speculates why Parsons went wrong.
Briefly, between writing *The Structure of Social Action* in 1937 and the
publication of *The Social System* and *Toward a General Theory of Action* in
1951, Parsons abandoned the stress on individual efficacy of his early work
(e.g., in his critique of positivism and behaviorism) in favor of treating the
individual as the effect of socialization that when successful produces social
order, and when unsuccessful produces social pathology. In *The Structure of
Social Action*, Parsons mentions the term "socialization" only once, writing
(pp. 400–401):

> Ultimate values of the individual members of the same commu-
> nity must be, to a significant degree, integrated into a system
> common to these members... not only moral attitudes but even
> the logical thought on which morality depends only develop as
> an aspect of the process of socialization of the child.

Moreover, in *The Structure of Social Action* Parsons uses this fact only
to show the impossibility of a "utilitarian" model of individual choice, by

which he means a model in which individuals fail to share a common moral dimension. He writes (p. 401), "This evidence confirms the negative proof of the impossibility of a truly utilitarian society." In 1951, in both *The Social System* and *Toward a General Theory of Action*, "socialization" is used constantly throughout. By contrast, the term "voluntarist," liberally dispersed throughout *The Structure of Social Action*, is replaced by "general" in later versions of the theory of action. The term "voluntarist" appears not at all in *The Social System* and only in the Index of *Toward a General Theory of Action*—doubtless left there by mistake. Moreover, by 1951 Parsons has come to treat the *demand* for agents to fill social roles, which is determined by the social division of labor, and the *supply* of agents to fill social roles, which is determined by the socialization process, as not simply interrelated, but in fact *identical*. He writes in *The Social System* (Parsons 1951, p. 142):

> The allocation of personnel between roles in the social system and the socialization processes of the individual are clearly the same processes viewed in different perspectives. Allocation is the process seen in the perspective of functional significance to the social system as a system. Socialization on the other hand is the process seen in terms of the motivation of the individual actor.

By the time he wrote *Economy and Society* with Neil Smelser in 1956, nothing is left of individual action at all, the economy being simply a system of intersectoral flows and boundary interchanges with other social subsystems. The individual becomes for Parsons like a cell in the metazoan body, having important work to do to maintain the organism (the social system), but either doing it well or poorly. It cannot affect the organization of the system itself.

The idea that the demand for agents to fill social roles and the supply of agents capable of and willing to fill these roles are identical is not simply false. It is preposterous. I cannot understand how Parsons could come to this conclusion, or why his close friends and colleagues did not call him on this. In terms of the general social equilibrium model developed in the previous chapter, Parsons' claim would take the form of asserting that markets for social roles are always in equilibrium. In fact, the ensemble of social roles follows quite a different logic from the ensemble of individuals with the motivations and capacities to fill these roles. In a dynamic society, the two are rarely if ever in equilibrium, although there may be strong tendencies towards equilibrium.

Parsons' possibly thought, following his assignment of "positive" to economics and "normative" to sociology, that economic theory could deal with the *skills and incentives* side of the supply and demand for role positions, leaving sociology to attend to the *normative* side of the equation. While it would be far-fetched to maintain that the supply and demand for various types of agent services are always in equilibrium, it is bordering on plausible that socialization could flexibly adjust to the motivational needs of society by suitably restricting which actions are allowed and which are not. Indeed, Parsons' positive vs. normative distinction between economics and sociology lends itself to this treatment. Then the economic subsystem could be in dynamic movement while the normative subsystem is in equilibrium. But this way of carving up the social world is not in fact tenable.

Two theoretical commitments appear to have led Parsons to identify economics vs. sociology with positive-rational action vs. normative-nonrational action. The first is his treatment of socialization as the internalization of a society's universal and pervasive culture. As we have seen, this treatment, as opposed to a more plausible construct in which socialization reflects and codifies individuals positions in the variety of social networks in which they participate, leads directly to a deeply functionalist view of the supply and demand of role positions. The second commitment is that rational choice is *instrumental* to the achievement of material goals. Viewed in this way, much of human action, including morality-motivated choices, appears to be nonrational.

Parsons' instrumental understanding of the rational actor model appears early in *The Structure of Social Action*. He writes (Parsons 1937, p. 44):

> An "act" involves logically the following: (1)... an agent ...(2)...an "end"...(3)...and a "situation"...This situation is in turn analyzable into two elements: those over which the actor has no control...and those over which he has such control. The former may be termed the "conditions" of action, the latter the "means"...Finally (4)...in the choice of alternative means to the end...there is a "normative orientation" of action.

It is challenging to cast this notion of an act (or what Parsons generally calls a "unit act") into the modern rational actor framework. Parsons defines the "end" as "a future state of affairs toward which the process of action is oriented." This concept is missing in the contemporary rational actor model. The reason for this is that rational choice theory as developed in this book, and as inspired by von Neumann and Morgenstern (1944) and Savage (1954), does not presume *instrumental rationality*—the notion that behavior is always oriented towards some specific goal and rationality takes the form of choosing the best action towards achieving that goal. Rather, we use the more restricted notion of *formal rationality*, which merely means that the Savage axioms, which say nothing about ends or goals, are obeyed.

The problem with the instrumental interpretation of rational choice is that often agents are in situations where they must make choices but they have no clear notion of what the goals of action are. For instance, if I see someone faint on a New York subway platform, I must choose how to react, but I have no obvious goal. Indeed several distinct considerations must be adjudicated in deciding how to act. Similarly, subjects in the behavioral economics laboratory may have no well-defined goals. They may have come with the goal of making money, but they often do not maximize their monetary rewards.

The "means" for Parsons include the agent's capacity to choose among alternatives according to his preferences and beliefs towards attaining the "end," and the "conditions" are the objective and observable personal and social relations that form the context of choice. A search through *Structure* fails to elucidate Parsons' notion of "conditions," except that he tends to attribute to "positivism" the notion that "conditions" determine choices; i.e., that the voluntaristic and subjective factors in behavior are absent.

What appears novel, and what Parsons contends is preeminently socio-logical, is element (4), involving a "normative orientation."

One might expect Parsons to devote some effort to explicate this model. Does not one need a "moral preference function" of some sort to evaluate al-ternative choices leading to a particular "end"? Will not alternative choices leading to the same "end" have additional costs or benefits that must be balanced against the normative value of the choice? Might we not actually choose our ends taking into consideration the normative costs and benefits of attaining these ends? Parsons never in his work directly addresses these obvious questions. He appears to hold that we cannot make utility cal-culations involving alternative material and moral aspects of our choices. Rather, for each given end there is a set of feasible choices leading to that end and normative considerations eliminate some of these choices while permitting others. The properly socialized individual will simply limit his choice set to those that are normatively permissible. The choice among what remains is then the business of economics, not sociology. Indeed, Parsons distinguishes economics from sociology precisely by defining the former as studying how agents rationally choose means to satisfy ends, while the latter studies the normative restrictions on choice that make the social order possible.

For instance, Parsons' study of the professions, as in his essay "The Motivation of Economic Activities" (Parsons 1949), supported the notion that physicians and other professionals internalize norms of social service leading them to consider the health interests of clients above their personal material interests. He writes in *The Social System* (p. 293):

> The "ideology" of the profession lays great emphasis on the obligation of the physician to put the "welfare of the patient" above his personal interests, and regards "commercialism" as the most serious and insidious evil with which it has to contend. The line, therefore, is drawn primarily vis-à-vis "business." The "profit motive" is supposed to be drastically excluded from the medical world. This attitude is, of course, shared with the other professions.

Moreover, the culture of the profession fosters a reputational system in which individual physicians are rewarded for ethical behavior. Thus while professionals are clearly motivated to enhance their income and wealth, they are obliged to subordinate such motivations when advising patients, and they normally do so even when the probability of being caught acting unprofessionally is close to zero. The efficacy of normative constraints on choice render it feasible to allow professionals extreme autonomy in their interactions with patients.

Hans Joas and Wolfgang Knöbl, in their popular exposition of Parsons' voluntarist theory of action, express Parsons' notion that norms are binding constraints on actions by asserting the impossibility of locating normative effects elsewhere (Joas and Knöbl 2009, p. 37–38):

> It is quite simply impossible to make our own values the subject of utility calculations...I cannot simply manipulate and overrule my own values...Value and norms themselves...are constitutive of every criterion underpinning such calculations.

In fact, as we have seen in virtually every chapter of this book, agents make such calculations all the time. Indeed, the experimental evidence presented in Section 7.1, a small slice of the literature, indicate that preference functions including both material and moral payoffs have the same properties as the preferences found in the neoclassical economics textbooks. Moreover, people trade off moral principle for material reward, and conversely, as part of daily life. It is thus simply incorrect to treat values as binding constraints on action.

The alternative, which I have embraced throughout this book, is that all decision making by socially situated individuals has a moral component, and that individuals are constantly involved in trading off between self-interest and a variety of moral values. Norms thus appear in the individual's preference ordering rather than as constraints on his choices.

These considerations suggest that Parsons' attempt to partition off economic theory from normative concerns is simply a failure. Economic concerns are both moral and material concerns, and the fact that this is not reflected in economics and sociology is a failure of theory, not of a plausible intellectual division of labor.

7.1   The Moral and Material Bases of Choice

There is nothing irrational about caring for others. But do preferences for altruistic acts entail transitive preferences as required by the notion of rationality in decision theory? Andreoni and Miller (2002) showed that in the case of the Dictator Game, they do. Moreover, there are no known counterexamples.

In the Dictator Game, first studied by Forsythe et al. (1994), the experimenter gives a subject, called the Dictator, a certain amount of money and instructs him to give any portion of it he desires to a second, anonymous, subject, called the Receiver. The Dictator keeps whatever he does not choose to give to the Receiver. Obviously, a self-regarding Dictator will give nothing to the Receiver. Suppose the experimenter gives the Dictator m points (exchangeable at the end of the session for real money) and tells him that the price of giving some of these points to the Receiver is p, meaning that each point the Receiver gets costs the giver p points. For instance, if $p = 4$, then it costs the Dictator 4 points for each point that he transfers to the Receiver. The Dictator's choices must then satisfy the budget constraint $\pi_s + p\pi_o = m$, where π_s is the amount the Dictator keeps and π_o is the amount the Receiver gets. The question, then, is simply, is there a prefer- ence function $u(\pi_s, \pi_o)$ that the Dictator maximizes subject to the budget constraint $\pi_s + p\pi_o = m$? If so, then it is just as rational, from a behavioral standpoint, to care about giving to the Receiver as to care about consuming marketed commodities.

Varian (1982) showed that the following generalized axiom of revealed preference (GARP) is sufficient to ensure not only rationality but that individuals have nonsatiated, continuous, monotone, and concave utility functions—the sort expected in traditional consumer demand theory. To de- fine GARP, suppose the individual purchases bundle $x(p)$ when prices are p. We say consumption bundle $x(p_s)$ is *directly revealed to be preferred* to bundle $x(p_t)$ if $p_s x(p_t) \leq p_s x(p_s)$; i.e., $x(p_t)$ could have been purchased when $x(p_s)$ was purchased. We say $x(p_s)$ is *indirectly revealed to be pre- ferred* to $x(p_t)$ if there is a sequence $x(p_s) = x(p_1), x(p_2), \ldots, x(p_k) = x(p_t)$, where each $x(p_i)$ is directly revealed to be preferred to $x(p_{i+1})$ for $i = 1, \ldots, k-1$. GARP then is the following condition: if $x(p_s)$ is in- directly revealed to be preferred to $x(p_t)$, then $p_t x(p_t) \leq p_t x(p_s)$; i.e., $x(p_s)$ could not have been purchased for the amount of money paid for $x(p_t)$ when $x(p_t)$ is purchased.

Andreoni and Miller (2002) worked with 176 students in an elementary economics class and had them play the Dictator Game multiple times each, with the price p taking on the values $p = 0.25, 0.33, 0.5, 1, 2, 3$, and 4, with amounts of tokens equaling $m = 40, 60, 75, 80$, and 100. They found that only 18 of the 176 subjects violated GARP at least once and that of these violations, only four were at all significant. By contrast, if choices were randomly generated, we would expect that between 78% and 95% of subjects would have violated GARP.

As to the degree of altruistic giving in this experiment, Andreoni and Miller found that 22.7% of subjects were perfectly selfish, 14.2% were per- fectly egalitarian at all prices, and 6.2% always allocated all the money so as to maximize the total amount won (i.e., when $p > 1$, they kept all the money, and when $p < 1$, they gave all the money to the Receiver).

We conclude from this study that, at least in some cases, and perhaps in all, we can treat altruistic preferences in a manner perfectly parallel to the way we treat money and private goods in individual preference functions.

A particularly clear example of including both material ends and moral principles in a preference ordering is reported by Gneezy (2005), who stud- ied 450 undergraduate participants paired off to play three games in which all payoffs were of the form (b, a), where player 1, Bob, receives b and player 2, Alice, receives a. In all games, Bob was shown two pairs of payoffs, $A{:}(x, y)$ and $B{:}(z, w)$ where x, y, z, and w are amounts of money with $x < z$ and $y > w$, so in all cases B is better for Bob and A is better for Alice. Bob could then say to Alice, who could not see the amounts of money, either "Option A will earn you more money than option B," or "Option B will earn you more money than option A." The first game was $A{:}(5,6)$ vs. $B{:}(6,5)$ so Bob could gain 1 by lying and being believed while imposing a cost of 1 on Alice. The second game was $A{:}(5,15)$ vs. $B{:}(6,5)$, so Bob could gain 1 by lying and being believed, while still imposing a cost of 10 on Alice. The third game was $A{:}(5,15)$ vs. $B{:}(15,5)$, so Bob could gain 10 by lying and being believed, while imposing a cost of 10 on Alice.

Before starting play, Gneezy asked the various Bobs whether they expected their advice to be followed. He induced honest responses by promising to reward subjects whose guesses were correct. He found that 82% of Bobs expected their advice to be followed (the actual number was 78%). It follows from the Bobs' expectations that if they were self-regarding, they would always lie and recommend B to Alice.

The experimenters found that, in game 2, where lying was very costly to Alice and the gain from lying was small for Bob, only 17% of Bobs lied. In game 1, where the cost of lying to Alice was only 1 but the gain to Bob was the same as in game 2, 36% of Bobs lied. In other words, Bobs were loath to lie but considerably more so when it was costly to Alices. In game 3, where the gain from lying was large for Bob and equal to the loss to Alice, fully 52% of Bobs lied. This shows that many subjects are willing to sacrifice material gain to avoid lying in a one-shot anonymous interaction, their willingness to lie increasing with an increased cost to them of truth telling, and decreasing with an increased cost to their partners of being deceived. Similar results were found by Boles et al. (2000) and Charness and Dufwenberg (2006). Gunnthorsdottir et al. (2002) and Burks et al. (2003) have shown that a socio-psychological measure of "Machiavellianism" predicts which subjects are likely to be trustworthy and trusting.

We conclude that moral choices are as much subject to trade-offs as are purely material choices, and the two strongly interact. Many philosophers will disagree with this. Typical is John Mackie (1977), who writes:

> A moral judgment is...is absolute, not contingent upon any desire or preference or policy or choice.

Perhaps a moral *judgment* has this character, but whether and how we act on this judgment surely has a relative character.

7.2   Carving an Academic Niche for Sociology

Talcott Parsons participated in the early stages of the formation of sociology as a discipline. His central preoccupation in writing his first and greatest book, *The Structure of Social Action* (1937), was how to retain neoclassical economic theory's stress on social outcomes as the aggregation of individual choices, but carve out an area outside of economics where sociology could live in neighborly harmony. His solution in *The Structure of Social Action* was to broaden the economist's decision model to the "voluntaristic theory of action," which retained the framework of choice subject to constraints. Parsons broadened the concept of instrumental choice to deal with normative concerns, and to suggest that the "utilitarian" rational actor model, by which he meant the rational actor model with purely self-regarding actors, could not "solve the problem of order" in society, because it embodies no principles leading self-regarding individuals to share enough expectations to cooperate effectively. Parsons called his alternative "voluntaristic," to stress the central effectivity of individual choice in his framework.

Parsons' voluntaristic theory of action is in sharp contrast to the devaluing of individual choice behavior in institutionalism (Veblen 1899; Berle and Means 1932), and the mechanistic conception of choice in the behaviorist psychology of the day (Watson 1913). Sociology, for Parsons, was to supply the theory behind important social parameters, including morality and ethical values, that economic theory ignores even though they are central to understanding economic activity.

For instance, economic theory assumes an adequate supply of entrepreneurship as a function of price. Where does this supply come from? Parsons' first step in answering this is was to turn to the greatest economist of his day, Alfred Marshall. Marshall (1930) dealt insightfully with the relationship between values and markets, expressing the view that a major attraction of the market economy is that it not only satisfies needs, but fosters morally valuable ethical principles in citizens. Parsons uses Marshall in *The Structure of Social Action* to show that economic theory is compatible with a serious analysis of social norms and personal ethics.

Parsons then turns to Max Weber's analysis of the role of Protestantism in the growth of capitalism (Weber 2002[1905]). Weber argued that these preferences come from a *cultural system* that had emerged in England that was appropriately geared towards hard work and material reward, as a symbol of proximity to God and salvation. Many studies since *The Structure of Social Action* attest to the fact that values underpin labor productivity. See, for instance, Gintis (1976), Akerlof (1982), and Fehr and Gächter (2000). Similarly, Parsons derived the taste for entrepreneurship on which growth and innovation depend from Weber's Protestant Ethic (Weber 2002[1905]; Schumpeter 1947; McCloskey 2010).

Parsons also was deeply impressed with the stress laid by Émile Durkheim on the "conscience collective"—a communality of beliefs among all members of society, even under circumstances of an organic division of labor that leads to social upheaval through extensive class differences. In *The Structure of Social Action*, Parsons stresses that modern economic theory takes class harmony for granted—there are no classes, only persons in neoclassical economics—but in fact, when there is class harmony in modern society, this is not an accident, but rather the product of specific social institutions and practices.

The above considerations were Parsons' material for creating an analytical core for sociological theory. However, two personal characteristics of Parsons limited his ability to turn his insights into a successful analytical foundation on which sociological theory might rest. One was his limited writing skills. Parsons is generally verbose and imprecise. Perhaps he inherited his approach to writing from his German teachers, for whom lucidity was often considered a sign of superficiality. The second characteristic was his lack of mathematical sophistication. When Parsons uses algebra (as he does at one point in *The Structure of Social Action*), it is awkward and sophomoric. Together, these weaknesses prevented Parsons from forging a general theory of action from which economic and sociological aspects would fall out naturally.

Parsons was also handicapped by the fact that game theory and rational decision theory had not been invented when *The Structure of Social Action* was published, and the literary and graphic economics of Marshall was being replaced by the high-tech mathematical approaches of Paul Samuelson and Léon Walras. Moreover, neoclassical economics in this period was more hostile to taking consideration of ethical values, social norms, and thick descriptions of social reality than at any point before or after in the history of economic theory. When Parsons and Neil Smelser wrote *Economy and Society* (1956), a second attempt at integrating sociology and economics, this time based on Parsons' structural-functional AGIL system, game theory had been invented but was in the doldrums, and it was the peak point of enthusiasm for the highly mathematical and abstract Walrasian general equilibrium model. Not surprisingly, the Parsons-Smelser effort was widely rejected by economists and sociologists alike.

It is curious as to why Parsons did not see immediately that the economist's standard decision model could be expanded into the basis for a sociological theory of action by enriching the model's treatment of "beliefs," and by introducing other-regarding objectives and character virtues into the individual preference function, thereby bringing ethical values and social norms directly into economic theory. I think an extended concept of "beliefs" was not open to Parsons because economic theory at the time only used a truncated form of the rational actor model in which preferences are defined directly over choice sets. With the advent of game theory, economics shifted to an enriched notion in which preferences are over "lotteries" whose outcomes are valued. As von Neumann and Morgenstern (1944) and Savage (1954) showed, this leads to the individual having a subjective prior that is naturally interpreted as the individual's "beliefs." A correct theory of action merely extends beliefs to an inter-subjective setting that is naturally affiliated with sociological theory.

As for the extension of preferences to include normative and other-regarding dimensions, economists at the time were so adamant in identifying "rationality" with "self-interest," a prejudice shared by the great economist/sociologist Vilfredo Pareto, that Parsons never had the temerity to think, much less to suggest, otherwise. The closest Parsons comes in *The Structure of Social Action* is to say that ethical issues could be a "constraint" on economic choice behavior. It was not until behavioral game theorists showed clearly that "rationality" is completely compatible with altruism and other forms of inherently ethical behavior that a theory of action that bridges economics and sociology could be productively contemplated (Fehr and Gächter 1998; Gintis et al. 2005), and structural-functionalism could be safely buried.

7.3 The Parsonian Synthesis

Talcott Parsons' social theory was at base very simple. Society is a highly differentiated nexus of "role positions" (for example, husband, worker, voter, hospital patient, subway rider, and so on). Each social role position is occupied by a person and a single individual can occupy many different role positions. Institutions are then structured networks of social roles. The rules and norms associated with a social role place specific requirements of how an individual behaves in a given role position, and the individual is motivated to behave in conformance with these requirements by virtue of material and moral incentives. Individuals respond to moral incentives, according to Parsons, because they are socialized to accept the norms and values associated with society in general and specific role positions in particular. We developed this approach in Chapter 6, extending Parsons by developing the theory in the context of the rational actor model, game theory, and general social equilibrium.

If we restrict this picture to the economy, we recreate modern economic theory, except that in Parsons' time it was assumed in economics the people are purely selfish and a process of socialization could not possibly induce rational individuals to forego behaving perfectly selfishly. The role structure of the economy is the Walrasian general equilibrium system of firms and households, plus the economic aspects of government. The actors are the employers and employees who occupy positions in the economy, as well as government actors who are involved in regulating and policing economic activity.

Economists argued that all of these positions could be efficiently filled by selfish agents, provided appropriate material incentives (rewards and penalties) were attached to the various economic roles. We now know that the assumption that the capitalist economy can operate effectively through purely material incentives applied to self-regarding agents is quite indefensible— see, for instance, my book *The Bounds of Reason* (2009), which supplies the appropriate empirical and theoretical references and, with Samuel Bowles, *A Cooperative Species* (2011), as well as the more anthropologically oriented Joseph Henrich et al., *Foundations of Human Sociality* (2004) for details. But when Parsons wrote, the rationality-equals-selfishness axiom was virtually universal in economic theory. Parsons was perfectly on the mark to stress that economics based on self-regarding agents could not solve the problem of order.   .

Parsons did say that economic theory is a "subset" of social theory, but he never managed to articulate how his vision of a theory of action would mesh with the economist's decision theoretic model. The answer, as developed in this book, is that we must extend the goals of action to include moral behavior and non-material ends such as reciprocity, empathy, considerateness, and justice, even in purely economic transactions, and more so in general social life. Role performance, as we argued in Chapter 6—see also Aoki (2010)—can be quite nicely modeled with such an expanded rational actor model, in which individuals have goals that they try to meet as best they can subject to their material and informational constraints, and in conformance with their beliefs (called subjective priors in economics) and where behaving morally may be among their goals.

Parsons' treatment of culture in his theory of action was far in advance of its treatment in economic theory. In economics each individual has beliefs in the form of a subjective prior, but there is no formal way to compare or adjudicate among the beliefs of distinct economic agents. Economic theory thus is often forced to assume *common priors* concerning the relative probability of various events in order to predict economic behavior, but with no serious suggestion as to the forces giving rise to these common priors. Economists also tended to assume *common knowledge* of certain social facts, without which behavior could not be predicted, without offering a plausible suggestion as to how this commonality of knowledge across agents might arise. The argument that all rational agents must have the same priors is not in the least plausible (Harsanyi 1968; Morris 1995) .This problem is especially severe in dealing with how individuals represent the internal states (beliefs and intentions) of others (Gintis 2009a, Ch. 8).

Parsons' alternative, which is really due to Émile Durkheim, was to posit the existence of a "common culture" independent from and above both social institutions and individual personalities. This common culture is reproduced by specialized institutions (rituals, schools, communications media, etc.) that ensure that culture remains common, changes only slowly under normal circumstances, and is internalized by the youth of each new generation. This common culture provides a common framework of assumptions and expectations that all social agents share, and provides the conditions needed by game theory to justify the assumption that agents will coordinate their activities appropriately (i.e., play Nash equilibria, in the language of game theory).

Parsons' treatment of culture as a set of society-wide mutually reinforcing principles that normal individuals internalize leads, as we have seen, directly to structural-functionalism. It has the additional weakness of precluding a dynamical theory of cultural change. For if culture determines individual consciousness, the only way culture can evolve endogenously in a social system is through the actions of deviants, whose actions will be rejected by the vast majority of properly socialized individuals. Our alternative, in which individuals are situated within overlapping social networks of minds over which their cognition is distributed immediately suggests a cultural dynamic based on resonances and clashes of principles based on the structure of social networks. Culture is thus an *effect* as much as a *cause* and the dynamics of cultural change require careful development, as in the biological-anthropological treatment of culture in the theory of gene-culture coevolution, as developed by Robert Boyd, Peter Richerson, Marcus Feldman, Luca Cavalli-Sforza, and others as explored in Chapter 1.

7.4   The Attempt to Separate Morality from Rationality

The Italian researcher Vilfredo Pareto (1848–1923) was a towering figure in both the sociology (*Trattato di Sociologia Generale*, 4 vols., 1916) and the economics (*Cours d'Economie Politique*, 2 vols. 1896, 1897) of his time. Pareto was a lecturer in economics at the University of Lausanne for most of his life. He was analytically powerful (having developed the concepts of Pareto optimality and the Pareto statistical distribution), but in his mature years, he despaired of fully understanding economic behavior using the standard rationality concept of economic theory. Pareto thus embraced *sociology* for deeper understanding. Much social action, he held, stems from mental "residues," by which he meant instincts and nonlogical sentiments, and their superficial appearance of logicality is due to "derivations," which are the individual's pseudo-logical justifications of residues.

While the university structure of the time did not then admit a separate discipline of "Sociology," the term had considerable academic currency in Pareto's time, having been widely used by the French philosopher Auguste Comte in 1838. Pareto defined his work on residues and derivations as being part of this field "Sociology," that encompassed but went beyond "Economics."

The first prominent sociology department was set up at Harvard University in 1931. The young Talcott Parsons, who had been teaching in Harvard's economics department, alienated by the turn to mathematical formalism in this department, moved to the new sociology department in 1931, and joined L. J. Henderson's influential Vilfredo Pareto study group. Other illustrious participants included George Homans and Crane Brinton.

Parsons, whose work defined the discipline of sociology in the post-World War II period, during which autonomous sociology departments were set up at most American and European universities, was preoccupied with maintaining a clear and defensible boundary between economics and sociology. In writing perhaps his greatest book, *The Structure of Social Action* (1937), Parsons retained the theoretical orientation of economics, while carving out an area of "non-economics," where Sociology could live in harmony with Economics. His accomplished this, in the spirit of methodological unity, by broadening the rational actor model so that it became a general "theory of action." Parsons' theory of action retained the centrality of choice under conditions of constraint. In the spirit of clear boundary-maintenance, however, as we have seen, Parsons drew on Paretian sociology by asserting that "sociological action" is based on *normative* and *ethical* concerns, in contrast with "economic action," which is founded upon self-regarding "utilitarian" choice. Parsons and his contemporaries thus developed a subtle and powerful theory of social values, social norms, the psychological internalization of norms, and the role-actor model to flesh out Pareto's theory of residues as motives to action.

Parsons' intellectual effort was so successful that his boundary criterion was virtually universally accepted without ever acquiring a name or being analyzed seriously in the literature by major social scientists. We may call this the *Parsons-Schumpeter criterion*, because Parsons and Joseph Schumpeter, the eminent Harvard economist who contributed to sociological theory (Schumpeter 1951), led a Harvard faculty group discussion of the concept of rationality in 1940, attended by several economists, including Abram Bergson, Gottfried Haberler, and Wassily Leontief. Following this, the Parsons-Schumpeter interpretation of the economics-sociology boundary was quickly embraced by leading economists of the time.

By fateful coincidence, the dominant position among philosophers of the time, even of very different theoretical persuasions, was that such moral concepts as "good" and "right" are logically without meaning, leading economists and sociologists alike to treat moral issues as "nonrational." Thus economics came to be seen as the study of self-interested rationality, leaving sociology to deal with the moral, normative, and other nonrational "residues" of human behavior. Thus in the *Foundations of Economic Analysis* (1947), Paul Samuelson writes:

> Many economists well within the academic field would separate economics from sociology upon the basis of rational or irrational behavior.

There is a subtle distinction between Pareto's analysis and that of Parsons and Schumpeter, the latter dominating social theory for the remainder of the twentieth century. Pareto argued that sociology *encompasses* economics while Parsons and his followers held that the two disciplines were *mutually autonomous*, meaning that one could carry out economic analysis without regard to sociological issues, and conversely, one could carry out sociological analysis without regard for economic issues. By contrast, in the *Trattato*, Pareto admitted that he at first held this view, but had come to appreciate its untenability; one could *not* understand either economic or sociological phenomena without a balanced use of the tools and theories of both. This point was not admitted by those who embraced the Parsons-Schumpeter criterion.

No sooner had the Parsons-Schumpeter criterion become accepted wisdom in economics, when serious cracks began appearing. In 1957 Anthony Downs published his *Economic Theory of Democracy*, which became the *de facto* standard for public economics, which was in the process of becoming an independent subfield of economics. According to Downs and the ascendant rational choice political theory school, political action is dominated by self-interested rational behavior. However Mancur Olson, in his *Logic of Collective Action* (1965), showed that self-interested rationality could rarely explain collective action. Indeed, only a handful of selfish rational individuals will even vote in elections. Thus collective action and voting appear to be "irrational," although the motives of activists and the polity are plausibly explained by principles of self-interested rationality! For instance, a peasant revolt can be rendered plausible by noting that rural taxes had risen to an oppressive level, but as Olson noted, a truly self-interested peasant will participate in only the very smallest of collective actions, if at all.

At about the same time, in 1957, Gary Becker published his *Economics of Discrimination*. which clearly illustrated the power of economic analysis when connected with a utility function in which the race of one's coworkers is an entry. The critical element in Becker's thought allowing this "breech of boundaries" was the recent and widely accepted interpretation of "rationality" as preference consistency and Bayesian updating by Leonard Savage (1954). This interpretation, which contrasts with the traditional notion of rational self-interest as implying maximizing some combination of income, leisure, and job satisfaction, would have been unavailable to Pareto, for whom racial or ethnic preference is a "residue," but was acceptable within the Parsons-Schumpeter framework, because parochial attitudes can be interpreted as individual preferences.

Becker followed with his analysis of human capital (Becker 1962, 1964). In keeping with the Parsons-Schumpeter criterion, economists interpreted human capital as technical and cognitive skills, thus locating human capital theory squarely in the purview of economic analysis, while sociologists failed to adopt human capital models for the same reason. In fact, as Gintis (1971) showed, and as stressed in Bowles and Gintis (1976) and more recently in the work of James Heckman, the economic productivity of education is about equally cognitive and affective, the affective side involving personal values and norms (Cameron and Heckman 1993; Heckman et al. 1996, 1999; Cawley et al. 2000; Heckman 2000; Heckman and Rubinstein 2001, 2003).

In succeeding years, Becker routinely analyzed primordially sociological problems using economic tools, while nevertheless largely respecting the Parsons-Schumpeter boundary by considering only self-interested behavior (Becker 1968, 1981; Becker and Murphy 1988). Becker taught a course for many years with the great sociologist James Coleman, one product of which was Coleman (1990), an attempt to ground sociological theory in the economic model of the self-interested rational actor. While widely reviewed in top sociological journals, Coleman's attempt at creating a sociology based on self-interest was universally criticized and rejected. I concur with this judgment. I recall while I was presenting a paper in his famous Chicago seminar several years ago, Professor Becker asked me what I thought of Coleman's *magnum opus*. I replied that with friends like Coleman, rational choice theory needs no enemies.

By 1990 it was clear that the Parsons-Schumpeter criterion had resulted in a scientifically untenable situation in which sociologists and economists often studied exactly the same phenomena with non-overlapping theoretical tools and incompatible models, members of each discipline categorically rejecting the analysis of the other, yet with no discernible effort being made to adjudicate this dispute (Gintis 2007c, 2009a). We can now locate the conditions leading to this situation: neither sociologists nor economists were accustomed to using controlled experiments to distinguish among otherwise plausible models.

However, by 1990, under the aegis of Vernon Smith, John Kagel, Raymond Battalio, Alvin Roth, and others, experimental laboratory methods had come of age (Battalio et al. 1981; Green and Kagel 1987; Roth et al.

1991; Smith and Williams 1992). The first notable experiment with a fully satisfactory methodology (using game theory in experimental design, using material rewards, and carefully constructing conditions of anonymity and game repetition) that bore on the adequacy of the rational actor model as outlined by Savage (1954), Anscombe and Aumann (1963), and others was Güth et al. (1982), which used the ultimatum game to show that real human subjects do not play Nash equilibria and do not maximize the game's material payoffs. In the ultimatum game, no Nash equilibrium involves rejection of positive offers, and the only subgame perfect equilibrium involves the proposer offering the minimum possible amount and the responder accepting. In fact, most proposers made offers above this amount, and responders often rejected positive offers of less than a third of the total amount.

Werner Güth, his coauthors, and indeed the whole economics profession did not realize how important this result was. The experiment had been suggested by his teacher, Reinhard Selten, who interpreted subject behavior as a "weakness of will" leading to an inability to play the rational strategy. Indeed, doubtless influenced by the pathbreaking work of Daniel Kahneman, Amos Tversky and their coworkers beginning in the early 1970s and influential in economics by the mid-1980s, it became fashionable to treat all deviations from the predictions of the rational actor model as due to "heuristics," and "bounded rationality."

7.5 Why Did Parsons Fail?

In the early 1950s Talcott Parsons was at his height of influence in the sociology profession. In his book *Toward a General Theory of Action* (1951) Parsons gathered some of the most influential social scientists of his time (Edward Shils, Gordon Allport, Clyde Kluckhohn, and Samuel Stouffer, among others) to endorse his "general theory of action," which was his attempt at an analytical core for sociology. Despite the august company and a promising opening chapter by Parsons himself, the book was deemed a failure by his contemporaries. Indeed, the various essays in this book, while insightful, never go beyond Parsons himself, and are of marginal contemporary interest. This failure of Parsons to consolidate and broaden his position shattered his dream of grounding sociology in a generally acceptable analytical framework.

Parsons opens his introductory essay in *Toward a General Theory of Action* with these words: "The present statement and the volume which introduces are intended to contribute to the establishment of a general theory in the social sciences." Brave words indeed! By 1950 Paul Samuelson's well-known *Foundations of Economic Analysis* was revolutionizing economics, as was Kenneth Arrow and Gérard Debreu's treatment of general equilibrium theory. Similarly, biology was unified around the fundamental synthesis of Ronald Fisher, Sewall Wright, and J. B. S. Haldane, bringing on board Ernst Mayr, Theodosius Dobzhansky, and the other luminaries of modern biology. This was clearly the Age of Analytical Foundations in the behavioral sciences. Talcott Parsons, who published his first synoptic works in economics journals, had always seen himself as a synthesizer who might do for sociology what Samuelson did for economics. Why did he fail, even in this exceptionally auspicious intellectual environment?

Part of the problem was the mind-set of sociologists of his time, which was to attach politically correct stories to rather pedestrian observational and statistical material, and to reject any notion that there might be an analytical core for sociological theory. Also part of the problem was Parsons' failure to articulate the close affinities between biological and economic theory on the one hand and sociology on the other. As for biology, we now know that *Homo sapiens* is one of many social species, and the biological and evolutionary analysis of human society is part of a more general scientific agenda, that of *sociobiology*, which is the study of the emergence and transformation of sociality in the biosphere (see Chapters 1 and 9). As for economics, it is clear that Parsons' theory of action is an elaboration on rational decision theory, in which beliefs are generalized beyond the subjective priors of standard decision theory (Savage 1954), and individuals hold other-regarding and universal preferences in addition to the standard preference for money and wealth.

We might also note that the Samuelson-Walras foundational theory was completely within-discipline, and consisted mainly of displacing alternative visions of economic theory—historicist, institutionalist, literary, Marxist, and the like. The analytical synthesis in biology placed more pressure for institutional change, for instance the consolidation of biology, zoology, and botany departments, but these were rather minor in comparison with Parsons' problem of getting agreement on the *unit act* as the analytical core of sociology, when the most natural exposition of the voluntaristic theory of action led inexorably to foundational principles of both biological and economic theory.

By lacking this broad perspective, sociologists of the time were burdened with a painfully short-sighted notion of the object of their studies. Indeed, as we have seen, the very idea of anthropology and sociology being distinct areas of study with virtually no theoretical overlap is a product of this myopic vision, where anthropologists studied primitive society and kinship, while sociologists studied modern industrial society and social stratification. As a result, young people became professors of anthropology mostly because of their love for pre-capitalist cultures, and of sociology mostly because they wanted to make a more just modern society. These may be noble goals, but they are not scientific goals.

Rather than envision the links between his sociological concerns on the one hand and biological and economic theory on the other, Parsons made the strategic decision to embrace a structural-functionalism that could be expressed with no dependence upon other behavioral disciplines. This was a serious error, because it led Parsons to take a highly incomplete elaboration of the unit act and the voluntaristic framework as satisfactory, and to devote all his research energy to the macrolevel issues.

Why did Parsons take this path? As we have seen, several key background factors were (a) his ignorance of mathematical modeling; (b) the extremely underdeveloped state of game theory until the mid-1970s; (c) the underdeveloped state of sociobiology in the 1950s and 1960s; and (d) the inordinate fear Parsons, like most sociologists, had of being "swallowed up" by economics. This led him to stress the moral side and systematically ignore the material incentive side of motivating role performance. Indeed, Parsons accepted Pareto's view that moral concerns cannot be incorporated into a model of rational choice. As we argued in Chapter 5 and above, we now know that this is simply incorrect (Gintis 2009a, Ch. 4). The importance of the normative side to role performance is perfectly compatible with actors caring about material incentives. In fact, the interaction among agents in role-performance can be modeled as strategic interactions in which agents attempt to find best responses to the behaviors of others (Camerer 2003; Gintis et al. 2005).

If sociology had been a mature discipline, researchers may well have seen the promise in the work of the early Parsons, and may have developed it appropriately independent of Parsons' later structural-functional concerns. In fact, the widespread respect for Parsons' work did not prevent him from being bitterly criticized and rejected by most sociologists. Some of Parsons' critics were highly politicized academics deeply involved in the social movements of the time. They include Alvin Gouldner, Barrington Moore, George Homans, Ralf Dahrendorf, C. Wright Mills, Tom Bottomore, and many others. These influential theorists railed against Parsons because he did not deal with social oppression and his language was not emancipatory. Parsons' unrelenting functionalism and his tendency to see all conflict as dysfunctional certainly lay him open to this charge (Coser 1956). To my mind, this is like criticizing Leonardo da Vinci's studies of human anatomy on the grounds that they fail to deal with cancer or the plague.

Other critics dwelt on Parsons' turn to functionalism in works of 1951 and beyond. I think these criticisms are well-taken because, in fact, a functional explanation is descriptively useful, but is not a substitute for a cogent model that develops the causal processes involved in a social phenomenon. The renown philosopher Jürgen Habermas criticized Parsons on the grounds that structural-functional theory is incompatible with the voluntarist theory of action (Habermas 1984). This is an interesting critique, but I think the general social equilibrium model shows its weakness: only in equilibrium are the functional requirements for system stability harmonious with individual choice behavior. And equilibrium never really happens.

7.6   The Flourishing of Middle-Range Theory

Sociological theory after the collapse of the Parsonian system moved in two directions. In 1957, in a highly influential book, *Social Theory and Social Structure*, Robert Merton suggested a strategic consolidation of Parsons' grand theory with his concept of *middle-range theory* (Merton 1968[1957]). Merton suggested that sociologists develop "special theories from which to derive hypotheses that can be empirically investigated." It was hoped that the accumulation of successful middle-range theories would merge into high theory "by evolving a progressively more general conceptual scheme that is adequate to consolidate groups of special theories" (Merton 1968[1957]).

Perhaps the most prominent example of middle-range theory in action was the illustrious career of James Coleman. His middle-range work includes the political structure of American labor unions, the culture of the American high school, the social structure of the American educational system, and the transition from adolescence to adulthood. Coleman established his reputation with the analysis of "burning issue" social problems. At a time when Americans were concerned about an antisocial turn in teenage culture, Coleman studied the social relations of ten Midwestern high schools, his findings summarized in *The Adolescent Society* (1981[1961]). He was the lead author of the famous "Coleman Report" (1966) on educational inequality in the United States, and in 1976, he analyzed the "white flight" from the inner city that followed enforced school desegregation. Coleman's work showcases the strengths of modern sociology—the use of statistics and common-sense middle-level social theory to analyze complex social problems. Coleman always maintained an interest in turning his middle-range insights into a complete high-level theory, which he offered in his *Foundations of Social Theory* (1990). Unfortunately, this work does not live up to Coleman's expectations.

Coleman worked on *The Foundations of Social Theory* on and off for two decades. Upon publication, *Foundations* was reviewed in several sociology journals by leading sociologists. Nearly all were bitterly critical of his efforts. The critiques are almost wholly centered on Coleman's championing of the economist's decision model in the sociologically relevant form forged by Nobel Prize-winning economist Gary Becker, with whom Coleman worked at the University of Chicago for many years (Becker 1957, 1968, 1981; Becker and Stigler 1974; Becker et al. 1994).

Coleman's advocacy of the rational actor model, the centerpiece of economic theory, was courageous and far-sighted, but he imported from economics only one aspect of the model, and the part he imported is wrong. Economists, until recent years, in practice identified rationality with selfishness and the capacity to calculate gains and losses without regard to the well-being of others, moral virtues, or the needs of the larger society. It is precisely this sociopathic conception of rationality that Coleman makes the centerpiece of *Foundations of Social Theory*.

The critique of Coleman by the sociology profession was well-deserved. The *Foundations of Social Theory*, which equates rationality with selfishness and then makes rationality the basis of individual behavior, has few redeeming qualities. Of course, economist Gary Becker achieved exemplary results by applying economic theory to traditional sociological problems, and Becker also, with some notable exceptions, equates rationality with selfishness. But unlike Coleman, Becker has chosen his subject matter very carefully, and his analysis is always both brilliant and plausible. Coleman, by contrast, applies the selfish/rational actor model willy-nilly to every possible social situation, and the result is at best awkward, and often simply bizarre, such as when Coleman wonders how workers can be talked into doing and believing things that are not in their material self-interest, or why mothers appear to love their children.

Unlike Becker, Coleman is forced by the nature of his discipline to deal with socialization, through which individuals are led to internalize important social values, so that they tend to conform to these values purely volitionally, even when there is no chance of being subject to external sanctions. For example, in many circumstances people are honest even when no one is looking, so they could easily cheat with impunity. Coleman never does explain why a rational/selfish individual would submit to internalization. Moreover his idea of internalization is that internalized values are *psychic constraints* on action. That is, people behave prosocially because they would feel guilty if they did not. In fact, people often attempt to avoid situations where they would be obliged to act charitably. The evidence indicates that internalized norms more often are integrated into the individual's preference function, so people often feel good when they act morally—see Gintis (2009a) and Chapter 6. Indeed, as stressed by virtue philosophers from the time of Aristotle to the present, virtuous people are not crippled by a domineering and repressive superego, but rather are happier and more complete than the sociopaths that populate Coleman's world of selfish rationality.

Coleman is clearly inspired by economic theory, but he did not have a serious grasp of decision theory. He never references the basic works in this area, such as Savage (1954) and di Finetti (1974), and he never bothers to mention that economic rationality is simply defined as preference consistency and Bayesian updating. Nowhere in the basic theory is it said that people are selfish, are indifferent to social concerns, or that they attempt to maximize anything. As we have seen, there is nothing irrational about voting, loving your *alma mater*'s lacrosse team, or giving to charity.

Coleman is not alone. The degree to which sociologists understand the rational actor model that they love to criticize is abysmal, and Coleman merely reinforces the standard sociological prejudices. His description of the rational actor model is absolutely ripe for caricaturing. "Actors have a single principle of action," he says, "that of acting so as to maximize their realization of interests" (p. 37). This sounds sociopathic, but in fact the rational actor theory does not say that people act to maximize anything, any more than light rays act to minimize transit time, and an individual care not only about his own interests, but those of others, and his honor and integrity as well.

Coleman's critics rarely fail to mention that he has no place for culture or symbolic communication in his approach, and they blame this on his reliance on rational action. The complaint is correct, but the reason is incorrect. When we recognize that social behavior involves strategic interaction based on social norms, and social norms are legitimated and interpreted only in the context of a group's cultural traditions and web of symbolic meanings, the interaction between rationality, morality, and culture can be properly modeled.

7.7 High Theory as Interpretation

The second direction taken by sociological theory after the collapse of Talcott Parsons' intellectual edifice was what might be called *interpretive theorizing*. Interpretive theorists may make extensive use of data (e.g., Pierre Bourdieu), but their arguments are judged hermeneutically by how they *prima facie* help us make sense of one or more facets of social life. Thus Karl Marx gave us a theory of class and social change (Marx 1998[1848]), Ferdinand Tönnies explored the shift from Gemeinschaft to Gesellschaft (Tönnies 2001[1887]), George Homans and Alvin Gouldner viewed social life through the lens of social exchange (Homans 1958; Gouldner 1960), Peter Blau analyzed power in the context of social exchange (Blau 1964), Mark Granovetter studied the network structure of social relations (Granovetter 1985), Jürgen Habermas explored the history and meaning of communicative discourse in modern society (Habermas 1991), Anthony Giddens (1986) integrate the insights of structuralism into a broader philosophical and sociological setting, Niklas Luhmann analyzed society as a system (Luhmann 2012[2004]), and Pierre Bourdieu gave us thick ethnographic, yet philosophically informed and motivated, descriptions of modern social life, as well as the psychological processes that lend stability to social structure (Bourdieu 1972/1977). These contributions, many of inestimable value, are themselves middle-level theories, in the sense that they take most of social life as given, and focus upon one facet, or several interrelated facets, of social life to analyze in depth. Often this task is accompanied by assertions that these facets are in fact the whole, as in Marx's single-minded historical materialism, Homans' insistence that there are no social structures above the individual, or Luhmann's contention that a theory of individual behavior is superfluous in social theory. These assertions, however, are scarcely plausible. Interpretive theory is useful precisely because society is a complex dynamical system, so any attempt to reduce thick description to simple analytical principles is doomed to failure.

Another approach at high-level theorizing is centrally motivated by particular continental philosophical schools. For instance, phenomenology inspired Alfred Schutz (Schutz 1932/1967), Peter Berger and Thomas Luckmann (Berger and Luckmann 1967), and Erving Goffman (Goffman 1959), structuralism influenced Michel Foucault (Foucault 1982), and postmodernism inspired Harrison White (White 2008). These approaches are also suggestive, but they stand and fall with their underlying philosophical positions, which are not generally accepted even by philosophers, much less by social scientists.

8

The Evolution of Property

> It is labor alone that is productive: it creates wealth and therewith lays the outward foundations for the inward flowering of man.
>
> Ludwig von Mises

> Every Man has a property in his own Person. This no Body has any Right to but himself. The Labour of his Body, and the Work of his Hands, we may say, are properly his.
>
> John Locke

> **C:**   When can property rights be effective without state enforcement?
> **HG:** When those who hold property are willing to fight harder to maintain it than a usurper is to seize it.
>
> Choreographer interview

Authors tracing back to the origins of political liberalism have treated property rights as a convention whose value lies in reducing conflict over incumbency (Schlatter 1973). In his *Leviathan*, Hobbes (1968[1651]), Chapter 29, argues:

> Every man has a Propriety that excludes the Right of every other Subject: And he has it onely from the Soveraign Power; without the protection whereof, every other man should have equall Right to the same.

We argue in this chapter that Hobbes is profoundly wrong. Property rights exist in human societies without states, including the hunter-gatherer societies in which we evolved as a species (see Chapter 2). Indeed, the predispositions of individuals to protect their rights without resort to courts of law strongly enhances human social life. Moreover, the well-known phenomena of *loss aversion* and the *endowment effect*, often considered irrational, are strongly fitness-enhancing predispositions in humans and they correspond to the motivations in other species that give rise to territorial behavior.

165

8.1   The Endowment Effect

According to the *endowment effect*, people value a good that they possess more highly than they value the same good when they do not possess it. Experimental studies have shown that subjects exhibit a systematic endowment effect (Kahneman et al. 1991). Because the endowment effect is an aspect of prospect theory (Kahneman and Tversky 1979), it can be modeled by amending the standard rational actor model to include an agent's *current holdings* as a parameter. The endowment effect gives rise to *loss aversion*, according to which agents are more sensitive to losses than to gains. We show here that the endowment effect can be modeled as respect for property rights without the need for legal institutions ensuring third-party contract enforcement (Jones 2001; Stake 2004). In this sense, pre-institutional "natural" property rights have been observed in many species in the form of *territorial possession*. We develop a model loosely based on the Hawk-Dove Game and the War of Attrition (Maynard Smith and Price 1973; Bishop and Cannings 1978) to explain the natural evolution of property rights.

We show that if agents in a group exhibit the endowment effect for an indivisible resource, then property rights for that resource can be established on the basis of *incumbency*, assuming incumbents and those who contest for incumbency are of similar perceived fighting ability.[1] The enforcement of these rights is then carried out by the agents themselves, so no third-party enforcement is needed. This is because the endowment effect leads the incumbent to be willing to expend more resources to *protect* his incumbency than an intruder will be willing to expend to *expropriate* the incumbent. For simplicity, we consider only the case where exactly one unit of the resource is useful to the incumbent (e.g., a homestead, a spider's web, or a bird's nest).

The model assumes the agents know the present value of the "good" state π_g of incumbency, as well as the present value of the "bad" state π_b of nonincumbency, measured in units of biological fitness. We assume utility and fitness coincide, except for one situation, described below. This situation explicitly involves *loss aversion*, where the disutility of loss exceeds the fitness cost of loss. When an incumbent faces an intruder, the

---

[1]The assumption of indivisibility is not very restrictive. In some cases it is naturally satisfied, as in a nest, a web, a dam, or a mate. In others, such as a hunter's kill, a fruit tree, a stretch of beach for an avian scavenger, it is simply the minimum size worth fighting over rather than dividing and sharing.

intruder determines the expected value of attempting to seize the resource, and the incumbent determines the expected value of contesting vs. ceding incumbency when challenged. These conditions will not be the same, and in plausible cases there is a range of values of π_g/π_b for which the intruder decides not to fight and the incumbent decides to fight if challenged. We call this a (natural) *property equilibrium*. In a property equilibrium, since the potential contestants are of equal power, it must be the case that individuals are *loss-averse*, the incumbent being willing to expend more resources to *hold* the resource than the intruder is to *seize* it.

Of course, π_g and π_b are generally *endogenous* in a fully specified model. Their values depend on the supply of the resource relative to the number of agents, the intrinsic value of the resource, the ease of finding an unowned unit of the resource, and the like.

In our model of decentralized property rights, agents contest for a unit of an indivisible resource, contests may be very costly, and in equilibrium, incumbency determines who holds the resource without costly contests. Our model, however, fills in critical gaps in the Hawk-Dove Game. The central weakness of the Hawk-Dove Game is that it treats the cost of contesting as exogenously given and taking on exactly two values, high for hawk and low for dove. Clearly, however, these costs are partially under the control of the agents themselves and should not be considered exogenous. In our model, the level of resources devoted to a contest is endogenously determined, and the contest itself is modeled explicitly as a modified War of Attrition, the probability of winning being a function of the level of resources committed to combat. One critical feature of the War of Attrition is that the initial commitment of a level of resources to a contest must be *behaviorally ensured by the agent*, so that the agent will continue to contest even when the costs of doing so exceed the fitness benefits. Without this precommitment, the incumbent's threat of "fighting to the death" would not be credible (i.e., the agent would abandon the chosen best response when it came time to use it). From a behavioral point of view, this precommitment can be summarized as the incumbent having a degree of *loss aversion* leading his utility to differ from his fitness.

Our fuller specification of the behavioral underpinnings of the Hawk-Dove Game allows us to determine the conditions under which a property equilibrium will exist while its corresponding antiproperty equilibrium (in which a new arrival rather than the first entrant always assumes incumbency) does not exist. This aspect of our model is of some importance because the inability of the Hawk-Dove Game to favor property over antiproperty is a serious and rarely addressed weakness of the model (but see Mesterton-Gibbons 1992).

8.2   Territoriality

The endowment effect, according to which a good is more highly prized by an agent who is in possession of the good than by one who is not, was first documented by the psychologist Daniel Kahneman and his coworkers (Tversky and Kahneman 1991; Kahneman et al. 1991; Thaler 1992).

Thaler describes a typical experimental verification of the phenomenon as follows. Seventy-seven students at Simon Fraser University were randomly assigned to one of three conditions, seller, buyer, or chooser. Sellers were given a mug with the university logo (selling for $6.00 at local stores) and asked whether they would be willing to sell at a series of prices ranging from $0.25 to $9.25. Buyers were asked whether they would be willing to purchase a mug at the same series of prices. For each price, choosers were asked to choose between receiving a mug or receiving that amount of money. The students were informed that a fraction of their choices, randomly chosen by the experimenter, would be carried out, thus giving the students a material incentive to reveal their true preferences. The average buyer price was $2.87, while the average seller price was $7.12. Choosers behaved like buyers, being on average indifferent between the mug and $3.12. The conclusion is that owners of the mug valued the object more than twice as highly as nonowners.

The aspect of the endowment effect that promotes natural property rights is known as *loss aversion*: the disutility of giving up something one owns is greater than the utility associated with acquiring it. Indeed, losses are commonly valued at about twice that of gains, so that to induce an individual to accept a lottery that costs $10 when one loses (which occurs with probability 1/2), it must offer a $20 payoff when one wins (Camerer 2003). Assuming that an agent's willingness to combat over possession of an object is increasing in the subjective value of the object, owners are prepared to fight harder to *retain* possession than nonowners are to *gain* possession. Hence there will be a predisposition in favor of recognizing property rights by virtue of incumbency, even where third-party enforcement institutions are absent.

We say an agent *owns* something, or *is incumbent*, if the agent has exclusive access to it and the benefits that flow from this privileged access. We say ownership (incumbency) is *respected* if it is rarely contested and, when contested, generally results in ownership remaining with the incumbent. The dominant view in Western thought, from Hobbes, Locke, Rousseau, and Marx to the present, is that property rights are a human social construction that emerged with the rise of modern civilization (Schlatter 1973). However, evidence from studies on animal behavior, gathered mostly in the past quarter-century, has shown this view to be incorrect. Various territorial claims are recognized in nonhuman species, including butterflies (Davies 1978), spiders (Riechert 1978), wild horses (Stevens 1988), finches (Senar et al. 1989), wasps (Eason et al. 1999), nonhuman primates (Ellis 1985), lizards (Rand 1967), and many others (Mesterton-Gibbons and Adams 2003). There are, of course, some obvious forms of incumbent advantage that partially explain this phenomenon: the incumbent's investment in the territory may be idiosyncratically more valuable to the incumbent than to a contestant or the incumbent's familiarity with the territory may enhance its ability to fight. However, in the above-cited cases, these forms of incumbent advantage are unlikely to be important. Thus, a more general explanation of territoriality is needed.

In nonhuman species, that an animal owns a territory is generally established by the fact that the animal has occupied and altered the territory (e.g., by constructing a nest, burrow, hive, dam, or web, or by marking its limits with urine or feces). In humans there are other criteria of ownership, but physical possession and first to occupy remain of great importance, as expressed by John Locke in the head quote for this chapter.

Since property rights in human society are generally protected by law and are enforced by complex institutions (judiciary and police), it is natural to view property rights in animals as a categorically distinct phenomenon. In fact, however, decentralized, self-enforcing types of property rights, based on behavioral propensities akin to those found in nonhuman species (e.g., the endowment effect), are important for humans and arguably lay the basis for more institutional forms of property rights. For instance, many developmental studies indicate that toddlers and small children use behavioral rules similar to those of animals in recognizing and defending property rights (Furby 1980).

How respect for ownership has evolved and how it is maintained in an evolutionary context is a challenging puzzle. Why do loss aversion and the endowment effect exist? Why do humans fail to conform to the smoothly differentiable utility function assumed in most versions of the rational actor model? The question is equally challenging for nonhumans, although we are so used to the phenomenon that we rarely give it a second thought.

Consider, for instance, the sparrows that built a nest in a vine in my garden. The location is choice, and the couple spent days preparing the structure. The nest is quite as valuable to another sparrow couple. Why does another couple not try to evict the first? If they are equally strong, and both value the territory equally, each has a 50% chance of winning the territorial battle. Why bother investing if one can simply steal (Hirshleifer 1988)? Of course, if stealing were profitable, then there would be no nest building, and hence no sparrows, but that heightens rather than resolves the puzzle.

One common argument, borrowed from Trivers (1972), is that the original couple has more to lose since it has already put a good deal of effort into the improvement of the property. This, however, is a logical error that has come to be known as the *Concorde fallacy* or the *sunk cost fallacy* (Dawkins and Brockmann 1980; Arkes and Ayton 1999): to maximize future returns, an agent ought to consider only the future payoffs of an entity, not how much the agent has expended on the entity in the past.

The Hawk-Dove Game was offered by Maynard Smith and Parker (1976) as a logically sound alternative to the sunk cost argument. In this game hawks and doves are phenotypically indistinguishable members of the same species, but they act differently in contesting ownership rights to a territory. When two doves contest, they posture for a bit, and then each assumes the territory with equal probability. When a dove and a hawk contest, however, the hawk takes the whole territory. Finally, when two hawks contest, a terrible battle ensues, and the value of the territory is less than the cost of fighting for the contestants. Maynard Smith showed that, assuming that there is an unambiguous way to determine who first found the territory, there is an evolutionarily stable strategy in which all agents behave like hawks when they are *first* to find the territory, and like doves otherwise.

The Hawk-Dove Game is an elegant contribution to explaining the endowment effect, but the cost of contesting for hawks and the cost of display for doves cannot plausibly be taken as fixed and exogenously determined. Indeed, it is clear that doves contest in the same manner as hawks, except that they devote fewer resources to combat. Similarly, the value of the ownership is taken as exogenous, when in fact it depends on the frequency with which ownership is contested, as well as on other factors. As Grafen

(1987) stresses, the costs and benefits of possession depend on the state of the population, the density of high-quality territories, the cost of search, and other variables that might well depend on the distribution of strategies in the population.

First, however, it is instructive to consider the evidence for a close association, as Locke suggested in his theory of property rights, between ownership and incumbency (physical contiguity and control) in children and nonhuman animals.

8.3   Property Rights in Young Children

Long before they become acquainted with money, markets, bargaining, and trade, children exhibit possessive behavior and recognize the property rights of others on the basis of incumbency (Ellis 1985). In one study (Bakeman and Brownlee 1982), participant observers studied a group of 11 toddlers (12 to 24 months old) and a group of 13 preschoolers (40 to 48 months old) at a day care center. The observers found that each group was organized into a fairly consistent linear dominance hierarchy. They then cataloged *possession episodes*, defined as situations in which a *holder* touched or held an object and a *taker* touched the object and attempted to remove it from the holder's possession. Possession episodes averaged 11.7 per hour in the toddler group and 5.4 per hour in the preschool group.

For each possession episode, the observers noted (a) whether the taker had been playing with the object within the previous 60 seconds (prior possession), (b) whether the holder resisted the take attempt (resistance), and (c) whether the take was successful (success). They found that success was strongly and about equally associated with both dominance and prior possession. They also found that resistance was positively associated with dominance in the toddlers and negatively associated with prior possession in the preschoolers. They suggest that toddlers recognize possession as a basis for asserting control rights but do not respect the same rights in others. Preschoolers, more than twice the age of the toddlers, use physical proximity both to justify their own claims and to respect the claims of others. This study was replicated and extended by Weigel (1984).

8.4   Respect for Possession in Nonhuman Animals

In a famous paper, Maynard Smith and Parker (1976) noted that if two animals are competing for some resource (e.g., a territory), and if there is some discernible asymmetry (e.g., between an owner and a later-arriving animal), then it is evolutionarily stable for the asymmetry to settle the contest conventionally, without fighting. Among the findings of the many animal behaviorists who put this theory to the test, perhaps none is more elegant and unambiguous than that of Davies (1978), who studied the speckled wood butterfly (*Pararge aegeria*), which is found in the Wytham Woods, near Oxford, England. Territories for this butterfly are shafts of sunlight breaking through the tree canopy. Males occupying these spots enjoyed heightened mating success, and on average only 60% of the males occupied the sunlit spots at any one time. A vacant spot was generally occupied within seconds, but an intruder at an already occupied spot was invariably driven away even if the incumbent had occupied the spot for only a few seconds. When Davies "tricked" two butterflies into thinking each had occupied the sunny patch first, the contest between the two lasted, on average, 10 times as long as the brief flurry that occurred when an incumbent chased off an intruder.

Stevens (1988) found a similar pattern of behavior among the feral horses occupying the sandy islands of the Rachel Carson Estuarine Sanctuary near Beaufort, North Carolina. In this case, it is freshwater that is scarce. After heavy rains, freshwater accumulates in many small pools in low-lying wooded areas, and bands of horses frequently stop to drink. Stevens found that there were frequent encounters between bands of horses competing for water at these temporary pools. If a band approached a water hole occupied by another band, a conflict ensued. During 76 hours of observation, Stevens observed 233 contests, of which the resident band won 178 (80%). In nearly all cases of usurpation, the intruding band was larger than the resident band. These examples, and many others like them, support the presence of an endowment effect and suggest that incumbents are willing to fight harder to maintain their positions than intruders are to usurp the owner.

Examples from nonhuman primates exhibit behavioral patterns in respecting property rights much closer to those of humans. In general, the taking of an object held by another individual is a rare event in primate societies (Torii 1974). A reasonable test of the respect for property in primates with a strong dominance hierarchy is the likelihood of a dominant individual refraining from taking an attractive object from a lower-ranking individual. In a study of hamadryas baboons (*Papio hamadryas*), for instance, Sigg and Falett (1985) handed a food can to a subordinate who was allowed to manipulate it and eat from it for 5 minutes before a dominant individual who had been watching from an adjacent cage was allowed to enter the subordinate's cage. A takeover was defined as the rival taking possession of the can before 30 minutes had elapsed. They found that (a) males never took the food can from other males; (b) dominant males took the can from subordinate females two-thirds of the time; and (c) dominant females took the can from subordinate females one-half of the time. With females, closer inspection showed that when the difference in rank was one or two, females showed respect for the property of other females, but when the rank difference was three or greater, takeovers tended to occur.

Kummer and Cords (1991) studied the role of proximity in respect for property in long-tailed macaques (*Macaca fascicularis*). As in the Sigg and Falett study, they assigned ownership to a subordinate and recorded the behavior of a dominant individual. The valuable object in all cases was a plastic tube stuffed with raisins. In one experiment, the tube was fixed to an object in half the trials and completely mobile in the other half. They found that with the fixed object, the dominant rival took possession in all cases and very quickly (median 1 minute), whereas in the mobile condition, the dominant rival took possession in only 10% of cases, and then only after a median delay of 18 minutes. The experiment took place in an enclosed area, so the relative success of the incumbent was not likely due to an ability to flee or hide. In a second experiment, the object was either mobile or attached to a fixed object by a stout 2- or 4-meter rope. The results were similar. A third case, in which the nonmobile object was attached to a long dragline that permitted free movement by the owner, produced the following results. Pairs of subjects were studied under two conditions, one where the rope attached to the dragline was 2 meters in length and a second where the rope was 4 meters in length. In 23 of 40 trials, the subordinate maintained ownership with both rope lengths, and in 6 trials the dominant rival took possession with both rope lengths. In the remaining 11 trials, the rival respected the subordinate's property in the short rope case but took possession in the long-rope case. The experimenters observed that when a dominant attempted to usurp a subordinate when other group members were around, the subordinate screamed, drawing the attention of third parties, who frequently forced the dominant individual to desist.

In *Wild Minds* (2000), Marc Hauser relates an experiment run by Kummer and his colleagues concerning mate property, using four hamadryas baboons, Joe, Betty, Sam, and Sue. Sam was let into Betty's cage while Joe looked on from an adjacent cage. Sam immediately began following Betty around and grooming her. When Joe was allowed entrance into the cage, he kept his distance, leaving Sam uncontested. The same experiment was repeated with Joe being allowed into Sue's cage. Joe behaved as Sam had in the previous experiment, and when Sam was let into the cage, he failed to challenge Joe's proprietary rights with respect to Sue.

No primate experiment, to my knowledge, has attempted to determine the probability that an incumbent will be contested for ownership by a rival who is, or could easily become, closely proximate to the desired object. This probability is likely very low in most natural settings, so the contests described in the papers cited in this section are probably rather rare in practice. At any rate, in the model of respect for property developed in the next section, we will make informational assumptions that render the probability of contestation equal to zero in equilibrium.

8.5   Conditions for a Property Equilibrium

Suppose that two agents, prior to fighting over possession, simultaneously precommit to expending a certain level of resources in the contest. As in the War of Attrition (Bishop and Cannings 1978), a higher level of resource commitment entails a higher fitness cost but increases the probability of winning. We assume in the remainder of this chapter that the two contestants, an incumbent and an intruder, are *ex ante* equally capable contestants in that the costs and benefits of battle are symmetric in the resource commitments s_o (owner) and s_u (usurper) of the incumbent and the intruder, respectively, and $s_o, s_u \in [0, 1]$. To satisfy this requirement, we let $p_u = s_u^n / (s_u^n + s_o^n)$ be the probability that the intruder wins, where $n > 1$. Note that a larger n implies that resource commitments are more decisive in determining victory. We assume that combat leads to injury $\beta \in (0, 1]$ to the losing party with probability $p_d = (s_o + s_u)/2$, so $s = \beta p_d$ is the expected cost of combat for both parties.

We use a territorial analogy throughout, some agents being incumbents and others being migrants in search of either empty territories or occupied territories that they may be able to occupy by displacing current incumbents. Let π_g be the present value of being a currently uncontested incumbent and let π_b be the present value of being a migrant searching for a territory. We assume throughout that $\pi_g > \pi_b > 0$. Suppose a migrant comes upon an occupied territory. Should the migrant contest, the condition under which it pays an incumbent to fight back is then given by

$$\pi_c \equiv (1 - p_u)\pi_g + p_d p_u (1 - \beta)(1 - c)\pi_b$$
$$+ (1 - p_d) p_u \pi_b (1 - c) > \pi_b (1 - c).$$

The first term in π_c is the product of the probabilities that the incumbent wins $(1 - p_u)$ times the value π_g of incumbency, which the incumbent then retains. The second term is the product of the probabilities that the incumbent loses (p_u), sustains an injury (p_d), survives the injury $(1 - \beta)$, and survives the passage to migrant status $(1 - c)$ times the present value π_b of being a migrant. The third term is the parallel calculation when the incumbent loses but sustains no injury. This inequality simplifies to

$$\frac{\pi_g}{\pi_b(1 - c)} - 1 > \frac{s_u^n}{s_o^n}s. \tag{8.1}$$

The condition for a migrant refusing to contest for the territory, assuming the incumbent will contest if the migrant does, is

$$\pi_u \equiv p_d(p_u \pi_g + (1 - p_u)(1 - \beta)(1 - c)\pi_b) \tag{8.2}$$
$$+ (1 - p_d)(p_u \pi_g + (1 - p_u)\pi_b(1 - c)) < \pi_b(1 - c). \tag{8.3}$$

This inequality reduces to

$$\frac{s_o^n}{s_u^n}s > \frac{\pi_g}{\pi_b(1 - c)} - 1. \tag{8.4}$$

A property equilibrium occurs when both inequalities obtain

$$\frac{s_o^n}{s_u^n}s > \frac{\pi_g}{\pi_b(1 - c)} - 1 > \frac{s_u^n}{s_o^n}s. \tag{8.5}$$

An incumbent who is challenged chooses s_o to maximize π_c and then contests if and only if the resulting $\pi_c^* > \pi_b(1 - c)$, since the latter is the value of simply leaving the territory. It is easy to check that $\partial \pi_c / \partial s_o$ has the same sign as

$$\frac{\pi_g}{\pi_b(1 - c)} - \left(\frac{s_o \beta}{2n(1 - p_u)} + 1 - s\right).$$

The derivative of this expression with respect to s_o has the same sign as $(n-1)\beta\pi_b/(1-p_u)$, which is positive. Moreover, when $s_o = 0$, $\partial\pi_c/\partial s_o$ has the same sign as

$$\frac{\pi_g}{\pi_b(1-c)} - 1 + \frac{s_u\beta(1-c)}{2},$$

which is positive. Therefore, $\partial\pi_c/\partial s_o$ is always strictly positive, so $s_o = 1$ maximizes π_c.

In deciding whether or not to contest, the migrant chooses s_u to maximize π_u and then contests if this expression exceeds $\pi_b(1-c)$. But $\partial\pi_u/\partial s_u$ has the same sign as

$$\frac{\pi_g}{\pi_b(1-c)} - \left(s - 1 + \frac{s_u\beta}{2np_u}\right),$$

which is increasing in s_u and is positive when $s_u = 0$, so the optimal $s_u = 1$. The condition for not contesting the incumbent is then

$$\frac{\pi_g}{\pi_b(1-c)} - 1 < \beta. \tag{8.6}$$

In this case, the condition (8.4) for the incumbent contesting is the same as (8.6) with the inequality sign reversed.

By an *antiproperty* equilibrium we mean a situation where intruders always contest and incumbents always relinquish their possessions without a fight.

THEOREM 8.1 *If $\pi_g > (1 + \beta)\pi_b(1 - c)$, there is a unique equilibrium in which a migrant always fights for possession and an incumbent always contests. When the reverse inequality holds, there exists both a property equilibrium and an antiproperty equilibrium.*

Theorem 8.1 implies that property rights are more likely to be recognized when combatants are capable of inflicting great harm on one another, so β is close to its maximum of unity, or when migration costs are very high, so c is close to unity.

Theorem 8.1 may apply to a classic problem in the study of hunter-gatherer societies, which are important not only in their own right but also because our ancestors lived uniquely in such societies until about 10,000 years ago, and hence their social practices have doubtless been a major environmental condition to which the human genome has adapted (Tooby and

Cosmides 1992). One strong uniformity across current-day hunter-gatherer societies is that low-value foodstuffs (e.g., fruits and small game) are consumed by the families that produce them, but high-value foodstuffs (e.g., large game and honey) are meticulously shared among all group members. The standard argument is that high-value foodstuffs exhibit a high variance, and sharing is a means of reducing individual variance. But an alternative with much empirical support is the *tolerated theft* theory that holds that high-value foodstuffs are worth fighting for (i.e., the inequality in theorem 8.1 is satisfied), and the sharing rule is a means of reducing the mayhem that would inevitably result from the absence of secure property rights to high-value foodstuffs (Hawkes 1993; Blurton-Jones 1987; Betzig 1997; Bliege Bird and Bird 1997; Wilson 1998a).[2]

The only part of theorem 8.1 that remains to be proved is the existence of an antiproperty equilibrium. To see this, note that such an equilibrium exists when $\pi_c < \pi_b(1-c)$ and $\pi_u > \pi_b(1-c)$, which, by the same reasoning as above, occurs when

$$\frac{s_u^n}{s_o^n} > \frac{\pi_g}{\pi_b(1-c)} - 1 > \frac{s_o^n}{s_u^n}s. \tag{8.7}$$

It is easy to show that if the incumbent contests, then both parties will set $s_u = s_o = 1$, in which case the condition for the incumbent to do better by not contesting is exactly what it is in the property equilibrium.

The result that there exists an antiproperty equilibrium exactly when there is a property equilibrium is quite unrealistic since few, if any, antiproperty equilibria have been observed. Our model, of course, shares this anomaly with the Hawk-Dove Game, for which this weakness has never been analytically resolved. In our case, however, when we expand our model to determine π_g and π_g, the antiproperty equilibrium generally disappears. The problem with the above argument is that we cannot expect π_g and π_b to have the same values in a property and in an antiproperty equilibrium.

8.6   Property and Antiproperty Equilibria

To determine π_g and π_b, we must flesh out the above model of incumbents and migrants. Consider a field with many patches, each of which is indivisible and hence can have only one owner. In each time period, a fertile

---

[2]For Theorem 8.1 to apply, the resource in question must be indivisible. In this case, the "territory" is the foodstuff that delivers benefits over many meals, and the individuals who partake of it are temporary occupiers of the territory.

patch yields a benefit $b > 0$ to the owner and dies with probability $p > 0$, forcing its owner (should it have one) to migrate elsewhere in search of a fertile patch. Dead patches regain their fertility after a period of time, leaving the fraction of patches that are fertile constant from period to period. An agent who encounters an empty fertile patch invests an amount $v \in (0, 1/2)$ of fitness in preparing the patch for use and occupies the patch. An agent suffers a fitness cost $c > 0$ each period he is in the state of searching for a fertile patch. An agent who encounters an occupied patch may contest for ownership of the patch according to the War of Attrition structure analyzed in the previous section.

Suppose there are n_p patches and n_a agents. Let r be the probability of finding a fertile patch and let w be the probability of finding a fertile unoccupied patch. If the rate at which dead patches become fertile is q, which we assume for simplicity does not depend on how long a patch has been dead, then the equilibrium fraction f of patches that are fertile must satisfy $n_p f p = n_p(1 - f)q$, so $f = q/(p + q)$. Assuming that a migrant finds a new patch with probability ρ, we then have $r = f\rho$. If ϕ is the fraction of agents that are incumbents, then writing $\alpha = n_a/n_p$, we have

$$w = r(1 - \alpha\phi). \tag{8.8}$$

Assuming the system is in equilibrium, the number of incumbents whose patches die must be equal to the number of migrants who find empty patches, or $n_a\phi(1 - p) = n_a(1 - \phi)w$. Solving this equation gives ϕ, which is given by

$$\alpha r\phi^2 - (1 - p + r(1 + \alpha))\phi + r = 0. \tag{8.9}$$

It is easy to show that this equation has two positive roots, exactly one lying in the interval $(0, 1)$.

In a property equilibrium, we have

$$\pi_g = b + (1 - p)\pi_g + p\pi_b(1 - c), \tag{8.10}$$

and

$$\pi_b = w\pi_g(1 - v) + (1 - w)\pi_b(1 - c). \tag{8.11}$$

Note that the cost v of investing and the cost c of migrating are interpreted as fitness costs and hence as probabilities of death. Thus, the probability of a migrant becoming an incumbent in the next period is $w(1 - v)$, and the probability of remaining a migrant is $(1 - w)$. This explains (8.11). Solving these two equations simultaneously gives equilibrium values of incumbency and nonincumbency:

$$\pi_g^* = \frac{b(c(1-w)+w)}{p(c(1-vw)+vw)}, \tag{8.12}$$

$$\pi_b^* = \frac{b(1-v)w}{p(c(1-vw)+vw)}. \tag{8.13}$$

Note that $\pi_g, \pi_b > 0$ and

$$\frac{\pi_g^*}{\pi_b^*(1-c)} - 1 = \frac{c + vw(1-c)}{w(1-v)(1-c)}. \tag{8.14}$$

By theorem 8.1, the assumption that this is a property equilibrium is satisfied if and only if this expression is less than β, or

$$\frac{c + vw(1-c)}{w(1-v)(1-c)} < \beta. \tag{8.15}$$

We have the following theorem.

THEOREM 8.2 *There is a strictly positive migration cost c^* and a cost of injury $\beta^*(c)$ for all $c < c^*$ such that a property equilibrium holds for all $c < c^*$ and $1 > \beta > \beta^*(c)$.*

To see this, note that the left-hand side of (8.15) is less than 1 precisely when

$$0 < c < c^* = \frac{w(1-2v)}{1+w(1-2v)}.$$

Because $v < 1/2$, this is a nonempty interval. We then set

$$\beta^*(c) = \frac{c + vw(1-c)}{w(1-v)(1-c)}.$$

This ensures that $\beta^*(c) < 1$.

This theorem shows that, in addition to our previous result, a low fighting cost and a high migration cost undermine the property equilibrium, a high probability w that a migrant encounters an incumbent undermines the property equilibrium, and a high investment v has the same effect.

Suppose, however, that the system is in an antiproperty equilibrium. In this case, letting q_u be the probability that an incumbent is challenged by an intruder, we have

$$\pi_g = b + (1-p)(1-q_u)\pi_g + (p(1-q_u) + q_u)\pi_b(1-c) \quad (8.16)$$

and

$$\pi_b = w\pi_g(1-v) + (r-w)\pi_g + (1-r)\pi_b(1-c). \quad (8.17)$$

Solving these equations simultaneously gives

$$\pi_g^* = \frac{b(c(1-r)+r)}{((p(1-q_u)+q_u))(vw+c(1-vw))}, \quad (8.18)$$

$$\pi_b^* = \frac{b(r-vw)}{(((p(1-q_u)+q_u))(vw+c(1-vw)))}. \quad (8.19)$$

Also, $\pi_g, \pi_b > 0$ and

$$\frac{\pi_g^*}{\pi_b^*} - 1 = \frac{c(1-r)+vw}{r-vw}. \quad (8.20)$$

Note that $r - vw = r(1 - v(1 - \alpha\phi)) > 0$. We must check whether a nonincumbent mutant who never invests, and hence passes up empty fertile patches, would be better off. In this case, the present value of the mutant, π_m, satisfies

$$\pi_m - \pi_b^* = (r-w)\pi_g^* + (1-r+w)\pi_b^*(1-c) - \pi_b^*$$
$$= \frac{bw(v(r-w) - c(1 - v(1-r+2)))}{(p(1-q_u)+q_u)(vw+c(1-vw))}.$$

It follows that if

$$v \leq \frac{c}{(r-w)(1-c)+c}, \quad (8.21)$$

then the mutant behavior (not investing) cannot invade, and we indeed have an anti-equilibrium. Note that (8.21) has a simple interpretation. The denominator in the fraction is the probability that a search ends either in death or in finding an empty patch. The right side is therefore the expected cost of searching for an occupied patch. If the cost v of investing in an empty patch is greater than the expected cost of waiting to usurp an already productive (fertile and invested in) patch, no agent will invest. We have the following theorem.

THEOREM 8.3 *There is an investment cost $v^* \in (0, 1)$ such that an antiproperty equilibrium exists if and only if $v \leq v^*$. v^* is an increasing function of the migration cost c.*

To see this, note that the right-hand side of (8.21) lies strictly between 0 and 1, and is strictly increasing in c.

If (8.21) is violated, then migrants will refuse to invest in an empty fertile patch. Then (8.9), which implicitly assumes that a migrant always occupies a vacant fertile patch, is violated. We argue as follows. Assume the system is in the antiproperty equilibrium as described above and, noting the failure of (8.21), migrants begin refusing to occupy vacant fertile patches. Then, as incumbents migrate from newly dead patches, ϕ falls, and hence w rises. This continues until (8.21) is satisfied as an equality. Thus, we must redefine an antiproperty equilibrium as one in which (8.9) is satisfied when (8.21) is satisfied; otherwise, (8.21) is satisfied as an equality and (8.9) is no longer satisfied. Note that in the latter case the equilibrium value of ϕ is strictly less than in the property equilibrium.

THEOREM 8.4 *Suppose (8.21) is violated when ϕ is determined by (8.9). Then the antiproperty equilibrium exhibits a lower average payoff than the property equilibrium.*

The reason is simply that the equilibrium value of ϕ is lower in the antiproperty equilibrium than in the property equilibrium, so there will be on average more migrants and fewer incumbents in the antiproperty equilibrium. But incumbents earn positive return b per period, while migrants suffer positive costs c per period.

Theorem 8.4 helps to explain why we rarely see antiproperty equilibria in the real world, If two groups differ only in that one plays the property equilibrium and the other plays the antiproperty equilibrium, the former will grow faster and hence displace the latter, provided that there is some scarcity of resources leading to a limitation on the combined size of the two groups.

This argument does not account for property equilibria in which there is virtually no investment by the incumbent. This includes the butterfly (Davies 1978) and feral horse (Stevens 1988) examples, among others. In such cases, the property and antiproperty equilibria differ in only one way: the identity of the patch owner changes in the latter more rapidly than in the former. It is quite reasonable to add to the model a small cost δ of ownership change, for instance, because the intruder must physically approach the patch and engage in some sort of display before the change in incumbency can be effected. With this assumption, the antiproperty equilibrium again has a lower average payoff than the property equilibrium, so it will be disadvantaged in a competitive struggle for existence.

The next section shows that if we respecify the ecology of the model appropriately, the unique equilibrium is precisely the antiproperty equilibrium.

8.7   An Antiproperty Equilibrium

Consider a situation in which agents die unless they have access to a fertile patch at least once every n days. While having access, they reproduce at rate b per period. An agent who comes upon a fertile patch that is already owned may value the patch considerably more than the current owner, since the intruder has, on average, less time to find another fertile patch than the current owner, who has a full n days. In this situation, the current owner may have no incentive to put up a sustained battle for the patch, whereas the intruder may. The newcomer may thus acquire the patch without a battle. Thus, there is a plausible antiproperty equilibrium.

To assess the plausibility of such a scenario, note that if π_g is the fitness of the owner of a fertile patch and $\pi_b(k)$ is the fitness of a nonowner who has k periods to find and exploit a fertile patch before dying, then we have the recursion equations

$$\pi_b(0) = 0, \tag{8.22}$$
$$\pi_b(k) = w\pi_g + (1-w)\pi_b(k-1) \qquad \text{for } k = 1, \ldots, n, \tag{8.23}$$

where r is the probability that a nonowner becomes the owner of a fertile patch, either because it is not owned or because the intruder costlessly evicts the owner. We can solve this, giving

$$\pi_b(k) = \pi_g(1 - (1-r)^k) \qquad \text{for } k = 0, 1, \ldots n. \tag{8.24}$$

Note that the larger k and the larger r, the greater the fitness of an intruder. We also have the equation

$$\pi_g = b + (1-p)\pi_g + p\pi_g(n), \tag{8.25}$$

where p is the probability the patch dies or the owner is costlessly evicted by an intruder. We can solve this equation, finding

$$\pi_g = \frac{b}{p(1-r)^n}. \tag{8.26}$$

Note that the larger b, the smaller p, the larger r, and the larger n, the greater the fitness of the owner.

As in the previous model, assume the intruder devotes resources $s_u \in [0, 1]$ and the incumbent devotes resources $s_o \in [0, 1]$ to combat. With the same notation as above, we assume a fraction f_o of incumbents are contesters, and we derive the conditions for an incumbent and an intruder who has discovered the owner's fertile patch to conform to the antiproperty equilibrium. When these conditions hold, we have $f_o = 0$.

Let π_c be the fitness value of contesting rather than simply abandoning the patch. Then we have

$$\pi_c = s(1 - p_u)\pi_g + (1 - s)((1 - p_u)\pi_g + p_u\pi_b(n)) - \pi_b(n),$$

which reduces to

$$\pi_c = \frac{\pi_g}{2}\left(\frac{s_u^2 + s_o(2 + s_u)}{s_o + s_u}(1 - r)^n - s_u\right). \tag{8.27}$$

Moreover, π_c is increasing in s_o, so if the owner contests, he will set $\sigma_o = 1$, in which case the condition for contesting being fitness-enhancing for the owner then becomes

$$\frac{s_u + 2/s_u + 1}{1 + s_u}(1 - r)^n > 1. \tag{8.28}$$

Now let $\pi_u(k)$ be the fitness of a nonowner who must own a patch before k periods have elapsed and who comes upon an owned fertile patch. The agent's fitness value of usurping is

$$\pi_u(k) = (1 - f)\pi_g + \\ f(sp_u\pi_g + (1 - s)(p_u\pi_g + (1 - p_u)\pi_b(k - 1))) - \\ \pi_b(k - 1).$$

The first term in this equation is the probability that the owner does not contest times the intruder's gain if this occurs. The second term is the probability that the owner does contest times the gain if the owner does contest. The final term is the fitness value of not usurping. We can simplify this equation to

$$\pi_u(k) = \pi_g\frac{s_o(1 - f) + s_u}{s_o + s_u}. \tag{8.29}$$

This expression is always positive and is increasing in s_u and decreasing in s_o, provided $f_o > 0$. Thus, the intruder always sets $s_u = 1$. Also, as one might expect, if $f_o = 0$, the migrant usurps with probability 1, so $\pi_u(k) = \pi_g$. At any rate, the migrant always contests, whatever the value of f_o. The condition (8.28) for not contesting, and hence for there to be a globally stable antiproperty equilibrium, becomes

$$2(1 - r)^n < 1, \tag{8.30}$$

which will be the case if either r or n is sufficiently large. When (8.30) does not hold, there is an antiproperty equilibrium.

The antiproperty equilibrium is not often entertained in the literature, although Maynard Smith (1982) describes the case of the spider *Oecibus civitas*, where intruders virtually always displace owners without a fight. More informally, I observe the model in action every summer's day at my bird feeders and bathers. A bird arrives, eats or bathes for a while, and if the feeder or bath is crowded, is then displaced, without protest, by another bird, and so on. It appears that, after having eaten or bathed for a while, it simply is not worth the energy to defend the territory.

8.8   Property Rights as Choreographer

Humans share with many other species a predisposition to recognize property rights. This takes the form of *loss aversion*: an incumbent is prepared to commit more vital resources to defending his property, ceteris paribus, than an intruder is willing to commit to taking the property. The major proviso is that if the property is sufficiently valuable, a property equilibrium will not exist (theorem 8.1).

History is written as though property rights are a product of modern civilization, a construction that exists only to the extent that it is defined and protected by judicial institutions operating according to legal notions of ownership. However, it is likely that property rights in the fruits of one's labor has existed for as long as humans have lived in small hunter-gatherer clans, unless the inequality in theorem 8.1 holds, as might plausibly be the case for big game. The true value of modern property rights, if the argument in this chapter is valid, lies in fostering the accumulation of property even when $\pi_g > (1 + \beta)\pi_b(1 - c)$. It is in this sense only that Thomas Hobbes may have been correct in asserting that life in an unregulated state of nature is "solitary, poor, nasty, brutish, and short." But even so, it must be recognized that modern notions of property are built on human behavioral propensities that we share with many species of nonhuman animals. Doubtless, an alien species with a genetic organization akin to that of our ants or termites would find our notions of individuality and privacy curious at best and probably incomprehensible.

9

The Sociology of the Genome

> In the evolution of life...there has been a conflict between selection at several levels...individuality at the higher level has required that the disruptive effects of selection at the lower level be suppressed.
>
> John Maynard Smith

> A Mendelian population has a common gene pool, which is its collective or corporate genotype.
>
> Theodosius Dobzhansky

> **C:** What do you think of the critique of Edward O. Wilson signed by 137 prominent population biologists and published in *Nature*?
> **HG:** When asked about the book *One Hundred Authors Against Einstein*, he replied "Why 100 Authors? If I were wrong, then one would have been enough."
> **C:** What is their error?
> **HG:** Organisms in social species do not maximize inclusive fitness and inclusive fitness, though a correct and important theory, does not explain social structure.
> **C:** How would you summarize this for non-experts?
> **HG:** Inclusive fitness is about a single genetic locus. Social behavior is never determined by a single gene.
>
> Choreographer Interview

Many animals interact in groups for at least a part of their lifecycle. Such groups may be called flocks, schools, nests, troupes, herds, packs, prides, tribes, and so on, depending on the species. There appears not to be a common term for these groups, so I will call them *animal societies*, or simply *societies*. Animal societies have at least rudimentary social structures governing the typical interactions among group members. Even animals that live solitary lives often have mating practices involving signaling and ritualistic interactions (Noe and Hammerstein 1994; Fiske et al. 1998).

Sociobiology is the study of the social structure of such species. Edward O. Wilson introduced the term in his pathbreaking book of the same name (Wilson 1975). Wilson is an expert on social insects, not humans, but the concluding chapter of his book addressed human sociobiology. At the time, the idea that biology had anything useful to say about human society had few proponents. Virtually all social scientists at the time believed that the only thing biologically distinctive about humans was *hypercognition* (see Chapter 2), and that human behavior was completely determined by social and cultural institutions (Cosmides et al. 1992). Biology, it was thought, simply had nothing to add.

Wilson's book, not surprisingly, generated some years of heated and indeed venomous criticism (Segerstrale 2001). However, science eventually won out over prejudice. We are now all sociobiologists. Indeed, the pendulum has perhaps swung too far in the other direction: all sorts of human behaviors are currently attributed, without much foundation, to our evolved dispositions (Gould and Lewontin 1979; Boyd and Richerson 2005).

Animal societies exist because living in a society enhances the fitness of its members. In economics this is called *increasing returns to scale*, the term applying perfectly to the aggregation of individuals in an animal society. While there may be some contingent and variable aspects of animal societies, with the exception of humans, the social structure of a given species is quite uniform across time and space. The social structure of animal societies is thus likely optimized, or close to optimized, for contributing to the fitness of members of the species, within the bounds set by the gene pool of the species. The same cannot be said of human society, given the massive effects of cumulative culture and technology, some of which is actually or potentially fitness-reducing (see Chapter 1).

The general social equilibrium model developed in Section 6.3 applies nicely to animal societies. There are social roles and social actors that fill these roles, the goal of social theory being to describe how actors are recruited to fill roles, and how roles interact to attain some degree of social efficiency. Sociobiology is part of sociology.

A basic principle of sociobiology is that behavior is conditioned by genes. In most species, age, sex, and caste condition the individual to assume a particular role. In highly social species, differential nurturing can create castes, such as worker vs. soldier vs. reproductive in eusocial bees and ants. In humans, of course, culture and socialization influence the allocation of individuals to social roles.

Basic evolutionary theory asserts that a gene for a particular behavior can persist in the population only if the behavior leads the gene's carrier (the individual) to contribute a sufficient number of copies of the gene to the next generation. The most straightforward way for this to occur is if the behavior enhances the fitness of the individual himself. Cooperation among social actors in this case is called *mutualistic* (Milinski 1996; Dugatkin 1997). Mutualistic interaction is particularly important in humans, and is called *collaboration* (Tomasello 2014). Genes for mutualism induce individuals to seek cooperative rather than solitary solutions to problems, and provide them with skills for effective collaboration.

Mutualism, however, is not enough to capture increasing returns to scale in social life. Often cooperation demands that participating individuals incur personal fitness costs. This is called *altruism*, and genes that code for altruistic behavior are called *altruistic genes*. Except in humans, this sort of biological altruism has no connection with moral sentiments, of course. Clearly, altruistic genes can spread only if the fitnesses of the beneficiaries of the altruistic act carrying the altruistic gene increase sufficiently to offset the sacrifice of the altruist. William Hamilton (1964a) was the first to fully develop this idea, culminating in *Hamilton's rule*. This rule says that if the altruist incurs fitness cost c, and confers fitness benefit b on another individual with relatedness r to the altruist, the altruistic gene will spread if $br > c$. The reason is that br is the expected number of copies of the altruism gene gained in the recipient and c is the number of copies lost in the donor. Calling $br - c$ the *inclusive fitness* of the altruist, the implications of Hamilton's rule are called *inclusive fitness* theory.

My aim in this chapter is to clarify the position of inclusive fitness theory in sociobiology, drawing on Gintis (2014). The issue is highly contentious. Edward O. Wilson, for instance, who strongly supported Hamilton's analysis in the years immediately following its appearance, has become a serious critic. He writes in his recent book, *The Social Conquest of Earth* (2012):

> The foundations of the general theory of inclusive fitness based on the assumptions of kin selection have crumbled, while evidence for it has grown equivocal at best.... Inclusive fitness theory is both mathematically and biologically incorrect.

To supporters of inclusive fitness theory, this statement is outrageous, striking a blow at population genetics itself. As Stuart West et al. (2007a) explain:

> The importance of Hamilton's work cannot be overstated—it is one of the few truly fundamental advances since Darwin in our understanding of natural selection.

Richard Dawkins' (2012) review of *The Social Conquest of Earth*, exclaims:

To borrow from Dorothy Parker, this is not a book to be tossed lightly aside. It should be thrown with great force.

Edward O. Wilson's critique culminated in a powerful paper, with coauthors Corina Tarnita and Martin Nowak (Nowak et al. 2010), that appeared in the high-profile journal *Nature*. The authors argue:

> Considering its position for four decades as the dominant paradigm in the theoretical study of eusociality, the production of inclusive fitness theory must be considered meagre…inclusive fitness theory…has evolved into an abstract enterprise largely on its own.

This paper drew the ire of a host of population biologists. *Nature* subsequently published several "brief communications" vigorously rejecting the claims of Nowak, Tarnita, and Wilson. One of these was signed by no fewer than 137 well-known biologists and animal behaviorists (Abbot 2011; Boomsma 2011; Strassmann 2011). In a leading biology journal article, Rousset and Lion (2011) accuse Nowak, Tarnita, and Wilson of saying nothing new and of using "rhetorical devices." They then attack the journal *Nature* itself, arguing that

> the publication of this article illustrates more general concerns about the publishing process.…*Nature*'s extravagant editorial characterization of the paper as "the first mathematical analysis of inclusive fitness theory" recklessly tramples on nearly 50 years of accumulated knowledge.

This controversy, a veritable clash of the titans (Gintis 2012a), has been avidly followed in the popular science literature, which has characterized the disagreement as to whether societies can be best modeled using concepts of *group selection* (with Nowak, Tarnita, and Wilson) or *individual selection* (with Dawkins and the signers of protest letters to *Nature* who argue that the notion that genes maximize inclusive fitness lies at the very *core* of evolutionary theory). For instance, West et al. (2011, p. 233) assert:

> Since Darwin, the only fundamental change in our understanding of adaptation has been Hamilton's development of inclusive fitness theory.…The idea [is] that organisms can be viewed as maximizing agents.

By contrast, opponents claim that *higher-level social organization is the driving force of evolutionary change*, and gene flows react by conforming to and promoting such higher-level social forms. For instance, Nowak et al. (2010) argue that eusocial species are successful because they develop social systems that suppress kin favoritism and promote generalized loyalty to the hive. Organisms that maximized inclusive fitness surely would not behave this way.

Prominent popular writers with solid academic backgrounds have strongly supported the inclusive fitness maximization position of Dawkins et al., yet do not seriously address the issues raised by Nowak, Wilson, and others (Pinker 2012; Coyne 2012).

I argue in this chapter that inclusive fitness theory is analytically valid, and is very important. However, it does not imply that individuals maximize inclusive fitness, and it fails to elucidate central driving forces in animal society formation and evolution. Nowak and Wilson correctly note the limitations of inclusive fitness theory, but they err in questioning its validity and in understating its contribution to sociobiology. Their critics correctly defend inclusive fitness theory, but they err in claiming that organisms in a social species maximize their inclusive fitness and that inclusive fitness theory explains social structure.

The conditions under which evolutionary dynamics leads to inclusive fitness maximization have been carefully studied by Alan Grafen and his associates, who have shown that Darwinian population dynamics entail inclusive fitness maximization at the individual and gene levels, but only assuming that fitness effects are *additive* (Grafen 1999, 2006; Gardner et al. 2011; Gardner and Welsh 2011). But if fitness effects were additive in general, then there would be no increasing returns to scale, and animal societies would not exist. Because societies are complex adaptive nonlinear systems, inclusive fitness is only *one tool* in the explanation of the social structure of animal societies.

Another way of expressing this point is that inclusive fitness theory applies to a *single* gene in the organism's genome, or to several *noninteracting genes*. But the evolutionary success of an organism depends on the way the various genes *interact synergistically*. Claiming that inclusive fitness theory explains societies is like claiming that the analysis of word frequency in a book is sufficient to comprehend the book's meaning.

9.1   The Core Genome

Social relations in nonhuman societies are coded in the genes of its members. The characteristic rules of cooperation and conflict, as well as the meaning of signals passed among individuals, are *shared by all members of an animal society*. We call this communality of genes the species' *core genome*. The core genome is the complex of genes that are broadly shared by all members of a species (Dobzhansky 1953). Section 9.11 develops this notion in greater detail. The core genome is like the computer code for a software program in an agent-based computer model. The core genome sets up the rules for social interaction and the conditions for individual social success, creates a heterogeneous set of agents, each of whom incorporates both the core genome and an idiosyncratic *variant genome* that defines its individuality. These agents interact according to the rules coded by the core genome, which rewards the more successful agents with more copies of itself in the future. In the case of human societies, additional rules and meanings are culturally specified, and as we explained in Chapter 1, human culture and the human core genome *coevolve*.

The core genome of a social species endows individuals with incentives to aggregate into social groups—packs, flocks, tribes, hives, and the like. The size and social structure of these groups coevolve with the genetic constitution of its members, as reflected in the evolution of the core genome over time. Group selection is not selection *among* groups, but rather *for* groups with a fitness-enhancing size and social structure. Selection for group characteristics requires individual selection because the social rules are inscribed in individuals who both instantiate the rules and are evolutionarily successful given these rules.

Societies are complex dynamical systems with emergent properties—properties that we cannot deduce from the DNA of the core genome, any more than we can deduce consciousness and mind from the chemical composition of the brain (Deacon 1998; Morowitz 2002).

Yet societies are effective because of the behaviors of its members, these behaviors are determined by the core genome, and an individual gene can evolve only if it *directly* enhances the fitness of its carriers, or it promotes *interactions among its carriers* that enhance its *inclusive fitness*—the sum of the increases in fitnesses of all carriers of the gene influenced by the behavior. In particular, a gene that leads its carrier to sacrifice its inclusive fitness certainly cannot evolve, except possibly in very small societies where random luck can temporarily outweigh systematic selective forces.

Although the concept of the core genome is somewhat new, I cannot conceive of there being any serious objection to the above paragraphs. Indeed, the danger is more that they are uncomfortably close to tautologies.

Why then this conflict between group and individual selection proponents? The participants themselves agree that whether one does the accounting on the level of the group, the individual, or the single gene, the answer must come out the same (Dugatkin and Reeve 1994). What then can account for Richard Dawkins' venom in attacking Edward O. Wilson (Dawkins 2012), or David Sloan Wilson's sense of triumph in observing that group selection has been resurrected from its status as an outcast of biological theory (Wilson 2008)? Must it not be simply a matter of personal preference and modeling ease which perspective one chooses in any particular situation?

I suspect the answer is that inclusive fitness theorizing leads researchers to think *atomistically*, while group selection theorizing leads researchers to think *structurally*. Inclusive fitness theory leads one to the beautiful Margaret Thatcher head quote of Chapter 2: "There is no such thing as society. There are only individual men and women, and there are families." Group selection theorizing, by contrast, leads researchers to the Martin Luther King head quote in that chapter: "We are caught in an inescapable network of mutuality, tied in a single garment of destiny." Of course, I am not suggesting that sociobiologists are embroiled in the ideologies of Left and Right, or any other political ideology. Nor are they closely connected to any particular set of moral or ethical principles. Rather, they are personal preferences—highly contrasting yet equally useful ways of thinking about society. The correct way of thinking is to embrace both atomistic and structural approaches and analyze the corresponding interplay of forces. This is the approach defended in this chapter.

There is, however, a certain asymmetry in the mutual criticism of the two schools of thought. Few supporters of group selection deny the importance of inclusive fitness theory, while virtually all its opponents regularly deny the importance of group selection theory. For instance, Steven Pinker writes in "The False Allure of Group Selection" (2012):

> Human beings live in groups, are affected by the fortunes of their groups, and sometimes make sacrifices that benefit their groups. Does this mean that the human brain has been shaped by natural selection to promote the welfare of the group in com-

petition with other groups, even when it damages the welfare
of the person and his or her kin?

The first problem with this quite disingenuous description is that group selection does not require "competition with other groups" any more than
individual selection requires "competition with other individuals." For instance, a mutant rabbit may be evolutionarily successful because it is more
adept at escaping the fox, not because it wins conflicts with other rabbits.
Similarly, a society may be evolutionarily successful because it better ex-
ploits its prey or contains its predators, not because it vanquishes other
societies in head-to-head competition.

   The more important problem with Pinker's critique is the notion that
group selection theory suggests that the group's success depends on behaviors that damage "the welfare of the person and his or her kin." This is
of course simply impossible. If the inclusive fitness of the gene for some
behavior is less than unity, that gene must in the long run disappear from the
population. No one disagrees with this. Only a reader who is quite ignorant
of sociobiology could take Pinker's argument seriously.

   Here is another rather randomly drawn, equally disingenuous, critique
from a prominent biologist (Coyne 2012):

> The idea that adaptations in organisms result from "group
> selection"…rather than from selection among genes them-
> selves…[is] in stark contrast to the views of most evolutionary
> biologists.

Of course, no group selection proponent sees group-level adaptations as an
alternative to selection among genes. Rather, they think of group selection
models as explanations of why particular gene are successful and others are
not. Why would so insightful a scientist like Coyne offer such a patently
deficient argument? Probably because, like Pinker, he could not think of a
good argument to support his position.

   In the first half of the twentieth century, most naturalists believed that
animal societies were effective because natural selection favors *altruism*, in
the form of individuals who sacrifice for the good of the species (Kropotkin
1989[1903]; Simpson 1941; Lorenz 1963). For instance, in times of food
scarcity, many believed that individuals would voluntarily restrict their reproductive activity (Wynne-Edwards 1962). This phenomenon was termed
group selection because the argument was that the altruist may have fewer
offspring, but its contribution to the success of the group would allow more

of these offspring to survive and reproduce. However, John Maynard Smith (1964), George Williams (1966), David Lack (1966), and others showed that virtually all apparent examples of animals sacrificing for the group could plausibly be explained by standard individual fitness maximization. Williams (1966) used the *principle of parsimony* to counsel that group selection be used only when the simpler principle of individual selection is incapable of explaining animal behavior. At that time no important examples of sacrifice for the good of the group were found.

As it stands today, there are *two* mechanisms of group selection. The first is the *evolutionary success of more effective collaboration* (Parsons 1964; Boyd and Richerson 1990; Bowles and Gintis 2011; Tomasello 2014). That is, social structures that effectively promote cooperation and punish antisocial behavior will tend to evolve. This mechanism works by an individual genetic mutation fostering a social structure mutation, the new social structure enhancing the fitness of social members, some of whom carry the mutant gene, which then is more frequently represented in the next generation. In this case it is the *social structure* that is favored by natural selection, and the genes that induce the behaviors given by the social structure are the beneficiaries of natural selection on the level of social structure.

Two forms of social organization are especially favored by this evolutionary process: *eusociality* and *extensive parental care*. In a eusocial species, one or very few individuals reproduce, and the remaining social members are sterile workers, soldiers, and foragers (Wilson 1975). Therefore a mutation in a reproductive will be inherited by a large fraction of her offspring, who will synergistically follow the principles of coordination, signaling, and task allocation indicated by the mutation. Not surprisingly, the eusocial insects have evolved into extremely complex and sophisticated societies— for instance the waggle dance in honeybees (Riley et al. 2005). A similar argument holds for animals that care for their young. Because there are at most only one or two individuals involved in mating and in nurturing offspring, a mutation in a male or female leading to a new social structure of mating can easily spread. Darwin called this *sexual selection*, an evolutionary process that has engendered sophisticated signaling and collaboration in many species (West-Eberhard 1983).

The second mechanism of group selection is exactly the altruistic behavior that had been discredited by Williams, Maynard Smith, Lack, and others, although now better understood in terms of game-theoretic models of social cooperation. Often the effectiveness of social cooperation is strongly enhanced when individuals are willing to incur personal costs to further collective goals. For instance, when a group of human hunters venture into the forest, they usually fan out in such a way that they are not visible to one another. Because the prey is shared irrespective of who killed the animal (Kaplan et al. 1984), and since the process of searching for prey is highly strenuous, each hunter has an incentive to shirk. Altruists do not. Successful groups foster altruism, which complements mutualistic collaboration in promoting efficient cooperation. Note that for species in general, this notion of biological altruism has nothing to do with either morality or psychology.

This sort of altruism was recognized by Darwin himself (Darwin 1871):

> An advancement in the standard of morality will certainly give an immense advantage to one tribe over another. A tribe including many members, who from possessing in a high degree the spirit of patriotism, fidelity, obedience, courage and sympathy, were always ready to aid one another, and to sacrifice themselves for the common good, would be victorious over most other tribes; and this would be natural selection.

The existence of altruism, the importance of which is not now widely disputed, nevertheless presents a serious problem for evolutionary theory: How can genes that promote altruistic behavior spread, since they disadvantage their carriers? Inclusive fitness theory provides the answer.

9.2   Inclusive Fitness and Hamilton's Rule

Classical genetics does not model cases in which individuals sacrifice on behalf of non-offspring, such as sterile workers in an insect colony (Wheeler 1928), cooperative breeding in birds (Skutch 1961), and altruistic behavior in humans (Darwin 1871). This problem was addressed by William Hamilton (1963, 1964ab, 1970), who noticed that if a gene favorable to helping others is likely to be present in the recipient of an altruistic act, then the gene could evolve even if it reduces the fitness of the donor. Hamilton called this *inclusive fitness* theory.

Hamilton developed a simple inequality, operating at the level of a single locus that gives the conditions for the evolutionary success of an allele. This rule says that if an allele in individual A, I will call it the *focal allele*, increases the fitness of individual B whose degree of relatedness to A is r, and if the cost to A is c, while the fitness benefit to B is b, then the allele will evolve (grow in frequency in the reproductive population) if

$$br > c. \tag{9.1}$$

We call $br - c$ the *inclusive fitness* of the focal allele. Subsequent research supported some of Hamilton's major predictions (Maynard Smith and Ridpath 1972; Brown 1974; West-Eberhard 1975; Krakauer 2005).

A critical appreciation of Hamilton's rule requires understanding when and why it is true. Rigorous derivations of Hamilton's rule (Hamilton 1964a; Grafen 1985; Queller 1992; Frank 1998) are mathematically sophisticated and difficult to interpret. For this reason, it is easy to assert implications of inclusive fitness that cannot be evaluated by a non-expert. I suspect that this accounts for the fact that non-experts have tended to support one or another in this debate without really understanding the technical issues involved. My goal in this chapter is to lay these issues bare, so that they can be appreciated by anyone willing to endure a bit of elementary algebra.

The usual popular argument (for instance, Bourke 2011) assumes that an altruistic helping behavior ($b, c > 0$) is governed by an allele at a single locus, and r is the probability that the recipient of the help has a copy of the helpful allele. The net fitness increment to carriers of the helpful allele is then $br - c$, so the allele increases in frequency if this expression is positive.[1]

However attractive, the popular argument has key weaknesses that render it unacceptable. First, the intuition behind Hamilton's rule is that r is the probability that the recipient has a copy of the helping gene, so br is the expected gain to the helping gene in the recipient, which must be offset by the loss c to the helper if the helping behavior is to spread. This argument, however, is clearly specious. Many have pointed this out, but perhaps none more elegantly than Washburn (1978, p. 415), who writes:

> All members of a species share more than 99% of their genes,
> so why shouldn't selection favour universal altruism?

Dawkins (1979) considers Washburn's argument the fifth of his "Twelve Misunderstandings of Kin Selection." Dawkins draws on Maynard Smith

---

[1] Hamilton's rule extends directly to behavior that is governed by alleles at multiple loci, provided that the interactions among the loci are frequency independent, or equivalently, that the effects at distinct loci contribute additively to the phenotypic behavior. Grafen (1984) calls such a phenotype a *p-score*. In this chapter I will use the term "single locus" even in places where the *p*-score generalization applies.

(1974) to show that Washburn's conclusion in favor of universal altruism is faulty. But he fails to explain what is wrong with Washburn's argument, except to say (correctly): "This misconception arises not from Hamilton's own mathematical formulation but from oversimplified secondary sources to which Washburn refers" (p. 191). One might, with Dawkins (1979), claim that r represents the probability of the identity of the helping gene in the two parties by *descent from a common ancestor*, but why should it matter whether it is identity by descent or otherwise? Descent is clearly beside the point. A copy is a copy, whatever its provenance.

In fact, part of the problem with the popular argument is rather subtle: it considers the conditions for an increase in the *absolute number* of copies of the helping allele in the population, but says nothing about its *relative frequency*, which is the quantity relevant to the evolutionary success of the helping allele. Indeed, $br - c$, the net increase in the number of copies of the helpful allele, is less than $b - c$, which is the net increase in the number of copies of all alleles at the locus, so the frequency of the helping allele in the next period will be *lower* than $br - c$, and *prima facie* may even decrease.

A second problem with the popular argument is that it makes sense if the relatedness r is a *probability*, so that br can be interpreted as the expected gain to the helping allele in the beneficiary. But in this case r must be *nonnegative*. By contrast, in a valid derivation of Hamilton's rule, r can be positive *or* negative. In the case $c > 0$ and $b, r < 0$, but with $br > c$, we call this *spite* (Hamilton 1970; Gardner et al. 2004). In fact, as we shall see, there is no simple relationship between the r in Hamilton's rule and genealogical coefficients of relatedness. The appropriate value of r in Hamilton's rule lies between plus and minus unity, but is generally a function of the social structure of the species in question, and can be positive or negative.

To address these deficiencies, we begin our study of inclusive fitness theory with a careful derivation of Hamilton's rule assuming, with Hamilton, that all interactions are dyadic. For simplicity, I will assume the species is haploid but sexual. That is, each new individual inherits a single gene from one of its two parents at each locus of the genome. A more general diploid treatment (individuals have two alleles at each genetic locus) is presented in Appendix A1, where we also drop the requirement that all interactions must be dyadic.

Our derivation of Hamilton's rule makes numerous simplifying assumptions. However, the argument can be extended to deal with heterogeneous relatedness, dominance, coordinated cooperation, local resource competition, inbreeding, and other complications (Uyenoyama and Feldman 1980; Michod and Hamilton 1980; Queller 1992; Wilson et al. 1992; Taylor 1992; Rousset and Billard 2007), with an equation closely resembling (9.1) continuing to hold. In general, however, the frequency q of the focal allele will appear in (9.1), and b and c may be functions of q as well, so the interpretation of r as relatedness becomes accordingly more complex (Michod and Hamilton 1980).

In general b and c will also depend on the frequency of alleles at other loci of the genome, and since the change in frequency q of the focal allele in the population will affect the relative fitnesses of alleles at other loci, inducing changes in frequency at these loci will in turn affect the values of b, c, and even r. For this reason, Hamilton's rule presupposes *weak selection*, in the sense that population gene frequencies do not change appreciably in a single reproduction period. Therefore Hamilton's rule does not imply that a successful allele will move to fixation in the genome. Moreover, alleles at other loci that are enhanced in inclusive fitness by the focal allele's expansion may undergo mutations that enhance the inclusive fitness of the focal allele, while alleles at other loci that are harmed by the expansion of the focal allele may develop mutations that suppress the focal allele. Such mutations can be evolutionarily successful and even move to fixation in the core genome.

Now to our derivation. Suppose there is an allele at a locus of the genome of a reproductive population that induces carrier A (called the *donor*) to incur a fitness change c that leads to a fitness change b in individual B (called the *recipient*). We will represent B as an individual, but in fact, the fitness change b can be spread over any number of individuals. If $b > 0$, A bestows a *gain* upon B, and if $c > 0$, A experiences a fitness *loss*. However, in general we make no presumption concerning the signs or magnitudes of b and c, except that selection is weak in the sense that b and c do not change, and the population does not become extinct, over the course of a single reproduction period. This assumption, which is extremely plausible, will be made throughout this chapter.

Suppose the frequency of the focal allele in the population is q, where $0 < q < 1$, and the probability that B has a copy of the allele is p. Then if the size of the population is n, there are qn individuals with the focal allele, they change the number of members of the population from n to $n + qn(b - c)$, and they change the number of focal alleles from qn to $qn + qn(pb - c)$. Thus the frequency of the allele from one period to the next will increase if

$$\Delta q = \frac{qn + qn(pb - c)}{n + qn(b - c)} - q = \frac{q(1 - q)}{1 + q(b - c)} \left(b\frac{p - q}{1 - q} - c \right) > 0. \quad (9.2)$$

The condition for an increase in the focal allele thus is

$$b \left(\frac{p - q}{1 - q} \right) > c. \quad (9.3)$$

To derive Hamilton's rule from (9.3), we must have

$$r = \frac{p - q}{1 - q}, \quad (9.4)$$

which can be rewritten as

$$p = r + (1 - r)q. \quad (9.5)$$

Provided the recipient is a single individual, equation (9.5) makes intuitive sense using the concept of *identity by descent* (Malécot 1948; Crow 1954), where r is the probability that both donor A and recipient B have inherited the same focal allele from a common ancestor. For instance, if A and B are full siblings, then $r = 1/2$ because this is the probability that both have inherited the focal allele from the same parent. Moreover, if the siblings have inherited the focal allele from different parents, then they will still be the same allele with a probability equal to the mean frequency q of the focal allele in the population, assuming no assortative mating. In general, r will then be the expected degree of identity by descent of recipients. This logic is developed in full by Michod and Hamilton (1980).

However, this cannot be the *general* argument because there is no reason for p to be greater than q; i.e., the recipient need not be more likely than average to carry the helping gene. But if $p < q$, then equation 9.4 shows that $r < 0$, so r cannot be interpreted as a genealogical relatedness coefficient.

Population biologists have generally responded to this problem by defining r as a beta coefficient in a least squares linear regression of the donor genotype on the recipient phenotype (Hamilton 1972; Queller 1992). This is an elegant approach, but rather unsatisfying. Why linear regression? Why least squares estimation? Why is it not just an approximation, as with standard linear regressions? Why is it a good approximation, given the strong nonlinear interactions of loci in the genome? It is comforting that the approach gives a reasonable result in many cases, but the conceptual foundations are wanting. Moreover, for an elementary exposition, like the present, where the reader should be able to follow perfectly what is going on, it is like the magician pulling a rabbit out of a hat. In fact, it is just that. The fact that the regression approach gives a satisfying answer does not justify it.

There is another way to explain negative relatedness while sticking to a rigorously correct logic. Each potential recipient B has a certain relatedness to the donor A. Therefore we can partition the population of potential recipients into groups $j = 1, \ldots, k$ such that all individuals in group j have the same genealogical relatedness r_j to the donor A. Let q_j be the mean frequency of the helping allele in group j, and let π_j be the probability that the donor encounters a recipient from group j, so $\sum_j \pi_j = 1$. Then the probability that a recipient in group j has a copy of the helping allele is, using the same reasoning as led to equation (9.5),

$$p_j = r_j + (1 - r_j)q_j. \tag{9.6}$$

Moreover, we have $p = \sum_j \pi_j p_j$, and if we define $r^* = \sum_j \pi_j r_j$ and $q^* = \sum_j \pi_j q_j$, we then have

$$p = \sum_j \pi_j (r_j + (1 - r_j)q_j)$$

$$= r^* + q^* - \sum_j \pi_j r_j q_j$$

$$= r^* + (1 - r^*)q^* - \left(\sum_j \pi_j r_j q_j - r^* q^* \right),$$

$$= r^* + (1 - r^*)q^* - \operatorname{cov}_\pi(r_j, q_j), \tag{9.7}$$

from the definition of the covariance of two variables. Note that if the recipient is a single individual with relatedness r to the donor, then $r^* = r$ and $q^* = q$, so equation (9.7) reduces to the standard equation (9.5).

One point is notable in equation (9.7). Now p can be smaller than q, so $r < 0$ is possible in (9.5). Indeed, this is more likely the smaller is q^* (the average frequency of the helping allele in the donor's potential beneficiaries) and the larger the covariance between relatedness and mean frequency of the helping allele. The latter effect enters because p will be higher if low-relatedness beneficiaries tend to have high average q_j because low r_j means the random allele will be chosen with high frequency. For example, if there is only one group ($k = 1$), the covariance term in 9.7 drops out and we can write

$$p - q = r^* + (1 - r^*)q^* - q = r(1 - q) + (1 - r)(q^* - q). \quad (9.8)$$

The first term on the right-hand side of (9.8) is positive but the second is negative for $q^* < q$, and the second term dominates when r is small; i.e., when the behavior attacks non-relatives that do not share the focal allele.

It is reasonable to call the array $\{\pi_j, r_j, q_j\}$ the *social structure* of the population with respect to the behavior induced by the helping allele. This array in general is not defined at the level of the helping locus, but at the *social level*, coded by the core genome. The core genome determines particular mating patterns, particular rituals and signals, certain patterns of offspring care, and social collaboration. *Inclusive fitness thus presupposes a general type of social structure* and does not elucidate this social structure.

While the simple inequality $br > c$ at first sight appears to connect genealogical relatedness, costs, and benefits at the level of a single locus, in fact a correct derivation of the inequality reveals a complex social structure underlying each of the three terms. This fact does not detract from the importance of Hamilton's rule. Indeed Hamilton's rule must be satisfied by any plausible social structure. But it is more like an accounting relationship than an explanatory model.

9.3   Kin Selection and Inclusive Fitness

William Hamilton's early work in inclusive fitness focused on the role of genealogical kinship in promoting prosocial behavior. Hamilton speculates, in his first full presentation of inclusive fitness theory (Hamilton 1964a, p. 19):

> The social behaviour of a species evolves in such a way that in each distinct behaviour-evoking situation the individual will seem to value his neighbours' fitness against his own according to the coefficients of relationship appropriate to that situation.

Because of this close association between inclusive fitness and the social relations among genealogical relatives, John Maynard Smith (1964) called Hamilton's theory *kin selection*, by which he meant that individuals are predisposed to sacrifice on behalf of highly related family members.

A decade after Hamilton's seminal inclusive fitness papers, motivated by new empirical evidence and Price's equation (Price 1970), Hamilton (1975, p. 337) revised his views, writing:

> Kinship should be considered just one way of getting positive regression of genotype... the inclusive fitness concept is more general than kin selection.

Hamilton is surely correct. Nevertheless the two concepts are often equated, even in the technical literature. For instance, throughout his authoritative presentation of sexual allocation theory, West (2009) identifies inclusive fitness with kin selection in several places and never distinguishes between the two terms at any point in the book. Similarly, in Bourke's (2011) ambitious introduction to sociobiology, we find:

> The basic theory underpinning social evolution [is] Hamilton's inclusive fitness theory (kin selection theory).

This curious identification of inclusive fitness theory, which models the dynamics at a single genetic locus and is equally at home with altruistic and predatory genes, as we explain below, with kin selection theory, which is a high-level behavioral theory of kin altruism, is a source of endless confusion. For most sociobiologists, kin selection remains, as conceived by Maynard Smith (1964), a social dynamic based on *close genealogical association*:

> By kin selection I mean the evolution of characteristics which favour the survival of close relatives of the affected individual.

The Wikipedia definition is similar:

> Kin selection is the evolutionary strategy that favours the reproductive success of an organism's relatives, even at a cost to the organism's own survival and reproduction.... Kin selection is an instance of inclusive fitness.

Moreover, while kin selection is a special case of inclusive fitness in the sense that Hamilton's rule applies generally, not just to situations where organisms favor their close genealogical kin, in another sense kin selection is far more general than inclusive fitness. This is because in all but the simplest organisms, kin selection does not describe the behavior at a single locus, or even at a set of independently contributing loci, but rather an inherently *high-level social behavior* in which individuals recognize their close relatives through complex phenotypic associations that require significant cognitive functioning and synergistic interactions among loci. Indeed, in general these phenotypic associations arise precisely to permit cooperation among close genealogical kin.

9.3.1   *Inclusive Fitness without Kin Selection*

A simple example shows that Hamilton's rule in principle has no necessary relationship with genealogy or kin selection, but rather is an expression of the social structure of the reproductive population. The model is based on Hamilton's analysis, but is more transparently presented than Hamilton (1975), which develops a similar model for the same purpose. For related models of positive assortment not based on kin selection see Koella (2000), Nowak (2006), Pepper (2007), Fletcher and Doebili (2009), and Smaldino et al. (2013).

Consider a population in which groups of size n form in each period. In each group individuals can cooperate by incurring a fitness cost $c > 0$ that bestows a fitness gain b that is shared equally among all group members. Individuals who do not cooperate (defectors) receive the same share of the benefit as cooperators, but do not pay the cost c and do not generate the benefit b. Let p_{cc} be the expected fraction of cooperating neighbors in a group if an individual is a cooperator, and let p_{cd} be the expected fraction of cooperating neighbors if the individual is a defector. Then the payoff to a cooperator is $\pi_c = bp_{cc} - c$, and the payoff to a defector is $\pi_d = bp_{cd}$.

The condition for the cooperative allele to spread is then $\pi_c - \pi_d = b(p_{cc} - p_{cd}) - c > 0$, or

$$b(p_{cc} - p_{cd}) > c. \tag{9.9}$$

Now p_{cc} is the probability that a cooperator will meet another cooperator in a random interaction in a group, so we can define the relatedness r between individuals, following (9.5), by

$$p_{cc} = r + (1 - r)q, \tag{9.10}$$

where q is the mean frequency of cooperation in the population. If we write $p_{dd} = 1 - p_{cd}$ for the probability that a defector meets another defector, then we similarly can write

$$p_{dd} = r + (1 - r)(1 - q), \tag{9.11}$$

since $1 - q$ is the frequency of defectors in the population. Then we have

$$p_{cc} - p_{cd} = r + (1 - r)q - (1 - (r + (1 - r)(1 - q))) \tag{9.12}$$
$$= r. \tag{9.13}$$

Substituting in (9.9), we recover Hamilton's rule, $br > c$.

Of course, if group formation is random, then $p_{cc} = p_{cd}$ so $r = 0$ and Hamilton's rule cannot hold. However, to illustrate the importance of social structure, suppose each group is formed by k randomly chosen individuals who then each raises a family of n/k clones of itself. We need not assume parents interact with their offspring, or that siblings interact preferentially with each other. There is no kin selection in the standard sense of Maynard Smith (1964). At maturity, the parents die and the resulting n individuals interact, but do not recognize kin. In this case a cooperator surely has $k - 1$ other cooperators (his sibs) in his group, and the other $n - k$ individuals are cooperators with probability q. Thus

$$p_{cc} = \frac{k - 1}{n - 1} + \frac{n - k}{n - 1}q = q + \frac{(k - 1)(q - 1)}{n - 1}.$$

Similar reasoning, replacing q by $1 - q$ gives

$$p_{dd} = 1 - q + \frac{q(k - 1)}{n - 1}.$$

Then

$$r = p_{cc} - p_{cd} = p_{cc} - 1 + p_{dd} = \frac{k - 1}{n - 1},$$

so Hamilton's rule will hold when

$$br = b\left(\frac{k - 1}{n - 1}\right) > c.$$

Note that the related recipients are all clones of the donor, with relatedness unity, although the r in Hamilton's rule is $(k - 1)/(n - 1)$. The inclusive fitness inequality is accurate here, but kin selection as defined above is inoperative in this model: the altruistic behavior is more likely to spread when the number of families n/k in a group is small.

This model suggests that the interesting question from the point of view of sociobiology is how the core genome of the species manages to induce individuals to aggregate in groups of size n and to limit family size to n/k, so that the benefits of cooperation $(b-c)$ can accrue to the population. This is a true miracle of Nature.

9.4   A Generalized Hamilton's Rule

When we think of Hamilton's rule in the context of an animal society, we must account for the possibility that the focal allele may impose a cost $(\beta > 0)$ or bestow a benefit $(\beta < 0)$ uniformly on all members of the population. We call this a *social fitness effect*. The case $\beta > 0$ may be termed a *pollution effect*. It occurs, for instance, in "tragedy of the commons" cases (Hardin 1968; Wenseleers and Ratnieks 2004), such as when the focal allele depletes a protein used in chemical processes by somatic cells in conferring the benefit b on others and incurring a cost c (Noble 2011). The benefit case $\beta < 0$ may be called a *public good effect* (West et al. 2007b). This follows the common use of the term in economic theory (Olson 1965). It occurs, for instance, in a parasite when the focal allele induces its carriers to suppress an alternative allele at the focal locus that induces carriers to grow so rapidly that it kills its host prematurely (Frank 1996). Equation (9.17) below shows *the degree of pollution or public good has no bearing on whether the allele can evolve.*

Hamilton's seminal paper (1964a) explicitly includes the pollution and public goods aspect of inclusive fitness, an aspect of his analysis that later writers have ignored. Hamilton called the public good/pollution effect the *dilution effect* because it affects the *rate* but not *direction* of change in the frequency of the focal allele. Hamilton also notes that the dilution effect can lead a successful allele to *reduce* population fitness. A streamlined presentation of Hamilton's argument, which is quite opaque in the original, is presented in Gintis (2014).

We will also consider the case where the focal allele imposes a cost α on all alleles *other than* the focal allele (Keller and Ross 1998). We may call α a *thieving effect*. This effect occurs, for instance, if A redirects brooding care from non-relative to relative larvae in an insect colony, and $\alpha < 0$ (stealing from one's kin to help others) can occur, for instance, if the focal allele helps other alleles at the focal locus that benefits carriers by avoiding possibly deleterious homozygosity at the focal locus. We can clearly treat

α as cost imposed on all alleles at the focal locus, plus a benefit of equal magnitude enjoyed by carriers of the focal allele. Thus if the population size is n in the current period, population size n' in the next period will include $n + qn(b + \alpha - c)$ individuals because of the behavior induced by the focal allele, but this will be reduced by $n(\alpha + \beta)q$ due to the effects on non-focal alleles. The number of relatives of the focal allele in the current period is qn, which is increased by the behavior by $qn(pr + \alpha - c)$, and decreased through lower efficiency by $qn(\alpha + \beta)q$. Thus the new population size is given by

$$n' = n(1 - (\alpha + \beta)q) + qn(b + \alpha - c), \tag{9.14}$$

and (9.2) becomes

$$\Delta q = \frac{qn(1 - (\alpha + \beta)q) + qn(pb - \alpha - c)}{n(1 - (\alpha + \beta)q) + qn(b - \alpha - c)} - q > 0, \tag{9.15}$$

which simplifies to

$$b(p - q) > (c - \alpha)(1 - q). \tag{9.16}$$

Substituting $p = r + (1 - r)q$, we get the generalized Hamilton's rule

$$br > c - \alpha. \tag{9.17}$$

The effect of an increase in the focal allele on population fitness is the sign of dn'/da, where $a = qn$ is the number of helping genes, which is given by

$$\frac{dn'}{da} = b - c - \beta. \tag{9.18}$$

Note that in the case of Hamilton's rule, which is the above with $\alpha = \beta = 0$, population fitness increases with the frequency of the focal allele in the case of altruism or cooperation, where $b > c$, and decreases in the case of spite ($b - c < 0$). In the case of the generalized Hamilton's rule, the fitness effect is indeterminate. As we explain below, Hamilton (1964a) included the $\beta \neq 0$ effect in his calculations, but he did not consider the case where the generalized fitness effects are unevenly distributed among the alleles at the focal locus ($\alpha \neq 0$).

It is useful to give descriptive names to the social interactions when α is nonzero. We may call the case $\alpha > 0$ *theft*, and the case $\alpha < 0$ as *charity*. Moreover, a thieving altruist ($b, c, \alpha > 0$) will always evolve, as will a thieving cooperative allele ($b, \alpha > 0 > c$). Finally, the producer of a public good will evolve only if it gains in inclusive fitness from so doing ($br > c$).

The most critical implication of the generalized Hamilton's rule is that neither social generosity nor pollution has any bearing on whether an allele will evolve, as seen in equation (9.17), despite the fact that a socially generous allele unambiguously enhances the population fitness, and a polluting allele unambiguously has the opposite effect, as seen in equation (9.18). In addition, a thieving allele does not directly affect the mean population fitness (see equation 9.18) but it allows the generalized Hamilton's rule to be satisfied even when $br - c < 0$ (see equation 9.17).

9.5   Harmony and Disharmony Principles

A rather stunning conclusion can be drawn from our exercise in elementary algebra and gene-counting. I call it the Harmony Principle. To state this principle succinctly, we say the allele a is *helpful* if its carriers enhance the fitness of other individuals that it encounters ($b > 0$), *altruistic* if it is helpful and incurs a fitness cost ($b, c > 0$), *predatory* if it is harmful to others but helps itself ($b, c < 0$), *mutualistic* if it helps itself and others ($b > 0, c < 0$), *prosocial* if it increases mean population fitness ($b - c > 0$), and *antisocial* if it reduces mean population fitness ($b - c < 0$). We then have:

Harmony Principle: An evolutionarily successful gene that is a helpful non-polluter is necessarily prosocial.

From equation (9.18), the allele is prosocial if $b - c - \beta > 0$. We can write $b - c - \beta = (br - c) + b(1 - r) - \beta$. Now $br - c > 0$ by Hamilton's rule, $b > 0$ by the assumption of helpfulness, since the probability p of the recipient having the helpful allele is nonnegative, $r < 1$ by equation (9.4), and $\beta <= 0$ because the allele is a non-polluter. Thus $b - c - \beta$, the net contribution per focal allele to the population, is strictly positive.

Because each individual gene is utterly selfish, the importance of this principle for sociobiology is *inestimable*, and mirrors similar assertions concerning the social value of selfishness in humans offered by Bernard Mandeville in his famous *Fable of the Bees* (1705), in which "private vices" give rise to "public virtues," and Adam Smith's (1776) equally famous dictum, "It is not from the benevolence of the butcher, the brewer, or the baker that we expect our dinner, but from their regard to their own interest." While economists have determined the precise conditions—they are far from universal—under which Mandeville and Smith are correct (Mas-Colell et al. 1995), the Harmony Principle is true under much broader conditions. While genes are utterly selfish according to inclusive fitness theory, evolutionarily successful genes that are helpful non-polluters are necessarily prosocial. Note that we have not assumed that $c > 0$, so this principle applies both to altruistic genes and mutualistic genes that help others as well as helping themselves.

However, what is the social status of genes that are *not* helpful? It is curious that this case appears never to have been treated in the literature. I cannot imagine why not. An alternative to the Harmony Principle is, indeed, *prima facie* equally possible. Suppose the focal allele is predatory. Then Hamilton's rule becomes $(-b)r < (-c)$, which can be satisfied even though the focal allele is antisocial. Indeed, this will be the case whenever $|b|(1 - r) > c - b > 0$. We have:

Disharmony Principle: A gene that is evolutionarily successful but predatory may be antisocial even if it is a non-polluter.

To see this, note that $b - c - \beta = (br - c) + b(1 - r) - \beta$, where $br - c > 0$ and $1 - r > 0$. Thus for sufficiently large (negative) b, we must have $b - c < 0$.

Note that the Disharmony Principle is distinct from the *spite* phenomenon (Hamilton 1970; Foster et al. 2001; Gardner et al. 2004), in which $r < 0$ and $c > 0$, which is well-developed in the literature. Indeed, it is a common occurrence that the interaction is costly but involves reducing the fitness of others, and (9.5) can hold with $r < 0$, while the focal allele is still altruistic (Bourke 2011). Examples are warfare in ants (Hölldobler and Wilson 1990) and humans (Bowles and Gintis 2011), as well as generally spiteful behavior in many species (Hamilton 1970; Foster et al. 2001; Gardner et al. 2004).

More generally, we have the taxonomy of Table 9.1, where if $b > 0 > c$, then the allele is *cooperative*, and since $b - c > 0$, the allele contributes unambiguously to the fitness of its carrier. A cooperative allele will always be selected, as in this case Hamilton's rule is always satisfied. The unnamed boxes in the table necessarily violate Hamilton's rule.

9.6   The Utterly Selfish Nature of the Gene

Hamilton's rule ensures that the gene is *selfish* in the sense described by Dawkins (1976). In particular, Hamilton's rule implies that *the conditions*

b	c	$r > 0$	$r = 0$	$r < 0$
> 0	> 0	Altruistic		—
> 0	< 0	Cooperative		
< 0	> 0	—		Spiteful
< 0	< 0	Predatory		

Table 9.1. Variety of behaviors that can satisfy Hamilton's rule

for the evolutionary success of a gene are distinct from the conditions under which the gene enhances the mean fitness of the reproductive population. The validity of the Disharmony Principle shows that inclusive fitness does *not* explain the appearance of design in nature or, in other words, why the genome of a successful species consists of genes that predominantly *collaborate* in promoting the fitness of its members (Dawkins 1996). Indeed, Hamilton's rule equally supports the evolutionary success of prosocial altruistic genes and antisocial predatory genes, whereas the former predominate in a successful species and account for the appearance of design.

It is common for sociobiologists who, with Dawkins, adopt the "gene's-eye point of view" to overlook this fact, despite its being a simple logical implication of Hamilton's rule. Indeed, many population biologists claim that the appearance of design in nature is explained by Hamilton's rule. For instance, in the protest letter to *Nature* mentioned above, 137 professional evolutionary biologists agreed with the following statement:

> Natural selection explains the appearance of design in the living world, and inclusive fitness theory explains what this design is for. Specifically, natural selection leads organisms to become adapted as if to maximize their inclusive fitness. (Abbot 2011)

In fact, as we shall see, organisms do *not* generally maximize inclusive fitness. Rather, organisms in a social species interact strategically in a complex manner involving collaboration, as well as enhancement and suppression of gene expression. Moreover, relatedness may play a *derivative* role in the dynamics of a species, especially a species that exhibits a complex division of labor involving the suppression of kin altruism.

Inclusive fitness theory, however, permits a formulation of the central problem of sociobiology in a particularly poignant form: *how do interac-*

tions among loci induce utterly selfish genes to collaborate, or to predispose their carriers to collaborate, in promoting the fitness of the organism? Inclusive fitness theory, because it ignores interactions among loci, does not answer this question. But it does provide important insights.

Fitness-enhancing collaboration among loci in the genome of a reproductive population requires suppressing alleles that decrease, and promoting alleles that increase the fitness of its carriers. Suppression and promotion are effected by *regulatory gene networks*, each member of which is itself utterly selfish. This implies that genes, and *a fortiori* individuals in a social species, do not generally maximize inclusive fitness but rather interact strategically in complex ways. It is the task of sociobiology to model these complex interactions.

9.7   Prosocial Genes Maximize Inclusive Fitness

Egbert Leigh (1971) famously compared the genome to a *parliament of genes*:

> Each acts in its own self-interest, but if its acts hurt the others, they will combine together to suppress it.

Leigh was concerned with the maintenance of Mendelian segregation, but the remark applies quite broadly. Certainly some such mechanism must account for the tendency of genes in the genome to cooperate. However, the mechanism does not operate through inclusive fitness maximization.

To see this, we return to the model explored in Section 9.2. Let q_a be the frequency of the focal allele in the population, and let p_a be the probability that the recipient shares a copy of this allele. Now let q_b be the frequency of some allele b at another locus of the genome, and let p_b be the probability that the recipient shares a copy of this allele with the donor. Note that the cost c imposed on the donor is imposed *equally* on the allele b assuming Mendelian segregation, because both allele a and allele b have probability $1/2$ of being passed on to each offspring. Similarly, allele b receives the same benefit b as the focal allele in all carriers of both alleles. Therefore if the size of the population is n in the current period, the size in the next period will be $n + qn(b - c)$ and the number of b alleles will be $nq_b + qn(bp_b - c)$.

Thus the change in frequency of allele b is given by

$$
\begin{aligned}
\Delta q_b &= \frac{n q_b + q n (b p_b - c)}{n + q n (b - c)} - q_b \\
&= \frac{q_a (1 - q_b)}{1 + q_a (b - c)} \left(b \frac{p_b - q_b}{1 - q_b} - c \right) \\
&= \frac{q_a (1 - q_b)}{1 + q_a (b - c)} (b r - c).
\end{aligned}
\tag{9.19}
$$

Note that we have used the equation $r = (p_b - q_b)/(1 - q_b)$, which is equation (9.4) for allele b.

Thus every allele at locus B benefits from the behavior induced by the focal allele, and hence a mutation at locus B that suppresses the focal allele will, *ceteris paribus*, be at a disadvantage as compared with the incumbent type alleles at this locus. Moreover, this is true whether the focal allele is prosocial or antisocial, so long as it satisfies Hamilton's rule. Therefore there is no intragenomic incentive for genes to evolve to suppress an antisocial allele.

This result must of course be qualified in the diploid case, both because meiotic drive can favor an allele at one locus that harms the other loci in the genome (Haig and Grafen 1991; Burt and Trivers 2006), and males and females may have distinct fitness enhancement conditions based on physiological differences (Haig 2002).

These and related situations aside, we can safely conclude that *even utterly selfish genes have common interests on the intragenomic level*. It follows that suppression of antisocial alleles must be a response to the joint reduced fitness of all alleles *at the society level*, through natural selection. It also follows that a prosocial allele that satisfies Hamilton's rule will provoke suppression responses on neither the intragenomic nor the intergenomic level. Hence prosocial genes *do* maximize inclusive fitness.

9.8   The Boundaries of Inclusive Fitness Maximization

In asserting that "natural selection leads organisms to become adapted as if to maximize their inclusive fitness," Abbot (2011) doubtless expresses a view with which at least 137 of the world's most prominent population biologists appear to agree. The main source cited in support of this statement is a series of papers written by Alan Grafen (1999, 2002, 2006). For instance in a paper devoted to exposing the "misconceptions" of others, West et al. (2011) write:

> Individuals should appear as if they have been designed to maximize their inclusive fitness. Grafen (1999, 2002, 2006a, 2007b) has formalised this link between the process and purpose of adaptation, by showing the mathematical equivalence between the dynamics of gene frequency change and the purpose represented by an optimisation program which uses an "individual as maximising agent" (IMA) analogy.

However, Grafen expressly declares in each of his papers on the subject that *additivity across loci*, or what is equivalent, *frequency independence*, is assumed. Others who have carefully studied the conditions under which a population genetics model of gene flow implies fitness maximization at the gene or individual level, including Metz et al. (2008), Gardner and Welsh (2011), and Gardner, West, and Wild (2011), require the same assumption. No careful researcher has *ever* claimed analytical support for the notion that individuals maximize inclusive fitness without making the frequency independence assumption.

If a gene is prosocial, we have seen that the behavior it fosters can be modeled as the maximization of inclusive fitness. But if the genome's success is based on a pattern of cooperation, promotion, and suppression of antisocial genes across loci, which will occur, for instance, if the production of a protein, RNA sequence, or social behavior requires the collaborative activity of many genes (Noble 2011), or if there are frequency dependent social interactions among individuals in a social species (Maynard Smith 1982), then neither genes nor individuals can be characterized as maximizing inclusive fitness.

9.9   The One Mutation at a Time Principle

Because genes code for proteins or RNA with very precise chemical functions, most mutations are fitness-reducing or fitness-neutral. The rate at which fitness-enhancing mutations occur is very low. Let us say that genes at two loci are *synergistic* if their joint presence in the genome of an individual is fitness-enhancing, but each alone is fitness-reducing. Clearly the rate at which two synergistic mutations occur in an organism is generally orders of magnitude less likely than single favorable mutations. Moreover, even when two such mutations are present, unless they are tightly linked so that they are not broken up by meiosis, they will only rarely and sporadically occur together. Bodmer and Felsenstein (1967) show that synergistic double mutants can survive if $1 < (1 - \rho) f_m / f_w$, where ρ is the recombination rate, f_w is the fitness of the wild type genome, and f_m is the fitness of the same genome with two relevant wild type alleles replaced with the mutants. Thus with no linkage ($\rho = 1/2$), the mutants would have to be twice as fit as the wild types to evolve. Except in the case of highly improbable macromutations, the linkage rate $1 - \rho$ would have to be very close to unity for the pair of mutants to survive. Moreover, in the case of extremely high linkage, it is a good approximation to treat the two genes as one. Indeed, a supergene can be treated as such, although normally with meiosis properties closer to haploid than diploid.

Therefore for most purposes we can assume that *only one favorable mutation occurs at a time*, and its success depends on the frequency distribution of alleles at other loci at the time the mutation appears. We call this the *one-gene-at-a-time principle*.

9.10   The Phenotypic Gambit

The genome of a multicellular organism includes a myriad of interdependent RNA-producing genes, protein-producing genes, and regulatory gene networks. The dynamics of gene interaction are poorly understood, to the point where it is normally impossible to isolate the exact role of a single gene in modulating a social behavior. Indeed, when we say that a certain allele produces or controls a certain phenotypic trait, what we really mean is that the *absence* of the allele entails the partial or complete *absence* of the trait. This is, of course, quite a weaker statement, merely asserting that the allele in question contributes in some more or less essential way to the production of the phenotypic effect.

One implication of this state of affairs is that there are few, if any, cases in which a social behavior can be attributed to the presence of an allele at a particular locus of the genome. This fact does not compromise Hamilton's rule, but without additional assumptions, it renders Hamilton's rule inapplicable to analytical models of social behavior. By far the most widely used such assumption is the so-called *phenotypic gambit* (Grafen 1984). The phenotypic gambit assumes that a behavior that may be extremely complex at the genetic level can be modeled as though it were the product of the choice of allele at a single locus. In the words of Alan Grafen (1984, p. 63),

The phenotypic gambit is to examine the evolutionary basis of a character as if the very simplest genetic system controlled it: as if there were a haploid locus at which each distinct strategy was represented by a distinct allele, as if the payoff rule gave the number of offspring for each allele, and as if enough mutation occurred to allow each strategy the chance to invade.

The haploid assumption is not necessary—there are many examples in the literature where the phenotypic gambit assumes that behavior is controlled by a diploid locus. Moreover, the assumption of a single locus is not necessary, as there is a research tradition in which the production of a phenotypic effect is controlled by two loci, one of which modulates the effects produced at the other locus (Liberman and Feldman 2005). Two-locus models, however, are generally extremely difficult to model and yield few additional insights.

The one-gene-at-a-time principle, however, often justifies the phenotypic gambit, especially in conjunction with the core genome concept. The latter suggests that most behavior-relevant genes will be either fixed in the genome or exist in such stable form that changes in the frequency of a mutant allele will not appreciably alter the frequency of other relevant genes in the genome. In that situation, the one-gene-at-a-time principle suggests that we are not likely to go wrong by considering an evolving behavior as the effect of an allele substitution at a single locus. For instance, whether an organism is altruistic or selfish in a particular situation may depend on the interaction of multiple loci, but so long as the behavior is subject to continuous variation, such as occurring with greater or lesser frequency or intensity, we can take the baseline degree of the trait as given and study the evolutionary success of a mutant allele that contributes to the frequency or intensity of the behavior using the phenotypic gambit.

9.11   The Anatomy of the Core Genome

If a gene has no social effects, that is if $b = \alpha = \beta = 0$ in the generalized Hamilton's rule (equation 9.17), then it obviously evolves only if it is prosocial ($c < 0$), in which case its increase in the population benefits all other loci in the genome. Moreover, if a gene that evolves is prosocial and non-polluting, it also benefits all genes both in the genome in which it is located, and in the population as a whole. These are strong harmony of interest principles that flow from inclusive fitness theory. But if a gene satisfies the generalized Hamilton's rule but is *antisocial* ($b - c - \beta < 0$) then, as we have seen in Section 9.7, it benefits all its co-resident genes, but it harms the population. Thus natural selection will favor the emergence of social forces that suppress such antisocial genes.

Enter *complexity*, the bitter enemy of classical systems theory. The gene pool of a species, consisting of many copies of long strings of DNA, interact *biochemically* to produce a metazoan organism whose cells manage to cooperate despite the evolutionary interest of each to ignore the others, and which interact *socially* through *emergent structural properties* that suppress defection and enhance cooperation sufficiently to ensure survival. We call these properties "emergent" because in our current state of knowledge, we are no more capable of explaining their provenance than we are in understanding how a sac of chemicals in the skull of a human being can give rise to consciousness.

The complex system of genes that gives rise to animal society is termed the *core genome*. The core genome of a sexually reproducing species is a subset of the loci in the genome that includes all loci that have certain key properties ensuring the general social character of the species. Included in the core genome are the *fixed loci* and *synonymous loci*. The fixed loci are those in which a single allele is shared by all members of the population, except for low-frequency mutations. The synonymous loci consist of loci in which all alleles, except for low-frequency mutations, produce identical biochemical and phenotypic effects. In addition certain non-synonymous alleles may have fitness-neutral, or near-neutral, phenotypic effects (e.g., tail length or eye color). The set of such *fitness-neutral gene sets* are stable across generations despite their somewhat labile internal composition, and are also part of the core genome. For instance, body size may be fitness independent over some range, and many genes interact to produce a phenotypic body size that is generally in the fitness-neutral range. The frequency distribution of these genes in the core genome is determined by natural selection and unchanged by meiosis and crossover.

In addition, if a set of alleles at a particular locus have equal fitness but distinct phenotypic effects, and if this set is preserved across generations, the alleles are likely to be equally fit alternative strategies in a Nash equilibrium among loci, each being a fitness-enhancing best response to the probability distribution of the other loci in the genome. We call such alleles *mixed strategy gene sets*, and we include these in the core genome. For example, a population equilibrium can sustain a positive fraction of

altruistic and selfish alleles, or alleles promoting aggressive *vs.* docile behavior, under certain conditions. Similarly, loci that protect carriers against
frequency-dependent variations in environmental conditions, including that
of bacterial and viral enemies, can be maintained in a polyallelic state as
a means of species-level risk reduction. These include the *immune system
gene sets* that maintain considerable heterogeneity to deal with a variety of
possible infectious agents.

Another example of a mixed strategy gene set is the interaction of suppressor genes and their targets, where the fitness of the suppressor depends
on a positive frequency of target genes. Leffler (2013) documents such a
set stabilized by balancing selection at least since the primate-hominin split.
Finally, heterozygote advantage involves a pair of alleles that maintain positive frequency despite the fitness cost to homozygous carriers. We may
call these *overdominance gene sets*. Additional features arise in dealing
with sex-linked genes, including maternal-paternal conflict, but these also
can be identified as characteristics of the species that are conserved across
many generations.

In species that recognize individuals, including many birds and mammals,
such recognition is based in part on the expression of alternative alleles
within a core genome gene set, as well as on genes *outside* the core genome,
which are shuffled and redistributed through meiosis and recombination,
accounting for the heterogeneity of phenotypes.

Finally, there are many complex phenotypes that are regulated by *supergenes*, that is, clusters of tightly linked loci. These supergenes segregate
as stable polymorphisms, and can constitute virtually entire chromosomes.
They arise through selection because reduced recombination between a pair
of synergistic genes is fitness-enhancing. Once a supergenome is formed,
mutations that render nearby genes synergistic with the supergene render
the inclusion of the mutant into the supergene additionally fitness-enhancing
(Rieseberg 2001; Bachtrog 2006; Schwander et al. 2014; Taylor and Campagna 2016). Many social species have two or more distinct phenotypes
that involve a combination of physical traits and linked behavioral patterns,
for instance, high coloration and aggression on the one hand and cryptic
coloration and low aggression on the other. The maintenance of these distinctions, which would be destroyed by meiosis and crossover, are often
explained as the result of supergenes.

Supergenes are maintained by several distinct mechanisms. One is simply
the fortuitous location of two synergistic genes sufficiently close, and per-

haps sufficiently close to the centromere of the chromosome on which they are located, that crossover is extremely unlikely. Another is through the formation of inversions, which are chromosomal rearrangements where large portions of the DNA molecule are flipped, suppressing local recombination.

In sum, the typical phenotypic characteristics of the species, including biochemistry, physiology, and behavioral predispositions, are conserved across generations due to the capacity of the core genome to self-replicate across generations. The non-core genes in the gene pool, largely accounting for the heterogeneity of individuals, may be called the *variant genome*—see Riley and Lizotte-Waniewski (2009) for an application to bacterial species. The core genome is subject to the laws of natural selection: replication, mutation, and selection of superior mutants. Individuals, their societies, and the social structure of these societies are the product of the evolution of the core genome.

While the core genome is an object of selection, it is not in any sense a *unit* of selection because it is specified by the frequency distribution of genomes in the population. Moreover, the very notion of units and objects of selection, while perhaps of use for a synthetic understanding of biological evolution, do not appear to play any role in modeling the social structure and dynamics of a reproductive population. However, recognizing the core genome as an object of selection is a useful heuristic in at least two ways. First, while not in any way undermining the insights of the gene's eye view of evolution, it captures the notion that precise combinations of gene interactions are adaptive and hence favored by natural selection. Second, the core genome allows us to conceptualize phenotypic effects that are located not in individuals, but in their social interactions. In other words, the core genome strongly predisposes a social species for certain forms of social behavior, including typical mating patterns, recognized forms of territoriality, and preferred forms of social grouping. The core genome also predisposes organisms to seek out particular natural environments, although there is natural variation in such environments that serve as epigenetic sources of social dynamics and social learning (Galef and Laland 2005; Goodnight et al. 2008; Smaldino et al. 2013).

The core genome is a *replicator* in the sense of Lewontin (1970). First, mutations in loci of the core genome give rise to *phenotypic heterogeneity*. Second, phenotypic differences can entail *fitness differences among members* of the reproductive population. Finally, such fitness differences are *heritable*. A mutation at a fixed locus, for instance, can lead to increased fitness of carriers of the mutated allele, leading to the increase in frequency of the new allele in the population. The focal locus then drops out of the core genome, but in the long run, with high probability, the mutation will either move to fixation or extinction, restoring the focal locus to the core genome.

Richard Dawkins (1982b) famously rejects the genome as an object of selection, arguing that because of meiosis and recombination, the genome dies with the body it inhabits. Dawkins concludes that the individual is but a *vehicle* for the transportation of genes across metazoan bodies, writing that a replicator must have a

> low rate of spontaneous, endogenous change, if the selective advantage of its phenotypic effects is to have any significant evolutionary effect.…too long a piece of chromosome will quantitatively disqualify itself as a potential unit of selection, since it will run too high a risk of being split by crossing over in any generation. (p. 47)

Cognizant of this important observation, I have defined the core genome so as to be impervious to meiosis and crossover. This is clear for fixed and synonymous loci, where no breaking up of synergistic genome interactions occur. Moreover meiosis creates as many heterozygote as it destroys, on average, and it does not alter the frequency distribution of mixed strategy or immune system gene sets in the population.

9.12 Explaining Social Structure

While inclusive fitness theory justifies selfish gene theory, neither inclusive fitness theory, nor any other plausible theory, supports the notion that genes or individuals in asocial species maximize inclusive fitness. We have shown that the maximization characterization is plausible for prosocial non-polluting genes that satisfy Hamilton's rule, but not otherwise.

The evolutionary process, from the first RNA molecules to advanced metazoans and complex social species, involves solving the problem of promoting cooperation among selfish genes (Maynard Smith and Szathmáry 1995). That genes generally contribute to the fitness of the individuals in which they reside is the result, not of inclusive fitness maximization, but of a complex evolutionary and intragenomic dynamic involving the suppression of antisocial and promotion of prosocial alleles (Leigh 1971; Buss 1987; Michod 1997; Frank 2003; Noble 2011).

The evolutionary forces that determine the complex interactions among loci in metazoans and among individuals in social species must be studied using, in addition to inclusive fitness theory, the phenotypic gambit (Grafen 1984), evolutionary game theory (Wilson 1977; Taylor 1992; Taylor 1996), agent-based modeling (Gintis 2009b), the physiology of suppressor and promoter genes (Leigh 1977; Noble 2011), as well as species-level systematics and ecology.

A1   Hamilton's Rule with General Social Interaction

This section presents a version of Hamilton's rule that assumes a diploid organism, and applies to sophisticated social species in which interactions are multi-adic, such as when there is a complex division of labor in hunting, defense, or rearing offspring. The resulting equations are similar to those deduced from the regression approach to Hamilton's rule (Queller 1992) but we have no need for least squares regression arguments. The most salient implication of this exercise is that Hamilton's rule holds with very great generality, although the three terms in the equation are reflections of the social structure of the reproductive population.

Consider a reproductive population X with individuals $\{X_i \in X | i = 1, \ldots, n\}$. Suppose the genome has a diploid autosomal locus with two alleles, s (selfish) which leads to a behavior that does not affect the fitness of other individuals, and a (altruistic) which leads its carrier X_i to incur an increased fitness cost c_i over that of the selfish allele, and to bestow fitness benefit b_i distributed over a subset Y_i of recipients. Suppose in addition that the altruistic allele has a fitness effect β (pollution when $\beta > 0$ or a public good when $\beta < 0$) on both alleles (see Section 9.4). This cost may be intragenomic, borne by the carrier, or intergenomic, distributed over the population in some arbitrary manner.

Hamilton (1964a) assumes the social fitness effect is distributed uniformly over the genome. This is a significant limitation of his analysis because intragenomically, meiotic drive and other forms of segregation distortion, and socially, altruistic acts that are purchased in part by reducing the fitness of non-relatives, which we call *thieving effects* (see Section 9.4), are important, although the Harmony Principle suggests that natural selection will limit their observed frequency. We can represent these thieving effects as transfers of fitness $\alpha > 0$ from non-relatives to relatives, and the reverse for $\alpha < 0$.

Standard expositions of Hamilton's rule take Y_i to be an individual. This, however, is a restrictive assumption because in many social species individuals interact in groups where it is difficult to apportion the benefit b_i among the various participants. For instance, agent i may play in an n-player public goods game in which the s allele promotes defection and the a allele promotes cooperation, or agent i may defend the nest against intruders, or punish a lazy coworker. As we shall see, Hamilton's rule does not depend on the assumption that the beneficiary is an individual.

The genotypic value X_g^i of X_i at the focal locus, the frequency of the focal allele at this locus, is 0, ½, and 1 for genotypes ss, sa, and aa, respectively. The phenotypic value X_p^i of X_i is 0, h, or 1 according as X_i is ss and never confers the benefit, is sa and confers the benefit with intensity h, or is aa and confers the benefit with intensity one. Here h can have any value, positive or negative, but if the allele effects are additive, then $h = 1/2$. Because there are $2n$ alleles at the focal locus in the population, the frequency of a is $q_a = \sum_i X_g^i / n$. Let Y_g^i be the mean genotype of members of Y_i.

The fitness cost to X_i in the current period is thus $c_i X_p^i$, and the fitness gain to the recipients Y_i is $b_i X_p^i$. The population size in the next period is then

$$n(1 - \beta q_a + (b - c)x_p) \tag{A9.1}$$

where $x_p = \sum_i X_p^i / n$ is the mean phenotype of the population, $b = \sum_i b_i X_p^i / x_p$ is the mean benefit, and $c = \sum_i c_i X_p^i / x_p$ is the mean cost. Note that because the thieving effect α is a within-population fitness transfer, it does not appear in (A9.1). The number of donor alleles in the next period is

$$nq_a(1 - \beta q_a + \alpha(1 - q_a)) + \sum_i b_i X_p^i Y_g^i - \sum_i c_i X_p^i X_g^i.$$

The increase in the frequency of the donor allele in the next period, writing the mean genotype of recipients as $q_a^y = \sum_i Y_g^i / n$, is then given by

$$\frac{nq_a(1 - \beta q_a + \alpha(1 - q_a)) + \sum_i b_i X_p^i Y_g^i - \sum_i c_i X_p^i X_g^i}{n(1 - \beta q_a + (b - c)x_p)} - q_a =$$

$$\frac{\left(\sum_i b_i X_p^i Y_g^i - nbx_p q_a^y\right) + nq_a\alpha(1-q_a)}{n(1-\beta q_a + (b-c)x_p)} -$$

$$\frac{\left(\sum_i c_i X_p^i X_g^i - ncx_p q_a\right) + nbx_p(q_a - q_a^y)}{n(1-\beta q_a + (b-c)x_p)} =$$

$$\frac{\mathrm{cov}(X_p^b, Y_g) - \mathrm{cov}(X_p^c, X_g) + \alpha\mathrm{var}(X_p) + bx_p(q_a^y - q_a)}{1-\beta q_a + (b-c)x_p}, \qquad \text{(A9.2)}$$

where X_p^b and X_p^c are the variables $b_i X_p^i$ and $c_i X_p^i$, respectively, and X_g is a binomial variable, so $\mathrm{var}(X_p) = nq_a(1-q_a)$. Note that the expression (A9.2) is positive, assuming weak selection, when

$$\frac{\mathrm{cov}(X_p^b, Y_g) + \alpha\mathrm{var}(X_p) + bx_p(q_a^y - q_a)}{\mathrm{cov}(X_p^c, X_g)} > 1. \qquad \text{(A9.3)}$$

This inequality is the most general form of Hamilton's rule, including both social fitness and thieving effects. If we assume donors distribute benefits that are, on average, independent from the allelic composition at the focal locus, i.e., $q_a^y = q_a$, then (A9.3) becomes

$$\mathrm{cov}(X_p^b, Y_g) + \alpha\mathrm{var}(X_p) > \mathrm{cov}(X_p^c, X_g). \qquad \text{(A9.4)}$$

Note that in the standard treatment, where the beneficiary is an individual, the condition $q_a^y = q_a$ necessarily holds. To see this, note that

$$q_a^y = [r + (1-r)q_a]q_a + [1 - (r + (1-r)(1-q_a))](1-q_a) = q_a, \quad \text{(A9.5)}$$

where r is the relatedness coefficient.

If we further assume that $b_i = b$ and $c_i = c$ for all individuals $i = 1, \ldots, n$, we get the expression:

$$\frac{b\,\mathrm{cov}(X_p, Y_g) + \alpha\,\mathrm{var}(X_p)}{\mathrm{cov}(X_p, X_g)} > c. \qquad \text{(A9.6)}$$

Finally, if the effect of the altruistic allele is additive, so $h = 1/2$, then (A9.6) becomes

$$b\frac{\mathrm{cov}(X_p, Y_g)}{\mathrm{var}(X_g)} > c - \alpha. \qquad \text{(A9.7)}$$

This is a standard expression for Hamilton's rule (Michod and Hamilton 1980), except we have taken into account the thieving effect α (and the pollution/public good effect β, which does not appear in Hamilton's rule). More generally, for arbitrary h, we have

$$br > cr^p - \alpha, \tag{A9.8}$$

where

$$r = \frac{\text{cov}(X_p, Y_g)}{\text{var}(X_g)}$$

is the regression coefficient of Y_g on X_p, and r^p is the regression coefficient of X_p on X_g:

$$r^p = \frac{\text{cov}(X_p, X_g)}{\text{var}(X_g)}.$$

It should be clear that, while we use mathematical terminology from statistical estimation theory, no statistical estimation is in fact involved.

To illustrate the increased generality of the form (A9.4) of Hamilton's rule, suppose the reproductive population is partitioned into *social castes* $\{Z^j \subset X | j = 1, \ldots, m\}$, where caste j has frequency z_j in the population, and suppose members of the same caste j have the same costs c_j and benefits b_j. Let Y^j be the weighted sum of $\{Y_i | X_i \in Z^j\}$, where each individual is weighted by the number of times the individual appears in the sum. Then we can write (A9.4) as

$$\sum_{j=1}^{m} \left(b_j \, \text{cov}(Z_p^j, Y_g^j) - c_j \, \text{cov}(Z_p^j, Z_g^j) \right) + \alpha \, \text{var}(X_p) > 0. \tag{A9.9}$$

Equation (A9.9) shows that in general the social structure of the population allows a caste to be *fundamentally altruistic* in the sense that its net costs of helping exceed the net benefits that the caste contributes to the population. Because the inclusive fitness of caste j is

$$b_j \, \text{cov}(Z_p^j, Y_g^j) - c_j \, \text{cov}(Z_p^j, Z_g^j) < 0 \tag{A9.10}$$

it is then clear that caste j members would maximize their inclusive fitness by simply refusing to contribute to the social process. This shows that *in a caste social structure, individuals do not necessarily maximize their inclusive fitness.* Of course, if castes are genetically determined, then the partition $\{z_j | j = 1, \ldots, m\}$ will be variable across periods and a fundamentally altruistic caste will become extinct in the long run. However, if castes are determined by developmental conditions (e.g., feeding in eusocial insects or socialization in humans), fundamentally altruistic castes can be maintained in the long run.

A1.1   The Sociobiological Dynamics of Hamilton's Rule

The mapping $X_i \rightarrow Y_i$, which we have taken as given, reflects the *social structure of the reproductive population*. This mapping does not presume any particular set of social relations of kinship, which is why we suggest that *kin selection* is in general an inappropriate description of inclusive fitness dynamics. Note that if the frequency of the *a* allele in the population does not affect the fitnesses of alleles at other loci in the genome, then the *a* allele will move to fixation in the population if Hamilton's rule is satisfied, and will become extinct if the reverse inequality is satisfied. Ultimately, the focal locus will be heterozygous with zero probability.

With frequency dependence, when the focal allele becomes prevalent in the population, if $b - c > 0$, so the allele is beneficial to its carriers, there will be no selection at the level of the genome for genes that suppress the *a* allele at the focal locus, so the *a* allele will still move to fixation in the population. When the focal allele is prevalent and $b - c < 0$, there will be natural selection at other loci for genes that either alter the sociobiological mapping $X_i \rightarrow Y_i$ or otherwise suppress the *a* allele at the focal locus, so that Hamilton's rule no longer holds for the antisocial allele. This is the essence of the Inclusive Fitness Harmony Principle. Of course there may be no likely mutation that suppresses an antisocial *a* allele, in which case the antisociality reflected in the behavior induced by the *a* allele will become ubiquitous in the population. Natural selection does not guarantee optimality.

This phenomenon also represents a plausible counterexample to Fisher's Fundamental Theorem   (Ewens 1969; Price 1972; Frank and Slatkin 1992; Edwards 1994; Frank 1997): as an antisocial allele moves to fixation, the average fitness of population members declines. Some population biologists save Fisher's theorem by calling this a *transmission effect*, and insisting that natural selection always produces fitness-enhancing gene frequency changes   (Edwards 1994; Frank 1997; Gardner et al. 2011). This interpretation of natural selection should be avoided because it is arbitrary and difficult to understand for those who are not experts in population biology.

It follows that Hamilton's rule is useful only in charting short-term genetic dynamics. Weak selection and additivity across loci are extremely powerful analytical tools, but in the long run changes in gene frequency at one locus are likely to induce compensatory and synergistic changes at other loci. Indeed, the very mapping $X_i \rightarrow Y_i$ on which Hamilton's rule is based is itself coded in the core genome of the reproductive population, and hence in the long run is modified in the course of evolutionary selection and adaptation.

A1.2   Altruism Among Relatives

A relative is a person "allied by blood...a kinsman" (Biology Online). The argument to this point has nothing to do with genealogy, and hence says nothing about altruism among family members. This is an attractive property of our exposition because in a highly social species, individuals interact frequently with non-relatives.

It remains to determine the exact relationship between the sociobiological conception (A9.6) and the genealogical conception of relatedness. We follow Michod and Hamilton (1980), except that we assume the population is outbred at the focal locus. Suppose that each Y_i is an individual recipient, and all recipients have the same genealogical relationship to their donors (e.g., Y_i is a sibling of X_i). Let $\{p_{xyzw}\}$ be the joint distribution of genotypes xy for donor and zw for recipient where $x, y, z, w \in \{s, a\}$. Let p^x_{ss}, p^x_{as}, and p^x_{aa} be the marginal distribution of the genotypes ss, sa, and aa for the donor (i.e., the fraction of these genotypes in the population), and similarly for p^y_{ss}, p^y_{as}, and p^y_{aa} for the recipient.

We have

$$x_p = h p^x_{as} + p^x_{aa},$$
$$y_p = h p^y_{as} + p^y_{aa},$$

because p^x_{as} is the fraction of sa genotypes, their phenotypic value is h, and p_{aa} is the fraction of aa genotypes, which have phenotypic value one. Also,

$$p^x_{as} = 2 q_n q_a \tag{A9.11}$$
$$p^x_{aa} = q^2_a. \tag{A9.12}$$

To derive (A9.11), note that either the paternal allele is s with probability $q_n = 1 - q_a$ and the second is a with probability q_a, or else the paternal allele is a with probability q_a and the second is s with probability q_n. The second equation is derived in a similar manner.

We thus have

$$x_p = 2 h q_n q_a + q^2_a \tag{A9.13}$$
$$y_p = 2 h q_n q_a + q^2_a \tag{A9.14}$$

Note that

$$x_g = \tfrac{1}{2}p_{as}^x + p_{aa}^x = q_a$$
$$y_g = \tfrac{1}{2}p_{as}^y + p_{aa}^y = q_a.$$

To derive $\mathrm{cov}(X_g, X_p)$, note that

$$\sum_i X_p^i X_g^i / n = h p_{as}^x / 2 + p_{aa}^x$$

$$= h q_n q_a + q_a^2$$

Given the values of p_{as}^x and p_{aa}^x from equations (A9.11) and (A9.12), and after algebraic simplification, we find

$$\mathrm{cov}(X_p, X_g) = q_n q_a \gamma / 2, \tag{A9.15}$$

where

$$\gamma = 2(h + q_a(1 - 2h)). \tag{A9.16}$$

Also,

$$\mathrm{cov}(y_g x_p) = h p_{sasa}/2 + h p_{saaa} + p_{aasa}/2 + p_{aaaa} - y_g x_p.$$

Now let p_{11} be the probability X_i and Y_i share both alleles at the focal locus identically by descent, let p_{10} be the probability the share one allele at the focal locus identically by descent, and let p_{00} be the probability they share neither allele identically by descent. then we have

$$p_{asas} = 2q_n q_a p_{11} + q_n q_a p_{10} + 4q_n^2 q_a^2 p_{00} \tag{A9.17}$$
$$p_{asaa} = q_a q_n^2 p_{10} + 2q_n q_a^3 p_{00} \tag{A9.18}$$
$$p_{aaas} = q_n q_a^2 p_{10} + 2q_n q_a^3 p_{00} \tag{A9.19}$$
$$p_{aaaa} = q_a^2 p_{11} + q_a^3 p_{10} + q_a^4 p_{00}. \tag{A9.20}$$

If we define f_{XY} as the probability that a random allele in X_i and a random allele in Y_i are identical by descent, then

$$f_{XY} = p_{11}/2 + p_{10}/4. \tag{A9.21}$$

Then a little algebra shows that the r in Hamilton's rule is given by

$$r = \frac{\mathrm{cov}(X_p, Y_g)}{\mathrm{cov}(X_p, X_g)} = 2 f_{XY}. \tag{A9.22}$$

Note that r is then the expected number of copies of the focal allele in the recipient.

Consider, for instance, the case of siblings. The two share the same allele from the father with probability $1/2$, and similarly for the mother. Therefore $p_{11} = 1/4$, $p_{10} = 1/2$, and $p_{00} = 1/4$. Substituting these values in (A9.17), we get

$$r = \frac{\text{cov}(Y_g, X_p)}{\text{cov}(X_g, X_p)} = \frac{1}{2}. \qquad (A9.23)$$

Thus the sociobiological definition of relatedness and the genealogical definition coincide.

10

Gene-Culture Coevolution and the Internalization of Norms

> The scientist does not study nature because it is useful. He studies it because it gives him pleasure.
>
> Henri Poincaré

> Mathematics, rightly viewed, possesses not only truth, but supreme beauty.
>
> Bertrand Russell

> The Lord God is subtle, but malicious he is not.
>
> Albert Einstein

> **C:** How is the internalization of norms compatible with inclusive fitness maximization?
> **HG:** It is not.
> **C:** Which is correct?
> **HG:** For humans and other social species inclusive fitness maximization fails.
>
> Choreographer interview

10.1 Norms and Internalization

A *norm* is a pattern of behavior enforced in part by internal sanctions, including shame, guilt, and loss of self-esteem, as opposed to purely external sanctions, such as material rewards and punishments. Humans internalize norms through *socialization* by parents (*vertical transmission*) and extra-parental conspecifics (*oblique* and *horizontal transmission*). The capacity to internalize norms is widespread among humans, although in some so-called "sociopaths," this capacity is diminished or absent (Mealey 1995). Human behavior generally conforms to the rational actor model. In these terms, the capacity to internalize norms means human agents have *socially programmable* preferences. Human behavior thus depends not only on *beliefs*, which concern the effects of action (choice X leads to result Y), but *values*, which are the very *goals* of action.

The capacity to internalize norms certainly has survival value for society because it permits rapid cultural adaptation towards novel conditions, whereas a purely genetic adaptive process takes orders of magnitude more time to become effective. The main puzzle from an evolutionary game-theoretic viewpoint is why internalization is *individually* adaptive. For were it not, then it would not evolve. Why cannot self-interested sociopaths out-compete norm-followers by mimicking their behavior when it suits their purposes and behaving selfishly when it does not?

Suppose there is one genetic locus that controls the capacity to internalize norms. I construct models of gene-culture coevolution, following the theory developed in Chapter 1 and Chapter 9. These models show that if a norm is individually fitness-enhancing, then the allele for the internalization of norms is evolutionarily stable. Moreover, if the fitness payoff to the internalized norm is sufficiently large, the allele for internalization is globally stable.

As argued in Chapter 6, social values are transmitted across generations through the internalization of norms (Parsons 1967; Grusec and Kuczynski 1997). Successful societies tend to foster norms that enhance personal fitness, such as the ability to defer gratification (Mischel and Ebbeson 1970), good personal hygiene, positive work habits, and control of emotions, as well as altruistic norms that subordinate the individual's needs to group welfare, fostering such behaviors as bravery, honesty, fairness, willingness to cooperate, and empathy for others (Bowles and Gintis 2011). People value norms for their own sake, in addition to, or despite, the effects the behavior the norms suggest have on perceived well-being. For instance, an individual who has internalized the value of speaking truthfully will do so even in cases where the net payoff to speaking truthfully would otherwise be negative. It follows that where people internalize a norm, the frequency of its occurrence in the population will be higher than if people follow the norm only instrumentally; i.e., when they perceive it to be in their selfish interest to do so.

Altruism is prosocial towards unrelated others (e.g., helping those in distress and punishing antisocial behavior) at personal cost. On the importance of altruism in humans see Gintis (2000), Bowles and Gintis (2011), and Wilson (2012). Adding an altruism norm allows us to model Herbert Simon's (1990) explanation of altruism. Simon suggested that altruistic norms could *hitchhike* on the general tendency of norms to be personally fitness-enhancing. Of course, norms may persist even if they are fitness-reducing both for individuals and the group (Edgerton 1992; Boyd and Richerson 1992). This chapter develops a gene-culture coevolutionary model elucidating the process whereby genes for norm internalization become evolutionary successful.

10.2   Socialization and Fitness-Enhancing Norms

Suppose there is a norm **C** that can be internalized by a new member of society. Norm **C** confers fitness $1 + t > 1$, while the normless phenotype, denoted by **D**, has baseline fitness 1. There is a genetic locus with two alleles, **a** and **b**. Allele **a**, which is dominant, permits the internalization of norms, whereas **b** does not. We assume that possessing at least one copy of **a** imposes a fitness cost $u \in (0, 1)$, on the grounds that there are costly physiological and cognitive prerequisites for the capacity to internalize norms.[1] We assume $(1+t)(1-u) > 1$, so the cost of the internalization allele is more than offset by the benefit of the norm **C**. There are five phenogenotypes, whose fitnesses are listed in Figure 10.1.

Individual Phenogenotype	Individual Fitness
aaC	(1-u)(1+t)
aaD	(1-u)
abC	(1-u)(1+t)
abD	(1-u)
bbD	1

Figure 10.1. Fitnesses of the five phenogenotypes. Here u is the fitness cost of possessing the internalization allele, and t is the excess fitness value of possessing the norm **C**. Note that **bbC** cannot occur.

Families are formed by random pairing, and offspring genotypes obey the laws of Mendelian segregation. Thus there are six familial genotypes, **aaaa**, **aaab**, **aabb**, **abab**, **abbb**, and **bbbb**. We assume also that only the phenotypic traits of parents, and not which particular parent expresses them, are relevant to the transmission process. Therefore, there are three familial phenotypes, **CC**, **CD**, and **DD**, and 18 familial phenogenotypes, of which only 14 can occur. The frequencies of familial phenogenotypes are as shown in Figure 10.2, where $p(i)$ represents the frequency of phenogenotype $i = $ **aaC**, ..., **bbD**.

The rules of gene-culture transmission are as follows. If familial phenogenotype is **xyzwXY**, where **x,y,z,w** $\in$ {**a,b**}, $X, Y \in$ {**C,D**}, an offspring is equally likely to inherit **xz**, **xw**, **yz**, or **yw**. An offspring whose

---

[1]This assumes that $u > 0$ is conservative, in that it biases the model against the global stability of the internalization allele. However, the contrasting assumption $u < 0$ is also plausible. I will point out the implications of $u < 0$ where appropriate.

Familial Phenogenotype	Relative Frequency in Reproductive Pool
aaaaCC	$p(\mathbf{aaC})^2(1-u)^2(1+t)^2$
aaaaCD	$2p(\mathbf{aaC})p(\mathbf{aaD})(1-u)^2(1+t)$
aaaaDD	$p(\mathbf{aaD})^2(1-u)^2$
aaabCC	$2p(\mathbf{aaC})p(\mathbf{abC})(1-u)^2(1+t)^2$
aaabCD	$2(p(\mathbf{aaC})p(\mathbf{abD})+p(\mathbf{aaD})p(\mathbf{abC}))(1-u)^2(1+t)$
aaabDD	$2p(\mathbf{aaD})p(\mathbf{abD})(1-u)^2$
ababCC	$p(\mathbf{abC})^2(1-u)^2(1+t)^2$
ababCD	$2p(\mathbf{abC})p(\mathbf{abD})(1-u)^2(1+t)$
ababDD	$p(\mathbf{abD})^2(1-u)^2$
aabbCD	$2p(\mathbf{aaC})p(\mathbf{bbD})(1-u)(1+t)$
aabbDD	$2p(\mathbf{aaD})p(\mathbf{bbD})(1-u)$
abbbCD	$2p(\mathbf{abC})p(\mathbf{bbD})(1-u)(1+t)$
abbbDD	$2p(\mathbf{aaC})p(\mathbf{aaD})(1-u)(1+t)$
bbbbDD	$2p(\mathbf{bbD})^2$

Figure 10.2. Relative frequencies of phenogenotypes. The absolute frequencies are the entries in this table divided by the sum of the entries. Note that **aabbCC**, **abbbCC**, **bbbbCC**, and **bbbbCD** are not listed, since **bbC** cannot occur.

genotype includes a copy of the **a** allele is equally likely to inherit **X** or **Y**.[2] But an offspring of genotype **bb** always has the normless phenotype **D**. The transition table is shown in Figure 10.3.[3]

The above accounts only for parental transmission. In addition, extraparental transmission is ubiquitous in human society, in the form of social pressure (rumor, shunning, and ostracism), rituals (dancing, prayer, marriage, birth, and death), and in modern societies, formalized institutions (schools and churches).[4] To account for extraparental transmission, let p_C be the fraction of the population carrying the **C** phenotype, and let

[2]Simulations show that the assumption that the **a** allele is dominant is not critical. The stability results described below continue to hold, and indeed more strongly, when **a** is less than fully dominant.

[3]Biased parental transmission, in which heterogeneous familial phenotypes are more likely to transmit one phenotype to offspring than the other (Cavalli-Sforza and Feldman 1981) is discussed below.

[4]Extraparental transmission is generally individually costly and the benefits accrue to unrelated others. Hence it is a form of altruistic behavior, and ideally should not be intro-

$\gamma \in [0, 1]$. We assume a fraction γp_C of **aa**-types and a fraction νp_C of **ab**-types who have not internalized **C** through parental transmission are influenced by extraparental transmission to switch to **C**. **bb** types are not affected by extraparental transmission.

Familial Type	Offspring Phenogenotypic Frequency				
	aaC	aaD	abC	abD	bbD
aaaaCC	1				
aaaaCD	1/2	1/2			
aaaaDD		1			
aaabCC	1/2		1/2		
aaabCD	1/4	1/4	1/4	1/4	
aaabDD		1/2		1/2	
aabbCD			1/2	1/2	
aabbDD				1	
abbbCD			1/4	1/4	1/2
abbbDD				1/2	1/2
ababCC	1/4		1/2		1/4
ababCD	1/8	1/8	1/4	1/4	1/4
ababDD		1/4		1/2	1/4
bbbbDD					1

Figure 10.3. Phenotypic inheritance is controlled by genotype. Note that **aabbCC**, **abbbCC**, **bbbbCC**, and **bbbbCD** are not listed, since **bbC** cannot occur.

The resulting system consists of four equations in four unknowns (**bbC** cannot occur, and one offspring phenogenotype is dropped, since the sum of phenogenotypic frequencies equals unity). It is straightforward to check that there are three pure strategy equilibria (i.e., equilibria in which the whole population bears a single phenogenotype). These are **aaC**, in which all agents internalize the fitness-enhancing norm, **aaD**, in which the internalization allele is present but the phenotype **C** is absent, and **bbD**, in which neither the internalization allele nor the norm is present.

A check of the eigenvalues of the Jacobian matrix of the dynamical system shows that the **aaD** equilibrium is unstable. Eigenvalues of the system at

duced until our analysis of altruism is completed. We introduce it now purely for expositional purposes.

the **aaC** equilibrium are given by

$$\left\{ 0, 1, \frac{1-\gamma}{2(1+t)}, \frac{1-\gamma}{1+t} \right\}.$$

The unit eigenvalue is semisimple,[5] so the linearization of the equilibrium **aaC**, in which the fitness-enhancing norm is internalized, is stable. However, we cannot conclude that the nonlinear model itself is stable. Extensive simulations fail to find a case in which the **aaC** equilibrium is unstable.[6] Moreover, in the case where the **a** allele is incompletely dominant, the unit root disappears, so stability is assured (this remark applies as well to all cases in which unit roots appear, some of which are discussed below).

The eigenvalues of the Jacobian matrix of the unnormed equilibrium **bbD** are given by

$$\left\{ 0, 0, 1-u, \frac{1}{2}(1+t)(1-u) \right\}.$$

Therefore this equilibrium, in which no internalization occurs, is locally stable if $(1+t)(1-u) < 2$, and unstable when the opposite inequality holds. There may exist equilibria involving more than one type of behavior, although the system is too complex to determine whether or not this is the case. Extensive simulations suggest that if such equilibria exist, they are not stable. I shall assume this is the case in this chapter. It follows that for $t > 2/(1-u) - 1$, **aaC** is a globally stable equilibrium.[7]

There are four plausible conditions that render the **bbD** equilibrium unstable, in which case **aaC** will be globally stable. The first is $u < 0$, which means that the apparatus upon which internalization depends has net positive (pleiotropic) fitness effects independent from its contribution to the internalization of norms. The second is that t is sufficiently large that

---

[5]An eigenvalue is semisimple if its algebraic and geometric dimensions are equal. Semisimple unit roots of linear dynamical systems are stable.

[6]The process of coding this and the other models presented in this chapter is tedious and error-prone. To ensure accuracy I wrote the simulations in two completely different languages, one Lisp-like (Mathematica) and the other procedural (C++), and verified that the results agreed to six decimal places over thousands of generations of simulation.

[7]The above result depends on our assumption of unbiased parental transmission. Suppose, however, that a fraction δ of offspring who would acquire norm **C** under unbiased transmission in fact acquire **D**. In this case, inspection of the eigenvalues of the Jacobian tells us that the **aaC** equilibrium is locally stable provided $\delta < t$ and the **bbD** equilibrium is stable provided $(1-\delta)(1+t)(1-u) < 2$. Thus parental transmission biased against the internalizable norm **C** is hostile to internalization.

$(1 + t)(1 - u) > 2$. Third, if parental transmission is sufficiently biased in favor of **C**, the internalization equilibrium is globally stable.

The fourth condition leading to the global stability of the **aaC** equilibrium is that there is some assortative mating that overcomes the tendency of the internalization allele to become "diluted." Suppose each type mates with another of its type with probability ζ, and with a random member of the population with probability $1 - \zeta$. Then the eigenvalues of the **bbD** equilibrium become

$$\left\{ \frac{1}{2}\zeta(1 - u), \frac{1}{2}\zeta(1 + t)(1 - u), 1 - u, \frac{1}{2}(1 + t)(1 - u)(1 + \zeta) \right\}. \quad (10.1)$$

Therefore there is always a degree of assortative mating that renders the **bbD** equilibrium unstable. Thus, it is plausible that some combination of assortative mating, parental transmission biased towards the norm, and high returns to the norm assures the global stability of the **aaC** equilibrium.[8]

10.3   Altruism

We now add a second dichotomous phenotypic trait with two variants. Norm **A** is altruistic in the sense that its expression benefits the group, but imposes fitness loss $s \in (0, 1)$ on those who adopt it. The normless state, **B**, is neutral, imposing no fitness loss on those who adopt it, but also no gain or loss to other members of the social group.

We assume **A** has the same cultural transmission rules as **C**: individuals who have a copy of allele **a** inherit their phenotypes from their parents, while **bb** individuals always adopt the normless phenotype **BD**. In addition, there is extraparental transmission, as before. There are now three genotypes and four phenotypes, giving rise to nine phenogenotypes that can occur, which we denote by **aaAC**, **aaAD**, **aaBC**, **aaBD**, **abAC**, **abAD**, **abBC**, **abBD**, and **bbBD**, and three that cannot occur, **bbAC**, **bbAD**, and **bbBC**. We represent the frequency of phenogenotype i by $p(i)$, for $i = $ **aaAC**, ..., **bbBC**.

We maintain the assumption that families are formed by random pairing and the offspring genotype obeys Mendelian segregation. We assume also that only the phenotypic traits of parents, and not which particular

---

[8]The same results hold for a haploid version of the model, except that there is no unit root in the **aC** equilibrium. Moreover, if **a** is not completely dominant in the diploid model, the unit root disappears and the equilibrium is unambiguously stable.

parent expresses them, are relevant to the transmission process. Therefore there are nine family phenotypes, which can be written as **AACC**,
AACD, **AADD**, **ABCC**, **ABCD**, **ABDD**, **BBCC**, **BBCD**, and **BBDD**. It
follows that there are 54 familial phenogenotypes, which we can write as
aaaaAACC,…,**bbbbBBDD**, only 36 of which can occur. We write the
frequency of familial phenogenotype j as $p(j)$, and we assume the population is sufficiently large that we can ignore random drift. For illustrative
purposes, here are a few of the phenogenotypic relative frequencies, where
we define $t^* = (1 + t)(1 - s)$:

$$p(\mathbf{aaaaAACC}) = p(\mathbf{aaAC})^2 t^{*2}(1 - u)^2$$
$$p(\mathbf{aaaaAACD}) = p(\mathbf{aaAC})\,p(\mathbf{aaAD})t^*(1 - s)(1 - u)^2$$
$$p(\mathbf{ababABCD}) = 2p(\mathbf{abAC})\,p(\mathbf{abBD})$$
$$+ p(\mathbf{abAD})(\mathbf{abBC}))t^*(1 - u)^2$$
$$p(\mathbf{bbbbBBDD}) = p(\mathbf{bbBD})^2$$

The rules of cultural transmission are as before. If familial phenogenotype
is **xyzwXYZW**, where **x,y,z,w** $\in \{\mathbf{a}, \mathbf{b}\}$, **X,Y** $\in \{\mathbf{A}, \mathbf{B}\}$, and **Z,W** $\in \{\mathbf{C}, \mathbf{D}\}$,
an offspring is equally likely to inherit **xz**, **xw**, **yz**, or **yw**. An offspring
whose genotype includes a copy of the **a** allele is equally likely to inherit **X**
or **Y**, and equally likely to inherit **Z** or **W**. Offspring of genotype **bb** always
have the normless phenotype **BD**. Extensive simulations show that if **a** is
incompletely dominant, the results described below continue to hold. The
transition table is shown in Figure 10.4 (continued on next page).

Familial Type	\multicolumn{9}{c}{Offspring Phenotypic Frequency}
	aaAC	aaAD	aaBC	aaBD	abAC	abAD	abBC	abBD	bbB
aaaaAACC	1								
aaaaABCC	1/2		1/2						
aaaaBBCC			1						
aaaaAACD	1/2	1/2							
aaaaABCD	1/4	1/4	1/4	1/4					
aaaaBBCD			1/2	1/2					
aaaaAADD		1							
aaaaABDD		1/2		1/2					
aaaaBBDD				1					
aaabAACC	1/2				1/2				
aaabABCC	1/4		1/4		1/4		1/4		
aaabBBCC			1/2				1/2		
aaabAACD	1/4	1/4			1/4	1/4			
aaabABCD	1/8	1/8	1/8	1/8	1/8	1/8	1/8	1/8	

Familial Type	Offspring Phenogenotypic Frequency								
	aaAC	aaAD	aaBC	aaBD	abAC	abAD	abBC	abBD	bbBD
aaabBBCD			1/4	1/4			1/4	1/4	
aaabAADD		1/2				1/2			
aaabABDD		1/4		1/4		1/4		1/4	
aaabBBDD				1/2				1/2	
aabbABCD					1/4	1/4	1/4	1/4	
aabbBBCD							1/2	1/2	
aabbABDD						1/2		1/2	
aabbBBDD								1	
abbbABCD					1/8	1/8	1/8	1/8	1/2
abbbBBCD							1/4	1/4	1/2
abbbABDD						1/4		1/4	1/2
abbbBBDD								1/2	1/2
ababAACC	1/4				1/2				1/4
ababABCC	1/8		1/8		1/4		1/4		1/4
ababBBCC			1/4				1/4		1/2
ababAACD	1/8	1/8			1/4	1/4			1/4
ababABCD	1/16	1/16	1/16	1/16	1/8	1/8	1/8	1/8	1/4
ababBBCD			1/8	1/8			1/4	1/4	1/4
ababAADD		1/4				1/2			1/4
ababABDD		1/8		1/8		1/4		1/4	1/4
ababBBDD				1/4				1/2	1/4
bbbbBBDD									1

Figure 10.4. Cultural and biological transition parameters

We assume both genotypic and phenotypic fitness, as well as their interactions, are multiplicative. Thus, the fitnesses of the nine phenogenotypes
that can appear with positive frequency are as shown in Figure 10.5. The
resulting system consists of eight equations in eight of the nine offspring
phenogenotypes. One offspring phenogenotype is dropped, since the sum
of phenogenotype frequencies must be unity.

It is straightforward to check that there are five pure equilibria. These
are **aaAC**, in which all agents internalize both the altruistic and fitness-
enhancing norms, **aaAD**, in which only the altruistic norm is internalized,
aaBC, in which only the fitness-enhancing norm is internalized, **aaBD**,
in which agents carry the gene for internalization of norms, but no norms
are in fact internalized, and **bbBD**, in which internalization is absent, and
neither altruistic nor fitness-enhancing norms are transmitted from parents
to offspring. A check of the eigenvalues of the Jacobian matrix shows that
the **aaAD** and the **aaBD** equilibria are unstable.

Individual Phenogenotype	Individual Fitness	Individual Phenogenotype	Individual Fitness
aaAC	(1-u)t*	aaAD	(1-u)(1-s)
aaBC	(1-u)(1+t)	aaBD	(1-u)
abAC	(1-u)t*	abAD	(1-u)(1-s)
abBC	(1-u)(1+t)	abBD	(1-u)
bbBD	1		

Figure 10.5. Payoffs to nine phenogenotypes

The Jacobian of the altruistic internalization equilibrium **aaAC** has eigenvalues

$$\left\{0, 1, \frac{1}{2(1+t)}, \frac{1}{1+t}, \frac{1-\gamma}{1-s}, \frac{1-\gamma}{2t*}, \frac{1-v}{4t*}, \frac{1-v}{4t*}\right\}.$$

It is easy to check that the linearization of this equilibrium is stable if $s < \gamma$, since the unit root is semisimple. We cannot conclude that the equilibrium itself is necessarily stable for all parameters s, t, u, γ, and v satisfying the above inequalities. However, many simulations under varying parameter sets have failed to turn up an instance of instability.[9]

The eigenvalues of the Jacobian for the **aaBC** equilibrium are

$$\left\{0, 1, \frac{1}{2(1+t)}, \frac{1}{1+t}, \frac{1}{2(1+t)^2(1-u)^2}, \frac{1}{2}(1-s), \frac{1-s}{4(1+t)}, 1+\gamma-s\right\}.$$

Thus **aaBC** is stable when $\gamma < s$, and unstable when the opposite inequality holds.

Finally, the nonzero eigenvalues of the Jacobian for the **bbBD** equilibrium are

$$\left\{1-u, \frac{1}{2}(1+t)(1-u), \frac{1}{4}t*(1-u), \frac{1}{2}(1-s)(1-u)\right\}.$$

As in the single phenotype case, this is unstable if $u < 0$ or $(1+t)(1-u) > 2$, and is stable if either of the opposite inequalities hold. Moreover, it can be shown that adding assortative mating leads to the instability of the **bbBD** equilibrium under the same conditions as in the single phenotype case, shown in equation (10.1).

[9]When allele **a** is incompletely dominant, the unit root disappears, so the **aaAC** equilibrium is unambiguously stable.

In sum, under plausible conditions, one internalization equilibrium is stable—the altruism equilibrium when $\gamma > s$ and the nonaltruism equilibrium when $s > \gamma$. Since we expect s to be small, whereas the ubiquity of extraparental transmission favors a high γ, the altruism equilibrium appears the more plausible of the two. Under not implausible conditions, either a high return to the norm, assortative mating of agents that internalize, or pleiotropism in the form $u < 0$, the only stable equilibrium of the system involves internalization. The dynamics, which we present below, support this conclusion.

10.4   Copying Phenotypes: The Replicator Dynamic

The above models of cultural transmission have been strongly criticized in the literature for suggesting that agents adopt norms *independent of their perceived payoffs*. In fact, people do not always blindly follow the norms that have been inculcated in them, but at least at times treat compliance as a strategic choice (Wrong 1961; Gintis 1975). The "oversocialized" model of the individual developed above may be improved by adding a phenotypic copying process reflecting the fact that agents shift from lower to higher payoff strategies. We represent this process as a *replicator dynamic* (Taylor and Jonker 1978; Samuelson 1997; Nowak and Sigmund 1998; Gintis 2000). In the current context, there are four phenotypes whose relative fitness ranks them as **BC** > **AC** > **BD** > **AD**, and only agents with a copy of the **a** allele will copy another phenotype, since only such types are capable of internalizing a norm, and noninternalizers will not desire to mimic internalizers.

We assume an agent with the **a** allele and phenotype **XY** meets an agent of type **WZ** with probability αp_{WZ}, where p_{WZ} is the fraction of the population with phenotype **WZ**, and switches to **WZ** if that type has higher fitness than **XY**. The parameter α is a measure of the strength of the tendency to shift to high-payoff phenotypes.

It is easy to see that adding a replicator dynamic does not change the single phenogenotype equilibria. Checking the eigenvalues of the Jacobian matrix we find that the **aaAD** and **aaBD** equilibria remain unstable, and the replicator dynamic does not affect the conditions for stability of the unnormed equilibrium **bbBD**. The condition $\gamma > s$ for stability of the altruism equilibrium **aaAC** now becomes

$$\alpha < \frac{\gamma - s}{1 - \gamma}, \tag{10.2}$$

so a sufficiently strong replicator dynamic can undermine the stability of the **aaAC** equilibrium.[10] The condition $s > \gamma$ for stability of the nonaltruism internalization equilibrium **aaBC** when the replicator dynamic is included now becomes

$$\alpha > \frac{\gamma - s}{1 + \gamma - s},$$

and this equilibrium is unstable when the reverse inequality holds. Thus in this case, $s > \gamma$ continues to ensure that **aaBC** is stable, but now for sufficiently large α this equilibrium is stable even when $\gamma > s$.[11]

In sum, adding a replicator dynamic changes the stability properties of the model in only one important way: a sufficiently strong replicator process can render the nonaltruistic yet internalized equilibrium **aaBC**, rather than the altruistic equilibrium **aaAC**, stable. Realistically, while the replicator process is key to understanding social change, in general we expect this to be a relatively weak force, and certainly too weak to undermine altruistic norms, unless they incur substantial fitness costs. Norms do not come labeled "altruistic norm," "instrumental practice." Rather, they are inextricably intermingled. It is quite common to believe that immoral acts lead to disease, for instance, just as does poor hygiene. Human psychology conditions us to be uncritical absorbers of hosts of beliefs and values, only a small fraction of which can be seriously questioned by an individual member of society.

10.5   Why is Altruism Predominantly Prosocial?

Norms may be either prosocial or antisocial. Indeed, there are many accounts of social norms that are severely socially costly, such as those involving invidious displays of physical prowess (Edgerton 1992). The reason for the feasibility of antisocial norms is that once the internalization

---

[10]This model also has a semisimple unit root, so stability was checked by extensive simulations. A similar result holds for the haploid model.

[11]This model also has a semisimple unit root, so stability was checked by extensive simulations. A similar result holds for the haploid model, except the relevant inequality for stability of **aaBC** becomes the much stronger, and hence implausible, inequality $\alpha > \frac{\gamma - s}{s(1 + \gamma - s)}$.

gene has evolved to fixation, there is nothing to prevent group-harmful phenotypic norms, such as our **A**, from also emerging, provided they are not excessively costly in comparison with the strength of the replicator process. The evolution of these phenotypes directly reduces the overall fitness of the population.

Yet as Brown (1991) and others have shown, there is a tendency in virtually all successful societies for cultural institutions to promote prosocial and eschew antisocial norms. The most reasonable explanation for the predominance of prosocial norms, as explained in Chapter 1, is *gene-culture coevolutionary selection*: societies that promote prosocial norms have higher survival rates than societies that do not (Parsons 1964; Cavalli-Sforza and Feldman 1981; Boyd and Richerson 1985,1990; Soltis et al. 1995). Note that the usual arguments against the plausibility of genetic group selection do not apply to our model. This is because altruism is (a) phenotypic, and (b) "hitchhikes" on the fitness-enhancing phenotypic norm. Because altruism is phenotypic, and because a high degree of cultural uniformity can be maintained within groups, a high ratio of between-group to within-group variance on the phenotypic trait is easily maintained, and hence high-payoff groups quickly outpace low-payoff groups. Because altruism hitchhikes, the mechanism that generally undermines group selection, a high rate of intergroup migration (Maynard Smith 1976; Boorman and Levitt 1980), does not undermine internalization, as long as altruistic individuals adopt the **A** norms of the groups to which they migrate.

To test this argument, I created an agent-based model of society with the following characteristics (the specific assumptions made are not critical, unless otherwise noted). The society consists of 256 groups, each initially comprising 100 members, arranged spatially on a torus (a 16×16 grid with the opposite edges identified). Each group was seeded with 10 **aaAC** types, 10 **aaBC** types, 74 **bbBD** types, and one each of the other possible types. In all groups, $t = 0.3$ and $u = 0.05$. Each group was then randomly assigned a value of γ between 0 and 0.60, a value of α between 0 and 0.5, a value of s between 0.01 and 0.10, and a value of ζ (the degree of assortative mating) between 0 and 0.70. Each group was also assigned an **A** phenotype with a fitness effect between -1 and 1, such that if a group's **A**-fitness effect is q, and if a fraction f of the group exhibit the **A** phenotype, then each member of the group has its fitness augmented by fq.

In each round, for each of the 256 groups, I simulated the model as described in the previous sections, and updated the frequencies of the various types in each group, according to the fitness effect of their **A** phenotype and the fraction of the group that exhibits this phenotype. A fraction of each group (typically 5%) then migrated to a neighboring group. If altruistic migrants adopt the **A** norm of their new groups, migration never undermines the stability of the altruism equilibria. To be conservative, we assume here that agents take their genes with them to a new group, and they take their **C** phenotype to this new group (since **C** is the same for all groups), but they abandon their **A** phenotype when they migrate (e.g., an **aaAC** type becomes a **aaBC** type in the new group). This assumption is maximally geared to undermine the altruism phenotype since immigrants never exhibit altruism. We then allow for some random drift in the individual group parameters s, α, γ, and ζ, as well as the payoff of altruism to the group, q.

Our final modeling assumption is that when group size drops below a minimum (generally I set this to zero or ten agents), it is replaced by a copy of a randomly chosen other group.

I ran this model many times with varying numbers of rounds, and varying the parameters described above. The system always stabilized by 100 periods, and the specific assumptions concerning the parameters were never critical. The following conclusions hold for these simulations:

a. Groups exhibiting the noninternalization equilibrium **bbBD** were quickly driven from the population, except under the joint assumptions that altruistic agents become nonaltruistic when they migrate, and the migration rate is over 20%;

b. The equilibrium fraction of groups for which the noninternalization equilibrium was stable was highly variable and dependent upon the specifics of the initial distribution of groups. Thus in equilibrium, though all groups were near the internalization equilibria (**aaAC** and **aaBC**), in some this was globally stable and in others, only locally;

c. The equilibrium fraction of the population exhibiting the altruistic phenotype was greater than 85% (the mean at the start of each simulation was 10%);

d. All but the highest prosocial **A** phenotypes were eliminated from the population, so that the mean fitness effect of the altruistic phenotypes was greater than 0.9 (the maximum possible was 1.0, and the mean at the start of the simulation was approximately zero). In particular, no groups with antisocial norms ever survived in equilibrium;

e. The mean level of extraparental socialization, γ was also very high, being at least 45% (the mean at the start of the simulation was 30%);

f. The mean strength of the replicator dynamic, α, was 9% (the mean at the start of the simulation was 25%). This shows that while a higher α helps individuals, because they are then more likely to move to high-fitness phenotypes, it hurts the groups they are in, and on balance lowers group fitness;

g. The altruistic equilibrium was attained as long as the initial average assortative mating probability ζ was at least 12.5%. An increase in the rate of assortative mating led to more globally stable altruism equilibria, but had no measurable effect on the equilibrium values of other variables;

h. The emergence of the **aaAC** agents and the elimination of antisocial norms were both due to population growth alone. The extinction and replacement of groups by more successful groups accounted only for the change in the frequency of γ and α (extinctions began to occur after 20 rounds, and more than 1500 extinctions typically occurred in a 100 round simulation).

These simulations thus strongly support the basic arguments of this chapter. In particular, a high level of migration does not undermine the altruistic equilibrium, since most of the effects occur on the cultural rather than the genetic level. Moreover, plausible patterns of population growth and migration account for the prosociality of the altruism phenotype **A**. The critical assumption that drives the model is simply that there is a fitness-enhancing effect of the selfish **C** phenotypic norm sufficiently strong to ensure that **C** can invade a population of **D** agents. The ability of the altruism phenotype **A** to "hitchhike" on **C** is quite robust.

10.6   The Power of Altruistic Punishment

Consider the following social dilemma. Each member of a group can either cooperate or shirk. Shirking costs nothing, but adds nothing to the payoffs of the group members. Cooperating costs $s^* > 0$, but contributes an amount $f^* > s^*$ shared equally by the other members. Selfish individuals will always shirk in this situation, so the potential gains from cooperating will be forgone. If the situation is repeated sufficiently frequently with the same players, and if the group is sufficiently small, cooperation can be sustained even with selfish players (Trivers 1971; Axelrod and Hamilton 1981). However, with large groups and/or infrequent repetition, universal shirking is virtually inevitable (Boyd and Richerson 1988), as has been confirmed repeatedly in experiments with humans (Ledyard 1995).

Given the potential gains to society of people internalizing the altruistic norm **A** = "always cooperate" in the above situation, the absence of this norm in society suggests that the cost s^* is simply too high to sustain as an equilibrium of the **aaAC** form. However, experimental results (Fehr and Gächter 2002) and ethnographic data (Boehm 1993, 1999), not to mention everyday observation, suggests that the threat of being punished by other group members for shirking may serve to sustain cooperation where the internalized value of cooperating does not.

Punishment can succeed where the norm of cooperation cannot because the expected cost per period of punishing shirkers is typically much smaller than the cost of cooperating. This is because (a) punishment such as shunning and ostracizing are inherently low cost, yet effective when directed by large numbers against a few transgressors; and (b) the punishment need be carried out only when shirking occurs, which is likely to be infrequent in comparison with the number of times cooperation must be carried out.

Nevertheless, altruistic punishment likely has strictly positive cost, so a selfish individual still will refrain from engaging in this activity. While a genetic group selection model can explain the evolutionary stability of altruistic punishment (Gintis 2000), such models are sensitive to group size and migration rates (Eshel 1972; Rogers 1990). The gene-culture coevolutionary model presented in this chapter, by contrast, suffers less from these problems.

To see this, suppose a fraction p of a group with n members consists of altruistic punishers. To prevent intentional shirking by selfish agents, each must be prepared to inflict a punishment s^*/pn on a shirker. Suppose a fraction q of the group nevertheless shirks (or perhaps is simply perceived to shirk under conditions of imperfect information). Then the total amount of punishment per altruistic punishment is $s = qs^*/p$. If p is large (as in our simulations) and q is small (as is likely to be the case except under extreme conditions, since no one has an incentive to shirk), then this value of s will be close to zero for each altruistic punisher. But then the altruism equilibrium will be stable according to equation (10.2), even when it would be violated for $s = s^*$.

Since the fitness costs of altruistic punishment are low, a replicator dynamic is unlikely to render the altruism equilibrium unstable in this case. Moreover, there is evidence that altruistic acts serve as costly signals of agent fitness (Gintis et al. 2001), in which case the altruistic phenotype cannot be undermined by the tendency to shift from lower to higher payoff phenotypes.

10.7   Final Considerations

We have developed a plausible model of altruistic cooperation and punishment that does not depend on repeated interactions, reputation effects, or multilevel genetic selection. The latter obtains because there is no net within-group penalty to either the altruistic gene or the altruistic norm, even though there is a penalty to individuals carrying the gene and behaving according to the norm.

One shortcoming of our model is that payoffs are assumed constant, whereas in many cases we would expect payoffs to be frequency dependent, as when group members are engaged in a noncooperative game. For instance, the payoff to being self-interested may increase when agents are predominantly altruistic. Gintis (2003b) shows that such a situation gives rise to a heterogeneous equilibrium, in which both altruists and self-interested types participate. Since the payoffs to the two types are equal in equilibrium, once again we can dispense with multilevel selection in specifying an equilibrium with a positive level of altruism.

There are two objections that biologists naturally raise to this model of altruism. First, if the **C** norm is individually fitness-enhancing while the **A** is not, why is there not a genetic mutation (for instance at another genetic locus) that allows the individual to distinguish between altruistic and fitness-enhancing behaviors, and hence to eschew the former? The answer is that **A**- and **C**-type behaviors are exhibited only on the phenotypic level, and hence have no clear inherent characteristics according to which such a gene could discriminate. Moreover, if such an inherent characteristic does exist for a particular **A**-type norm, that type would be driven to extinction. But there are so many varieties of cultural norms that others, unaffected by this mutation, would arise to replace the one to which people have become "immune." Finally, we should note that generally the *degradation* of the genetic capacity to discriminate is much more likely than the *emergence* of such a capacity, since the latter, being complex in the case of **A**-type norms, requires the existence of a sequence of one-point mutations that are each fitness-enhancing, finally leading to the capacity to discriminate. This is implausible in the current context.

A second objection to our model of altruism is that we have as-sumed rather than provided an explanation of why the internalization of norms—having a programmable objective function—is individually fitness-enhancing. Why would an agent gain from an altered objective function when he always has the option of obeying a norm when it is his interest to do so, and violating the norm when it is not? However, agents do not maximize fitness, but rather an objective function that is itself subject to selection. In a constant environment, this objective function will track fitness closely. In a changing environment, natural selection will be too slow, and the objective function will not track fitness closely. Cultural transmission and the ensuing increase in social complexity produced such a rapidly changing environment in human groups. Imitation (the replicator dynamic) will not correct this failure, because agents copy objective-function-successful, not fitness-successful, strategies. In this situation, there are large fitness payoffs to the development of a *non-genetic mech-anism for altering the agent's objective function*, together with a *genetic mechanism for rendering the individual susceptible to such alteration*. Internalization of norms, which may be an elaboration upon imprinting and imitation mechanisms in nonhuman animals, doubtless emerged by virtue of its ability to alter the human objective function in a direction conducive to higher fitness. There is not to my knowledge a confirmed instance of internalization in nonhuman animals. This may in part be due to the fact that the relevant research has not been carried out. Yet there are obvious reasons to doubt that internalization might be important, because cultural transmission in nonhuman animals is relatively rudimentary.

It would be a serious mistake to conclude that the socialization process in humans is sufficiently powerful to permit *any* pattern of norms to be promulgated by internalization. For instance, many have suggested that it would be better if people acted on the principle of contributing to society according to one's ability, and taking from society according to one's needs. Whatever the moral standing of such a principle, no society has lasted long when its incentives have been based on it. Our model suggests one reason why such a principle might fail: the operation of the replicator dynamic. In this case, the payoff to defectors from the norm is simply too high to prevent its erosion.

There may be other criteria determining what types of altruistic norms are likely to emerge from the gene-culture coevolutionary process described in this chapter. For instance, behaviors that are altruistic, but very similar to ones that are personally fitness-enhancing may be relatively easy to internalize; e.g., since it is generally fitness-enhancing to speak truthfully, it may be relatively easy to move the decision to speak truthfully from the realm of instrumental calculation to that of the moral realm of right and wrong. Similarly, altruistic punishment may be widespread because it is generally prudent to develop a reputation for punishing those who hurt us, and it is a short step to turning this prudence into a moral principle.

11

The Economy as Complex Dynamical System

> I have no idea whether Adam Smith's invisible hand holds for the real world. But then no one else does either. No mathematical theory exists to justify it. Even simple models…can exhibit dynamical behavior far more complex than anything found in classical physics or biology.
>
> Donald Saari

> Subtle is the Lord, but malicious He is not.
>
> Albert Einstein

> **C:** How did you discover that the general equilibrium model is stable?
>
> **HG:** I thought I would simulate the model on my computer and show that it exhibits chaotic price movements as the standard theory says.
>
> **C:** What happened?
>
> **HG:** It always converged to market equilibrium.
>
> **C:** What did you do wrong?
>
> **HG:** Rather than having an auctioneer, I allowed all traders to set their own prices.
>
> Choreographer interview

Walras (1874) developed a general model of competitive market exchange. In the 1950s several researchers contributed to showing that a market-clearing equilibrium for Walras' model exists under quite general conditions (Debreu 1954; Arrow and Debreu 1954; Gale 1955; Nikaido 1956; McKenzie 1959).

Walras was aware that his model required a price adjustment mechanism that would ensure stability of equilibrium. He considered the key force leading to equilibrium to be *market competition*, which he thought would result in the continual updating of prices by traders until equilibrium was attained. However, he believed that a model where economic agents individually update their prices would be analytically intractable, whereas a simple centralized model of price adjustment, which he called the *tâtonnement* process, would more easily lend itself to a proof of the stability of equilibrium. He was wrong.

The stability of the Walrasian economy became a central research focus in the years following the existence proofs (Arrow and Hurwicz 1958,

1959, 1960; Arrow et al. 1959; Nikaido 1959; McKenzie 1960; Nikaido and Uzawa 1960). Following Walras' tâtonnement process, these models assumed that there is no production or trade until equilibrium prices are attained, and out of equilibrium, there is a price profile shared by all agents, with the time rate of change of prices being a function of excess demand.

These efforts at proving stability were unsuccessful (Fisher 1983). Indeed, Scarf (1960) and Gale (1963) complicated the situation by providing simple examples of unstable Walrasian equilibria under a tâtonnement dynamic. Moreover, Sonnenschein (1973), Mantel (1974, 1976), and Debreu (1974) showed that any continuous function, homogeneous of degree zero in prices, and satisfying Walras' Law, is the excess demand function for some Walrasian economy. These results showed that no general stability theorem could be obtained based on the tâtonnement process. Subsequent analysis showed that *chaotic price movements* are the generic case for the tâtonnement adjustment processes (Saari 1985; Bala and Majumdar 1992). This explains Donald Saari's quote at the head of this chapter.

A novel approach to the dynamics of large-scale social systems, evolutionary game theory, was initiated by Maynard Smith and Price (1973), and adapted to dynamical systems theory in subsequent years (Taylor and Jonker 1978; Friedman 1991; Weibull 1995). The application of these models to economics involved the shift from *biological reproduction* to *behavioral imitation* as the mechanism for the replication of successful agents (Mandel and Gintis (2014, 2016).

This chapter begins with a compact exposition of general market equilibrium, and provides evolutionary foundations for competitive market exchange. We treat a decentralized market economy as the *stage game* of an evolutionary process in which each agent in each period produces some mix of goods that he must trade to obtain the various goods he consumes. An agent's trade strategy consists of a set of *private prices* for the goods he produces and the goods he consumes, a trade being acceptable if the value of goods received is at least as great as the value of the goods offered in exchange according to these private prices.

11.1 The General Equilibrium Model Explained

Consider an economy consisting of individuals and firms. Individuals own labor and capital goods, which they supply to firms. These firms produce marketable goods which they sell to individuals. Individuals consume whatever they purchase from firms. We could allow firms to own capital goods as well, but this complicates the equations and adds nothing to the analysis. The economy is in *market equilibrium* when the vector of prices for labor, capital, and goods equates supply and demand in the labor market, the capital market, and all goods markets, assuming firms maximize profits and individuals maximize their consumption utility.

We model a highly simplified version of this economy in which there are only two goods, which we will call *apples a* and *nuts n*, and two individuals, *x* and *y*. We also assume there is a single type of labor (l). Suppose l_x, l_y and k_x, k_y are the amounts of labor and capital owned by individuals *x* and *y*, l_a^x and k_a^x are the amounts of labor and capital supplied by individual *x* to firms producing apples, l_a^y and k_a^y are the amounts of labor and capital supplied by individual *y* to firms producing apples, and l_n^x, k_n^x and l_n^y, k_n^y are the labor and capital supplied to firms producing nuts by individuals *x* and *y*. Then we have the four equations

$$l_a^x + l_n^x = l_x \tag{11.1}$$

$$l_a^y + l_n^y = l_y \tag{11.2}$$

$$k_a^x + k_n^x = k_x \tag{11.3}$$

$$k_a^y + k_n^y = k_y, \tag{11.4}$$

which say that the total amount of labor and capital goods demanded by firms to use in production equals the total amount of these factors supplied by individuals.

Now suppose the wage rate is w and the interest rate, which is the price for renting one unit of the capital good for one production period, is r. Also, suppose the price of the market goods are p_a for apples and p_n for nuts, and individual *x* consumes x_a of apples and x_n of nuts, while individual *y* consumes y_a of apples and y_n of nuts. Finally, we assume individual *x* owns share α_a of the net profit of apple-producing firms and share α_n of the net profit of nut-producing firms, while individual *y* owns shares $1 - \alpha_a$ and $1 - \alpha_n$, respectively.

Then if m_x and m_y are the incomes of individuals *x* and *y* from supplying labor and capital goods to firms, we have the following two equations:

$$m_x = \alpha_a \pi_a + \alpha_n \pi_n + w l_x + r k_x \tag{11.5}$$

$$m_y = (1 - \alpha_a)\pi_a + (1 - \alpha_n)\pi_n + w l_y + r k_y, \tag{11.6}$$

where π_a and π_n are the net profits of apple-producing and nut-producing firms.

The next equations are production functions for the individual goods and market goods firms. They say that each good is produced by using capital goods and labor.

$$g_a(l_a^x + l_a^y, k_a^x + k_a^y) = x_a + y_a, \tag{11.7}$$
$$g_n(l_n^x + l_n^y, k_n^x + k_n^y) = x_n + y_n, \tag{11.8}$$

where $g_a(l_a, k_a)$ and $g_n(l_n, k_n)$ are the aggregate production functions for the apples and nuts sectors, respectively, and x_a, x_n, y_a, and y_n are the apples and nuts purchased by individuals x and y.

We assume firms maximize profits, given by

$$\pi_a = p_a(x_a + y_a) - (w(l_a^x + l_a^y) + r(k_a^x + k_a^y)) \tag{11.9}$$
$$\pi_n = p_n(x_n + y_n) - (w(l_n^x + l_n^y) + r(k_n^x + k_n^y)). \tag{11.10}$$

Profit maximization gives four first-order conditions

$$\frac{\partial g_a}{\partial l_a} = \frac{w}{p_a} \qquad \frac{\partial g_n}{\partial l_n} = \frac{w}{p_n} \tag{11.11}$$

$$\frac{\partial g_n}{\partial k_a} = \frac{r}{p_a} \qquad \frac{\partial g_n}{\partial k_n} = \frac{r}{p_n}. \tag{11.12}$$

We assume individuals have utility function $u^x(x_a, x_n)$ and $u^y(y_a, y_n)$, which they maximize subject to their income constraints (11.5) and (11.6). Maximizing utility given these income constraints gives four additional equations

$$\frac{1}{p_a}\frac{\partial u^x}{\partial x_a} = \frac{1}{p_n}\frac{\partial u^x}{\partial x_n} \tag{11.13}$$

$$\frac{1}{p_a}\frac{\partial u^y}{\partial y_a} = \frac{1}{p_n}\frac{\partial u^y}{\partial y_n} \tag{11.14}$$

$$m_x = p_a x_a + p_n x_n \tag{11.15}$$

$$m_y = p_a y_a + p_n y_n. \tag{11.16}$$

Finally, we can normalize the nominal price level p_n to unity, and we assume that competition among firms equates profits in the two sectors. This gives

$$p_n = 1, \qquad (11.17)$$

$$\pi_a = \pi_n. \qquad (11.18)$$

We call a situation in which the price structure (p_a, p_n, w, r) is such that all these equations hold simultaneously and all prices are positive a *market equilibrium*. In this system, α, l_x, l_y, k_x, and k_y are parameters representing the structure of ownership in the economy. There remain twenty variables to be determined: l_a^x, l_a^y, l_n^x, l_n^y, k_a^x, k_a^y, k_n^x, k_n^y, x_a, y_a, x_n, y_n, π_a, π_n, m_x, m_y, p_a, p_n, w, and r. There are also twenty equations, expressed in (11.1)–(11.18). The equality in the number of equations and unknowns generically determines a locally unique equilibrium, but there is no general guarantee that prices and quantities will be nonnegative in this solution. However, the appropriate assumptions concerning the shape of the production function and utility functions will guarantee the existence of a market equilibrium, along the lines of Debreu (1952) and Arrow and Debreu (1954). The conditions that make this possible, roughly speaking, are that consumers have concave preferences (declining marginal utility) and firms have convex production functions (declining marginal productivity).

11.2   The Fundamental Theorems of Welfare Economics

To state the famous First Fundamental Theorem of Welfare Economics we say an individual is *locally nonsatiated* if he would be slightly worse off if his income were slightly lower. An allocation of labor and capital to firms and goods to individuals is called *Pareto optimal* if any reallocation that makes one individual better off must make at least one individual worse off. This is of course the standard definition of efficiency in economic theory (Mas-Colell et al. 1995). We then have

THEOREM 11.1 **First Theorem of Welfare Economics***: If all agents are locally nonsatiated, then every market equilibrium is Pareto optimal.*

The proof of this theorem is extremely simple, but not worth going through here. The interested reader can find a proof on Wikipedia. The Second Fundamental Welfare Theorem says that every Pareto-optimum can be achieved by an initial redistribution of ownership followed by market exchange. The main conditions for this are also quite reasonable but the proof uses rather sophisticated separating hyperplane or fixed point theorems. By the way, and to avoid confusion, the term "welfare" in economics does not mean subsidies to the poor, but rather individual and social well-being in general.

The Second Fundamental Theorem is of no practical use because it is not politically or ethically feasible to redistribute ownership arbitrarily to attain equity goals. Moreover, a one-time egalitarian redistribution would quickly return to its naturally inegalitarian state in subsequent periods because in any realistic economy, a market equilibrium is an absorbing state of an ergodic Markov process, as described in our analysis below. Of course, continual arbitrary redistributions are possible (if neither ethical nor politically feasible), but these interfere with incentives so lead to severe economic inefficiencies.

The First Fundamental Theorem, by contrast, is extremely important not because it accurately reflects actual economic conditions, but rather because it is instructive to understand when it does not, and why. The general equilibrium model is practically relevant because, as we show below, the economy is normally rather close to equilibrium, although it can experience rather dramatic excursions far from equilibrium for some number of time periods, as in the case of financial bubbles and persistent stagnation.

The basic question of when unregulated markets of sort described above are the most effective instruments of economic efficiency was worked out in the post-World War II period, and remains valid today, a half century later (Musgrave 1959; Atkinson and Stiglitz 1980). This theory is called the *theory of market failure*. There are four types of market failure:

1. **Increasing Returns to Scale:** In some production sectors, the optimal efficient firm size is so large that either efficiency or market competition must be sacrificed. For instance, a city's water supply may be most efficiently supplied using one large reservoir and a unified system of delivery and waste removal. There is simply no room for multiple firms to compete, so the service is supplied by the local government. The problem can sometimes be handled alternatively requiring firms to share the resource that accounts for increasing returns, such as railroad tracks or an electric grid.

2. **Public Goods:** Some goods are *non-exclusionary*—they are consumed equally by many or all individuals, although they may be *valued* differently by different individuals. For instance, national defense protects all equally, and many forms of public health measures affect the incidence of diseases for the entire population. Public goods must be publicly provided in most cases.

3. **Market Externalities:** Some goods are produced using technologies that release waste products into the environment at zero or low cost to the producer but that impose high costs on everyone else. Economic theory suggests that the costs imposed by these effluents be charged to producers, an example being carbon taxes.

4. **Quality Goods:** There are many market sectors in which the quality of a product cannot be ascertained until after purchase, and in which reputation effects are not sufficient to ensure a minimum quality level. We analyzed this situation in Section 4.6. In such sectors, quality can be maintained only by legal regulation. For instance, most countries have health standards for restaurants and quality standards for hotels that prevent an upstart from profiting at the expense of consumers and the high-quality firms. Similarly, professionals may be licensed (e.g., medical and legal services), and pharmaceuticals may be regulated for safety and effectiveness.

Of course, it is not feasible to correct all market failures through government intervention because regulation may be excessively costly. More important, *state failure* is quite as endemic as market failure. That is, the political dynamics that give rise to particular forms of state intervention are governed by forces that may place little value on public welfare. Small but powerful special interest groups, for instance, may agitate for interventions that benefit their members at the expense of the public. Similarly, some forms of regulation invite corruption by officials.

The general market equilibrium model is attractive not because it is realistic, but rather because it allows us to state and understand clearly the nature of market failures and the possible measures to correct them.

11.3 The Market Economy as a Dynamic Game

We construct a finite m-player game, where m is the number of agents in the economy. Each agent i's strategy π_i is his private price vector. There is an *exchange process* ξ such that in each period the strategy profile $\pi = (\pi_1, \ldots, \pi_m)$ leads to a feasible allocation $\xi_i(\pi)$ to each agent i. Under rather mild conditions on the exchange process ξ, general equilibrium market clearing equilibria are precisely the strict Nash equilibria of the game. If we then add a learning process whereby agents update their strategies by adopting those of other agents who have had superior trading success, we get a standard replicator dynamic as described in Section 10.4.

In this dynamic, the general market equilibrium prices are precisely the stable equilibria of the dynamical system.

Our stability result does not indicate how fast we can expect the convergence to equilibrium to take place, or how robust the market economic is to shocks. To address this problem Section 11.8 presents a Markov model of a simple exchange economy with many goods and many agents. The model is too complex to solve analytically, but a computer implementation of this model exhibits strong convergence to equilibrium and considerable robustness to shocks. There is good reason to believe that if we added a financial sector and other real-world economic institutions, the system would exhibit severe fragility to shocks under some realistic conditions (Gintis 2007a; Mandel et al. 2015).

11.4   The Walrasian Economy

We consider an economy with goods $G = \{1, \ldots, n\}$, and agents $A = \{1, \ldots, m\}$. Each agent i has consumption set $\mathcal{X} = \mathbf{R}^n_+$, where $\mathbf{R}_+$ is the set of nonnegative real numbers, utility function $u_i : \mathcal{X} \to \mathbf{R}_+$, and in each period produces a set of goods $e_i = (e_{i1}, \ldots, e_{in}) \in \mathcal{X}$. We denote this economy by $\mathcal{E}(u, e)$, where $u = (u_1, \ldots, u_m)$ and $e = (e_1, \ldots, e_m)$.

Agent i's production set is $E_i = \{g \in G \,|\, e_{ig} > 0\}$ and his consumption set is $C_i \subset G$, disjoint from E_i. The agent's utility u_i depends only on the goods in C_i. Formally, we have:

Assumption 1 (**Goods**): For all $i \in A$, there exists $C_i \subset G$ such that $C_i \cap E_i = \emptyset$ and $(\forall x, y \in \mathcal{X})(\forall g \in C_i, \ x_g = y_g) \Rightarrow u_i(x) = u_i(y)$.

In other words, each agent is capable of producing one set of goods and consumes a distinct set of goods. We then define the set of *buyers* of good g as $B_g = \{i \in A \,|\, g \in C_i\}$, and the set of *sellers* of good g as $S_g = \{i \in A \,|\, g \in E_i\}$.

We say an allocation $x \in \mathcal{X}^m$ of goods is *feasible* if for all $g \in G$:

$$\sum_{i=1}^m x_{ig} \leq \sum_{i=1}^m e_{ig}.$$

That is, the maximum amount of consumption of a good is the amount produced. We write $\mathcal{A}(e_1, \ldots, e_m) = \mathcal{A}(e) \subset \mathcal{X}^m$ for the set of feasible allocations. The *price space* is defined to be $\mathcal{P} = \mathbf{R}^{n-1}_+ \times \{1\}$. Good n is called the *numeraire* good and has the nominal value of unity.

Agent i's *demand* is $d_i(p, w)$, where p is his private price vector and w is his income. We assume $x_i = d_i(p, w)$ maximizes utility $u_i(x_i)$ where $\{x_i \in X | p \cdot x_i \leq w\}$; i.e., subject to his income constraint.

A feasible allocation $x \in \mathcal{A}(e)$ is an *equilibrium allocation* if there exists a price vector $p \in \mathcal{P}$ such that for all i, $x_i = d_i(p, p \cdot e_i)$. The price p is then called an *equilibrium price*. We denote the set of equilibrium price vectors by $\mathcal{P}_{equi}(u, e)$.

A feasible allocation $x \in \mathcal{A}(e)$ and a price vector $p \in \mathcal{P}$ form a *quasi-equilibrium* if for all agents i and for all $y \in X^m$, $u_i(y_i) > u_i(x_i)$ implies $p \cdot y_i > p \cdot x_i$. In other words, if an agents prefers bundle y to bundle x but he chooses x, then purchasing y must exceed his budget constraint.

We assume that utility functions satisfy the following standard conditions.

Assumption 2 (**Utility**): For all $i \in A$, u_i is continuous, its restriction to R^{C_i} is strictly concave and locally non-satiated, and $u_i(x) > 0$ only if for all $g \in C_i$, $x_g > 0$.

This assumption suffices to guarantee the existence of a quasi-equilibrium. Moreover, the strict concavity assumption implies that demand mappings are single-valued, which proves useful below. The last condition implies that the consumption of each consumption good is necessary and is fairly standard in general equilibrium theory (Mas-Colell et al. 1995).

To ensure that every quasi-equilibrium is an equilibrium allocation, it suffices to assume that at a quasi-equilibrium, agents do not receive the minimal possible income (Hammond 1993; Florenzano 2005). This condition is satisfied when all initial endowments are in the interior of the consumption set, as well as in settings with corner endowments such as those investigated in Scarf (1960) and Gintis (2007a). Formally, the assumption can be stated as follows.

Assumption 3 (**Income**): For every quasi-equilibrium (p, x), there exists an allocation y such that $p \cdot x_i > p \cdot y_i$ for all $i \in A$.

The Utility Assumption (2) and the Income Assumption (3) then imply that the economy $\mathcal{E}(u, e)$ has at least one equilibrium (Florenzano 2005). We assume they hold throughout the chapter and will further restrict attention to the generic case where the set $\mathcal{P}_{equi}$ of equilibria is finite (Balasco 2009).

11.5 Exchange Processes with Private Prices

We represent the exchange process in the economy $\mathcal{E}(u, e)$ as a game in which agents set prices in a decentralized manner and the resulting distribution of prices determines the agents' allocations. More precisely, we consider that each agent i has a strategy consisting of a *private price vector* $\pi_i \in \mathcal{P}_i = \mathbf{R}_+^{h_i}$, where h_i is the number of goods he produces or consumes. This private price vector represents the prices at which he is willing to sell the goods he supplies to the market and the maximum prices he is willing to pay for the goods he demands. We then associate to the economy $\mathcal{E}(u, e)$ the game $\mathcal{G}(u, e, \xi)$ such that:

- Each agent i has the set of prices $\mathcal{P}_i$ as strategy set, so that the set of strategy profiles for the game is $\Pi = \mathcal{P}_1 \times \cdots \times \mathcal{P}_m$.

- The game form is defined by an *exchange mechanism* $\xi \colon \Pi \to \mathcal{A}(e)$ that associates to a profile of private prices $\pi \in \Pi$ a feasible allocation $\xi(\pi) = (\xi_1(\pi), \dots, \xi_m(\pi)) \in \mathcal{A}(e)$.

- The payoff of player i is $u_i(\xi_i(\pi))$.

Many exchange processes can be represented in this way including central clearing systems such as double auctions, simultaneous multilateral exchanges as usually considered in general equilibrium models with out-of-equilibrium features such as Grandmont (1977), and sequential bilateral exchanges such as those considered by Gintis (2007a, 2012b).

In our analysis of the game $\mathcal{G}(u, e, \xi)$, we use the following notions:

- A strategy profile $\pi \in \Pi$ is *p-uniform* for price $p \in \mathcal{P}$ if, for all $i \in A$ and $g \in G_i$, one has $\pi_{ig} = p_g$.

- A p-uniform strategy profile π is a *general equilibrium price profile* if $p \in \mathcal{P}_{\text{equi}}(u, e)$.

- A strategy profile π is *p-seller uniform* if for all $g \in G$ and all $i \in S_g$, we have $\pi_{ig} = p_g$.

- For $\pi \in \Pi$, $i \in A$, $g \in G$, and $q_{ig} \in \mathbf{R}_+$ we denote by (π_{-ig}, q_{ig}) the strategy profile π' such that $\pi'_{ig} = q_{ig}$ and for $(j, h) \in A \times G_j$, $(j, h) \neq (i, g)$, we have $\pi'_{jh} = \pi_{jh}$.

- For $\pi \in \Pi$, $i \in A$, and $p \in \mathcal{P}_i$ we denote by (π_{-i}, p) the strategy profile π' such that $\pi'_i = p$ and for $j \in A$, $j \neq i$, $\pi'_j = \pi_j$.

11.6   Strict Nash Equilibria and Stability

We will specify a broad class of exchange processes ξ for which general equilibrium price profiles coincide with strict Nash equilibria of the game $\mathcal{G}(u, e, \xi)$. A strategy profile $\pi \in \Pi$ is a *strict Nash equilibrium* of $\mathcal{G}(u, e, \xi)$ if for all $i \in A$ and all $p \in \mathcal{P}_i$ such that $p \neq \pi_i$, we have $u_i(\xi_i(\pi)) > u_i(\xi_i(\pi_{-i}, p))$.

Our focus on strict Nash equilibrium is motivated by its central role in the theory of learning in games. In evolutionary game theory (Weibull 1995; Gintis 2009b) most learning processes, including the replicator dynamic, are monotone dynamics in the sense that players tend to switch from worse to better performing strategies. A key result is that for monotone dynamics, strict Nash equilibria of the stage game are precisely the asymptotically stable equilibria (Weibull 1995). Hence, if they can be identified with strict Nash equilibria of the game $\mathcal{G}(u, e, \xi)$, general market equilibria of the economy $\mathcal{E}(u, e)$ will be precisely the asymptotically stable states for any monotone dynamic. In particular, this sheds light on the convergence to general equilibrium observed in the simulations in Gintis (2007a) and Gintis (2012b) where agents update their prices by imitating those of more successful peers. Indeed there is a well-established relationship between these stochastic imitation models and the replicator dynamic (Benaim and Weibull 2003). We explore this relationship in Section 11.8.

11.7   The Characterization of Stable Exchange Processes

To analyze the relationships between the behavior of the agents in the economy $\mathcal{E}(u, e)$ and the game $\mathcal{G}(u, e, \xi)$, we first extend the notion of demand to our framework with private prices. An agent's demand may depend not only on his own private prices, but also on the prices posted by his potential transactors. For instance, a buyer of good g may sample a subset of sellers of good g who post their selling prices, and choose his demand for g based on the information received. To subsume all possible cases under a common framework, we assume that each agent i has an *extended demand function* $\delta_i : \Pi \times \mathbf{R}_+ \to \mathcal{X}$, that associates to a strategy profile $\pi \in \Pi$ and an income $w \in \mathbf{R}_+$, a demand $\delta_i(\pi, w)$.

Because trade takes place out of equilibrium, realized income, which depends on a particular trading history, cannot be determined *ex ante*. So we assume that agents choose strategies based on *expected* income, which depends on the agent's information concerning the distribution of private prices for the economy. For instance, an agent may sample the selling and buying prices of a number of other agents before specifying his expected income. Hence, we represent the income expected by seller $i \in S_g$ on good g market as a function $w_{ig} : \Pi \to \mathbf{R}_+$ and his total expected income by $w_i = \sum_{g \in E_i} w_{ig}$.

When the strategy profile is π, we refer to $\delta_i(\pi, w(\pi))$ as the *extended demand* (if there is no ambiguity) of agent i. We say there is *excess demand* or *excess supply* for good g at strategy profile π if $\sum_{i \in B_g} \delta_{ig}(\pi, w(\pi)) > \sum_{j \in S_g} e_{jg}$, or $\sum_{i \in B_g} \delta_{ig}(\pi, w(\pi)) < \sum_{j \in S_g} e_{jg}$, respectively.

We assume that when the strategy profile is uniform and the extended demand and income functions coincide with their general equilibrium counterparts:

Assumption 4 (**Price Consistency**): For all agents i and all $\pi \in \Pi$,

1. $w_{ih}(\pi) \leq \pi_{ih} e_{ih}$, with equality when π is p-uniform;
2. if π is p-uniform, then $\delta_i(\pi, w) = d_i(p, w)$ for all w.

Second, we assume that at a general equilibrium price profile, the corresponding equilibrium allocation prevails:

Assumption 5 (**Demand Rationality**): If $\pi \in \Pi$ is a p-uniform general equilibrium price profile, then for all agents i and all prices $q \in \mathcal{P}_i$, $\xi_i(\pi) = d_i(p, p \cdot e_i)$.

This assumption is a minimal efficiency requirement on the exchange process without which no relation could be established between the equilibrium strategy profiles of $\mathcal{G}(u, e, \xi)$ and the equilibria of the economy $\mathcal{E}(u, e)$.

Assumption 6 (**Constraint**): If $\pi \in \Pi$ is a p-uniform general equilibrium price profile, then for all agents i and all prices $q \in \mathcal{P}_i$, then if $q > p_i$ and $w_i(\pi_{-i}, q) \leq p \cdot e_i$, then $u_i(\xi_i(\pi_{-i}, q)) < u_i(d_i(p, p \cdot e_i))$.

This assumption ensures that, at a general equilibrium price profile, the exchange mechanism is consistent with the agent's demand. Namely, there can be a profitable deviation from a general equilibrium price profile only if there is an increase of income or a decrease in prices.

We now turn to the constraints private prices place on trading out of equilibrium. A private price represents the maximum price a buyer is willing to pay for one unit of good and conversely the minimum price a seller is willing to accept for one unit of good. Hence, the least constraint one can impose on the trading mechanism is that buyers whose private price for a good g is below that of the lowest seller price do not receive any allocation of good g. Conversely sellers whose private price for good g is above the highest buyer price cannot raise any income on the market for good g. That is, we have:

Assumption 7 (**Price Compatibility**): For any strategy profile $\pi \in \Pi$ and any good g, we have:

1. For any buyer $i \in B_g$, if for all sellers $j \in S_g$, $\pi_{jg} > \pi_{ig}$, then $\xi_{ig}(\pi) = 0$.
2. For any seller $j \in S_g$, if for all buyers $i \in B_g$, $\pi_{ig} < \pi_{jg}$, then $w_{jg}(\pi) = 0$.

We must assume that the exchange mechanism implements some form of competition both among buyers and among sellers. To account for competition among buyers, we assume that when there is positive excess demand for good g, a buyer has incentive to outbid his competitors.

Assumption 8 (**Buyer Competition**): Let $\pi \in \Pi$ be such that there is excess demand for good g. Then there exists a buyer $i \in B_g$ and a price $q_{ig} > \pi_{ig}$ such that $u_i(\xi_i(\pi_{-ig}, q_{ig})) \geq u_i(\xi_i(\pi))$.

For competition among sellers, we assume that when there is excess supply for good g, a seller of good g can raise his market share and hence his income by undercutting his competitors. That is:

Assumption 9 (**Seller Competition**): Let $\pi \in \Pi$ be such that there is weak excess supply for good g and there are sellers $i, j \in S_g$ with $\pi_{ig} > \pi_{jg}$. Then there is a price $q_{ig} < \pi_{ig}$ such that $w_{jg}(\pi_{-ig}, q_{ig}) \geq w_{jg}(\pi)$.

This assumption will have an actual impact on behavior only if, all other things being equal, an agent's allocation is a non-decreasing function of income. Therefore we posit:

Assumption 10 (**Income Monotonicity**): Let $\pi, \pi' \in \Pi$, and suppose that for some agent i, we have $\pi_{-i} = \pi'_{-i}$ and for all $g \in C_i$, $\pi_{ig} = \pi'_{ig}$. Then if $w_i(\pi) \geq w_i(\pi')$, we have $u_i(\xi_i(\pi)) \geq u_i(\xi_i(\pi'))$.

Last, in order to prevent buyers from indefinitely increasing their prices, one should assume that there is some form of counterweight to the Buyer Competition Assumption (8). That is, in case of excess supply, buyers incur losses if their private price for a good is above the highest selling price (for instance, because price dispersion hampers coordination).

Assumption 11 (**Homogeneous Pricing**): Let $\pi \in \Pi$ be such that there is weak excess supply for good g and for some buyer $i \in B_g$, $\pi_{ig} > \max_{j \in S_g} \pi_{jg}$. Then there is a price $q_{ig} < \pi_{ig}$ such that $u_i(\xi_i(\pi_{-ig}, q_{ig})) \geq u_i(\xi(\pi))$.

We now have structural assumptions (4) to (7) that ensure the consistency of the exchange game with the underlying economy and behavioral assumptions (8) to (11) that ensure the exchange mechanism implements competition. of Walrasian equilibrium. Together, these assumptions give:

THEOREM 11.2 *Let ξ be an exchange mechanism for economy $\mathcal{E}(u, e)$ such that Assumptions (1) to (11) hold. Then a strategy profile π is a strict Nash equilibrium of the game $\mathcal{G}(u, e, \xi)$ if and only if π is p-uniform for some market equilibrium price vector p.*

Proof: Suppose π is a p-uniform general equilibrium price profile. We must show that π is a strict Nash equilibrium. According to the Demand Rationality Assumption (5), for all agents i, $\xi_i(\pi) = d_i(p, p \cdot e_i)$. Suppose agent i deviates to a price vector $q \neq p_i$.

For $g \in E_i$, we have:

- If $q_g > p_g$, then according to the Price Compatibility Assumption (7), we have $w_{ig}(\pi_{-i}, q) = 0$.
- If $q_g < p_g$, then according to the Price Consistency Assumption (4), we have $w_{ig}(\pi_{-i}, q) \leq q_g e_{ig} < p_g e_{ig} = w_{ig}(\pi)$.

Hence we have

$$w_i(q, \pi_{-i}) \begin{cases} < w_i(\pi) & \text{if } q_g \neq p_g \text{ for some } g \in E_i \\ = w_i(\pi) & \text{otherwise.} \end{cases}$$

(11.19)

Then, for $g \in C_i$, we have three possibilities:

- Case 1: There is some $g \in C_i$ such that $q_g < p_g$. Then according to the Price Compatibility Assumption (7), we have $\xi_{ig}(\pi_{-i}, q) = 0$. This implies, by the Goods Assumption (1), that $u_i(\xi_i(\pi_{-i}, q)) < u_i(d_i(p, p \cdot e_i)) = u_i(\xi_i(\pi))$.

- Case 2: For all $g \in C_i$, $q_g \geq p_g$ with the inequality being strict for some g. Equation (11.19) gives $w_i(q, \pi_{-i}) \leq w_i(\pi)$, and the Constraint Assumption (6) that $u_i(\xi_i(\pi_{-i}, q)) < u_i(\xi_i(\pi))$.
- Case 3: $q_g = p_q$ for all $g \in C_i$. Then, as $q \neq p$, there is a good $h \in E_i$ such that $q_h \neq p_h$. This implies according to (11.19) that $w_i(\pi_{-i}, q) < w_i(\pi)$ and hence, using the Constraint Assumption (6), that $u_i(\xi_i(\pi_{-i}, q)) < u_i(\xi_i(\pi))$.

This proves that every general equilibrium price profile is a strict Nash equilibrium.

Conversely, let us show that if $\pi \in \Pi$ is not a general equilibrium price profile then π is not a strict Nash equilibrium of $\mathcal{G}(u, e, \xi)$. To prove this result by contradiction, suppose π were a strict Nash equilibrium.

- According to the Buyer Competition Assumption (8), if there is excess demand for good g for the strategy profile π, then there is a buyer $i \in B_g$ and a price $q_{ig} > \pi_{ig}$ such that $u_i(\xi_i(\pi_{-ig}, q)) \geq u_i(\xi(\pi))$ and hence π cannot be a strict Nash equilibrium. By contraposition, this implies that there can be no excess demand for any good at the price profile π.
- Note that if π is a p-uniform strategy profile, then by the Price Consistency Assumption (4) we have that for all $i \in A$, $\delta_i(\pi, w_i(\pi)) = d_i(p, p \cdot e_i)$. Walras' Law then implies that, unless p is an equilibrium price, there is excess demand for at least one good. This would contradict the preceding point. If the only good for which there is excess demand is the numeraire good, there must be excess supply for another good and a similar argument applies.

  Hence π is not a p-uniform strategy profile.
- We now have that π is not p-uniform and that there is weak excess supply for every good. We can then distinguish the following cases. In each of the cases, good g cannot be the numeraire good because there exist two different prices for good g in the population.:

  - Case 1: π is not p-seller uniform for any price $p \in \mathcal{P}$. There then exists a good g and sellers $i, j \in S_g$ such that there is weak excess supply for good g and $\pi_{jg} < \pi_{ig}$. Then by the Seller Competition Assumption (9), there exists $q_{ig} < \pi_{ig}$ such that $w_{ig}(\pi_{-ig}, q_{ig}) \geq w_{ig}(\pi)$. Therefore, using the Income

Monotonicity Assumption (10), we have $u_i(\xi_i(\pi_{-ig}, q_{ig})) \geq u_i(\xi_i(\pi))$ and π is not a strict Nash equilibrium.

- Case 2: π is p-seller uniform for some $p \in \mathcal{P}$. We then have:

  * Case 2a: There exists $i \in B_g$ such that $\pi_{ig} < p_g$. Then by the Price Compatibility Assumption (7), we have $\xi_{ig}(\pi) = 0$ and hence $u_i(\xi_i(\pi)) = 0$. Therefore, it is clearly profitable for agent i to shift to price p_g and π is not a strict Nash equilibrium.

  * Case 2b: For all $i \in B_g$, $\pi_{ig} \geq p_g$ and, because π is not uniform, for at least one $k \in B_g$, one has $\pi_{kg} > p_g$. As there is weak excess supply for good g, we can apply the Homogeneous Pricing Assumption (11). Hence there exists $q < \pi_{kg}$ such that $u_i(\xi_i(\pi_{-ig}, q)) \geq u_i(\xi(\pi))$ and π is not a strict Nash equilibrium.

We have thus shown that whenever π is not a p-uniform general equilibrium strategy profile, it is not a strict Nash equilibrium. □

11.8 A Markov Implementation of Walrasian Dynamics

While analytical solutions for the replicator dynamics exist (Kemeny and Snell 1960; Gintis 2009b), they are too complex to be solved symbolically or estimated numerically. Yet, the link between stochastic imitation models and deterministic replicator dynamics is well described in the literature (Helbing 1996; Benaim and Weibull 2003). Therefore, it is possible to construct a discrete version of our model of price dynamics in an exchange economy and study its behavior as a Markov process.

We initialize an exchange economy with m agents and n goods, by drawing randomly for each agent a single production good and a set of consumption goods. The initial endowment of each agent is set to one unit of its production good. The utility function of each agent is rendered unique by randomly setting several parameters of a hybrid CES (constant elasticity of substitution) utility function. More precisely, for each consumer, we partition the set of consumer goods into k segments of randomly chosen sizes. We randomly assign goods to the various segments, and for each segment, we generate a CES consumption with random weights and elasticity. Total utility is the product of the k CES utility functions to random powers f_j such that $\sum_j f_j = 1$. In effect, no two consumers have the same utility function. These utility functions, which are generalizations of a functional form widely used in economic models, do not satisfy the gross substitutability assumption (Arrow et al. 1959), so stability in the tâtonnement dynamic does not hold.

Private price vectors are initialized by choosing each price from a uniform distribution on the open unit interval, then normalizing so that the price of the last good is unity. The state variable of an agent consists in its private price vector and its inventory of goods (consumption and production), which is initially set equal to its initial endowment and can include any good acquired through trade.

In each period, the agents in the economy are randomly ordered and permitted one-by-one to initiate trades. When agent i is the currently active agent, for each good g for which i is a buyer, he is randomly matched with a seller j who also is a buyer of i's production good h_i. Provided their private price ratios are compatible, that is, agent i's relative price for good g in terms of good h_i is at least equal to that of j, the agents actually trade good h_i against good g. The amount traded is the maximal compatible with the agents' demands (evaluated at their private prices) and their inventories. If i fails to trade with j, he still might secure a trade giving him good g when he will be on the receiving end of trade offers from g producers at some point during the period. After each trading period, traders consume their whole inventories, and replenish them with their initial endowment. Moreover, each agent updates his private price vector on the basis of his trading experience over the period, raising the price of a consumption or production good by 0.05% if his inventory is empty (that is, if he failed to purchase any of the consumption good or sell all of his production good), and lowering the price by 0.05% otherwise (that is, if he succeeded in obtaining his consumption good or sold all of his production inventory). We allow this adjustment strategy to evolve endogenously according to an imitation process.

After ten trading periods, the population of agents is updated using a discrete approximation of the standard replicator dynamic, in which agents who have gained high utilities by trading and consuming have a high probability of reproducing, while unsuccessful traders are eliminated from the economy. In all cases, the new agents inherit the price vector of its parent, perhaps mutated a bit. The resulting updating process is a discrete approximation of a monotonic dynamic in evolutionary game theory. The reader should note that in differential equation systems, all monotonic dynamics have the same properties as the simplest, which is the replicator dynamic (Taylor and Jonker 1978; Samuelson and Zhang 1992). Other monotonic approximations, including choosing a pair of agents and letting the lower-scoring agent copy the price vector of the higher-utility agent, produce similar dynamical results.

The result of the dynamic specified by the above conditions is the change over time in the distribution of private prices. The general result is that the system of private prices, which at the outset are randomly generated, in rather short time evolves to a set of *quasi-public* prices with very low inter-agent variance. Over the long term, these quasi-public prices move toward their equilibrium, market-clearing levels.

We assume that there are equal numbers of producers of each good from the outset, although we allow migration from less profitable to more profitable sectors, so in the long run profit rates are close to equal in all sectors. The complexity of the utility functions do not allow us to calculate equilibrium properties of the system perfectly, but we will assume that market-clearing prices are approximately equal to unit costs, given that unit costs are fixed, agents can migrate from less to more profitable sectors, and utility functions do not favor one good or style over another, on average. Population updating occurs every ten periods, and the number of encounters per sector is 10% of the number of agents in the sector. The mutation rate is $\mu = 0.01$ and the error correction is $\epsilon = 0.01$.

We illustrate this dynamic assuming there are 54 goods and 300 producers of each good, amounting to a total of 16200 agents in the economy. The results of a typical run of this model is illustrated in Figures 11.1 and 11.2. Figure 11.1 shows the passage from private to quasi-public prices over the first 20,000 trading periods of a typical run in terms of the average of the standard errors for all goods. The passage from private to quasi-public prices is quite dramatic, the standard error of prices across individuals falling by an order of magnitude within 300 periods, and falling another order of magnitude over the next 8500 periods. The final value of this standard error is 0.029, as compared with an initial value of 6.7.

Figure 11.2 shows the movement of the absolute value of excess demand over 50,000 periods for fifty-four goods. Using this measure, after 1500 periods excess demand has decreased by two orders of magnitude, and it decreases another order of magnitude by the end of the run.

The distinction between low-variance private prices and true public prices is significant, even when the standard error of prices across agents is ex-

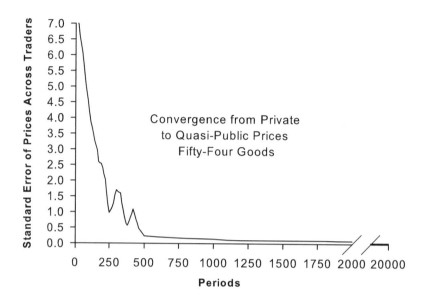

Figure 11.1. Convergence of private prices to quasi-public prices in a typical run with fifty-four goods

tremely small, because stochastic events such as technical changes propagate very slowly when prices are highly correlated private prices, but very rapidly when all agents react in parallel to price movement. In effect, with private prices, a large part of the reaction to a shock is a temporary reduction in the correlation among prices, a reaction that is impossible with public prices, as the latter are always perfectly correlated.

There is nothing special about the parameters used in the above example. Of course adding more goods increases the length of time until quasi-public prices become established, as well as the length of time until market quasi-equilibrium is attained. Increasing the number of agents increases the length of both of these time intervals.

11.9   Complex Dynamics

We have shown that the general equilibrium of a Walrasian market system is the only strict Nash equilibrium of an exchange game in which the requirements of the exchange process are quite mild and easily satisfied. Assuming producers update their private price profiles periodically by adopting the strategies of more successful peers, we have a multipopulation game in

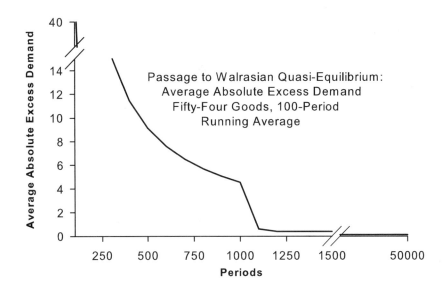

Figure 11.2. The path of aggregate excess demand over 50,000 periods.

which strict Nash equilibria are asymptotically stable in the replicator dynamic. Conversely, all stable equilibria of the replicator dynamic are strict Nash equilibria of the exchange process and hence Walrasian equilibria of the underlying economy.

The major innovation of our model is the use of private prices, one set for each agent, in place of the standard assumption of a uniform public price faced by all agents, and the replacement of the tâtonnement process with a replicator dynamic. The traditional public price assumption would not have been useful even had a plausible stability theorem been available using such prices. This is because there is no mechanism for prices to change in a system of public prices—no agent can alter the price schedules faced by the large number of agents with whom any one agent has virtually no contact.

The private price assumption is the only plausible assumption for a fully decentralized market system not in equilibrium, because there is in fact no natural way to define a common price system except in equilibrium. With private prices, each individual is free to alter his price profile at will, market conditions alone ensuring that something approximating a uniform system of prices will prevail in the long run.

Macroeconomic models have been especially handicapped by the lack of a general stability model for competitive exchange. The proof of stability of course does not shed light on the *fragility* of equilibrium in the sense of its susceptibility to exogenous shocks and its reactions to endogenous stochasticity. These issues can be studied directly through Markov process simulations, and may allow future macroeconomists to develop analytical microfoundations for the control of excessive market volatility.

12

The Future of the Behavioral Sciences

> Each discipline of the social sciences rules comfortably within its own chosen domain. . . so long as it stays largely oblivious of the others.
>
> Edward O. Wilson

> While scientific work in anthropology, sociology and political science will become increasingly indistinguishable from economics, economists will reciprocally have to become aware of how constraining has been their tunnel vision about the nature of man and social interaction.
>
> Jack Hirshleifer

> C: What do you call it when two behavioral disciplines disagree fundamentally and do not address their differences?
> HG: A scandal.
>
> Choreographer interview

I have found that when I attack problems concerning human behavior restricting myself to knowledge from a single academic discipline leaves me partially blind. I find I do much better by combining insights and models from a variety of behavioral disciplines, letting my research wander about in whatever direction seems fruitful at the moment. This book is the result of my efforts in this direction.

Before writing the professional journal articles on which this book draws, I honed my transdisciplinary skills working closely with economists, biologists, anthropologists, psychologists, and decision theorists. We produced numerous articles and books, including Gintis et al. (2005), Henrich et al. (2004), and Boyd et al. (2010). Cross-disciplinary collaboration works well.

Collaboration in the natural sciences is often interdisciplinary, scientists with distinct areas of expertise combining their ideas towards solving problems inaccessible to each alone. For instance, designing and running CERN's Large Hadron Collider required the cooperation of physicists, chemists, geologists, and engineers of all stripes. In the behavioral sciences, by which I mean the social sciences plus sociobiology (the biological study of the social behavior of living organisms), interdisciplinary collaboration has been rare indeed. Why is this?

Interdisciplinarity works in the natural sciences because the natural science disciplines are *mutually compatible*. Physicists, chemists, geologists, astronomers, and engineers have distinct areas of expertise, but when two natural science disciplines focus on a specific natural phenomenon, they offer complementary but non-conflicting theories. The same is unfortunately not true for the behavioral sciences. When experts from distinct behavioral science disciplines come together, they bring widely divergent theories and highly contrasting world views. In this situation, serious interdisciplinary work is virtually doomed.

This situation is curious, even scandalous. While it is not uncommon for scientists to disagree, there is only one truth in science and standard scientific protocols dictate that disagreements be adjudicated until some resolution is achieved. By contrast, distinct paradigms in the behavioral sciences have persisted interminably with no attempt by researchers to address their incompatibilities.

It is not hard to explain how this lamentable situation has come about. Almost all behavioral research is done in academia. The university structure in the twentieth century solidified disciplinary boundaries in the behavioral sciences around *communality of practices* rather than *scientific principles*. Thus economics became the study of money, profit-maximizing firms, and market exchange, sociology became the study of daily life in modern societies, anthropology became the study of small-scale pre-modern societies, political science became the study of government and political power, and psychology became the study of the mind. There is no serious justification for this particular carving up of the behavioral sciences.

As a result, the behavioral disciplines have become feudal fiefdoms. Each discipline has its own journals, its own conferences and hiring practices, and its own standards for accreditation and advancement. If economists generally accept one model of a social problem, such as drug addiction, while sociologists accept a completely different model and psychologists espouse yet a third, no one really cares about the discrepancies because the economists, sociologists, and psychologists are rewarded solely on the basis of intradisciplinary criteria, and what happens outside the discipline is of no account whatever. Such starkly contrasting explanations without the attempt to adjudicate the differences suggest that we are dealing with ideology, not science.

This situation has become untenable because it has become clear in many areas of social policy that no single discipline has the answer, but that several disciplines have important insights that might be integrated into an answer, were the proper theoretical tools carefully developed. For instance, successfully combating communicable diseases requires not only epidemiology, but also social network theory and models of belief formation and organizational behavior. Similarly, economic development depends intimately on cultural transformation, education and training, capital accumulation, and the containment of predatory elites.

Given this bizarre state of affairs, it would be an interesting exercise in the sociology of knowledge to explain how behavioral scientists, almost all of whom are wedded to standard canons of scientific method, justify their apathy. The incentive to accept the *status quo* is dictated by futility, prudence, ignorance, and self-selection. *Futility* because it appears hopeless to protest such incongruities, there being nothing a single researcher, or even a group of researchers, having neither the power nor the influence to affect what happens in another discipline, can do about it. *Prudence* because there is no reward for adjudicating cross-disciplinary conflicts. *Ignorance* because a researcher in one discipline generally has only the most superficial understanding of the theoretical structures of the other disciplines and the evidence on which these structures are based. And *self-selection* because individuals choose to work in fields with whose world-view they are most sympathetic, and tend to feel alienated in the foreign territory of another discipline.

To accommodate this state of affairs, *disciplinary provincialism* has come to dominate the thinking of behavioral scientists. The first principle of disciplinary provincialism is *rigorous policing of disciplinary boundaries*: each behavioral scientist belongs squarely in one discipline, and is not accorded the right to pronounce authoritatively on the principles of another discipline. For instance, even in multidisciplinary research gatherings, an economist is considered arrogant when venturing to speak authoritatively about biology, sociology, or anthropology. Similarly, it is considered rude to dispute a statement made authoritatively by an expert in another field.

That I routinely violate disciplinary boundaries is evident in every chapter of this book. For this I am often bitterly criticized not for the content of my ideas, but simply for having stated them without the usual diffidence required of those who would comment on a discipline while lacking the appropriate professional degree in that discipline. I recall one amusing event while walking to lunch in Budapest after having given a talk on the unification of the behavioral sciences. A young researcher approached and said to me, "You must be one of those arrogant New York Jews." I replied that I was from Philadelphia.

The second principle is *disciplinary incommensurability*: cross-disciplinary dialogue is in principle impossible because distinct disciplines deploy mutually untranslatable languages. *A fortiori* there is no common language underlying distinct disciplinary discourses, and hence it is a formidable task to develop models that successfully synthesize the insights of various disciplines.

The third principle is *interdisciplinary tolerance*: researchers judge other disciplines as irrelevant but non-threatening. For the tolerant provincialist, the various disciplines are at best like the seven proverbial blind men, each inspecting a different part of the elephant. At worst, the other disciplines are valuable, if at all, only when they deal with phenomena outside the purview of one's own discipline. When two disciplines deal with the same phenomena, each simply ignores the other's analysis, good taste advising against direct confrontation.

The chapters of this book, each in its own way, have had to overcome disciplinary provincialism.

The conditions that reproduce disciplinary provincialism are in serious decline. Three recent developments have made it possible, even easy, to develop expertise in two or more behavioral disciplines. The first is the emergence of a universal language of science, English, used throughout the world. The second is the prevalence of low-cost and high-speed digital information storage and transmission. The third is the Internet as a social medium. Any behavioral scientist with solid training in mathematics, statistics, and standard research methodology can learn a new field in a few years by (a) downloading reading lists from top universities; (b) ordering books online with practically overnight delivery; (c) accessing journal articles immediately through the cyberspace library system; (d) making use of Wikipedia, the online *Stanford Encyclopedia of Philosophy*, and a host of other Internet information sources; (e) requesting working and published papers from fellow researchers; (f) seeking advice by email from experts in the field, and asking such experts for comments on one's early attempts to work in allied disciplinary fields; and finally (g) working with experts in other fields, in other research institutions, and in countries spread across the world, as coauthors on research papers, communicating and coordinating from cyberspace.

The problem of interdisciplinary incommensurability may be a social problem, but it is not a scientific problem. The most important social impediment to cross-disciplinary dialogue is *skill-imbalance*. Most of the behavioral disciplines foster competence in statistical analysis and hypothesis testing, but some, especially economics and sociobiology, require a high level of mathematical sophistication in model building, whereas others require a high level of conceptual sophistication in dealing with ethnographic and historical data, as well as a deep feeling for the less formally modeled aspects of social life. The simple fact is that following an integrated approach to the behavioral sciences requires the researcher to be fluent in both analytical model building and the thick description of social behavior. The quest to build transdisciplinary research will draw the ire of the many researchers unwilling or unable to handle both analytical and synthetic reasoning capably and fluently.

There is a certain reality-imposed hierarchy in the natural sciences. Physics underlies all natural science disciplines, physics and chemistry underline the remaining natural science disciplines, and physics, chemistry, and biology underlie the life sciences. The natural sciences are *consilient* in the sense that whenever two disciplines study the same object of knowledge, their models and theories agree where they overlap.

The behavioral sciences are much younger than the natural sciences, and their subject matter is inherently more difficult for reasons discussed in this book. The future of the behavioral sciences also lies in consilience, with secure analytical foundations. In the place of hierarchy, we want to move towards a set of *common core principles* applicable to all fields, complemented by a set of *disciplinary core principles* for each of the key behavioral research areas. The common core and the disciplinary cores should be learned by all behavioral scientists as part of their early training. This chapter suggests some steps toward this future.

12.1   What are Analytical Foundations?

A *research area* in science is a linked set of questions, studied by one or more networks of researchers who have developed a common vocabulary for the communication of ideas and a common set of communication instruments, including journals, workshops, and conferences. Research areas can be telescoped, so for instance inner-city poverty research is part both of community research and social policy research, both of which are part of sociological research.

Maximal research areas, by which I mean research areas that are not part of larger research areas, are commonly called *disciplines*. Examples are sociology, economics, physics, chemistry, and mathematics. In the natural sciences, the disciplines coalesced following the development of stable research areas as a result of many years of scientific activity, and were relatively mature prior to the emergence of the modern system of higher education after World War II. A partial exception is biology, which developed later than chemistry and physics. The lack of an analytical foundation for biology often led universities to establish separate departments of biology, botany, and zoology. The modern evolutionary synthesis, uniting the evolutionary theory of Charles Darwin with the genetic theory of Gregor Mendel, was completed in the years around World War II (Mayr 1982). The various strands of biological research were then largely consolidated in to a single discipline in the years after World War II.

The history of the natural sciences suggests that maximal research areas coincide with disciplines, which roughly coincide with analytical foundations. This holds, for instance, for physics, chemistry, geophysics, and astronomy, the central disciplines in the natural sciences. Note, of course, that the analytical foundations of physics are included in all of the other natural science disciplines and the analytical foundations of chemistry are included in geophysics and astronomy. Accordingly, we might expect the analytical foundations of some behavioral science disciplines to be included in other disciplines. However, I have argued in this book that in the behavioral sciences, the basic principles and major results of each discipline are integral to successful research in the other disciplines. There is no natural hierarchy.

The behavioral science disciplines, including sociology, economics, anthropology, political science, sociobiology, and psychology, have had a very different social history. These disciplines were created at the time of the formation of the modern system of higher education by identifying disciplines with maximal research areas, even though no behavioral discipline had, at the time, a widely accepted analytical foundation. The analytical foundations for economic theory were forged in the two decades after World War II and conformed quite nicely to its original disciplinary boundaries. Sociobiology served as the analytical foundation for evolutionary biology, augmented by John Maynard Smith's game theoretic approach to animal behavior (Maynard Smith 1982), William Hamilton's inclusive fitness concept (Hamilton 1963, 1964), the sociobiological theory of Edward O. Wilson (Wilson 1975), and the gene-culture coevolutionary theories of Marcus Feldman, Luca Cavalli-Sforza, Robert Boyd, Peter Richerson, Robin Dunbar, William Durham, and others (Cavalli-Sforza and Feldman 1982; Boyd and Richerson 1985; Durham 1991; Dunbar 1996). Political science has no analytical foundations of its own, but rather borrows heavily from economic theory. This appears to cause no problems for the discipline, which defines itself not by a core theory, but by a common object of knowledge: political activity and government.

The same cannot be said for other behavioral science disciplines: sociology, anthropology, and psychology. These disciplines have nothing remotely approaching an analytical foundation. Anthropology and sociology suffer from not accepting the rational actor model and game theory. Chapter 2 was my attempt to treat anthropology as a branch of sociology, and Chapter 6 was my attempt to ground sociological theory in game theory and the rational actor model. At present only a small minority of researchers in these disciplines are interested in analytical social theory. Most anthropologists are dedicated to the thick description and preservation of small-scale societies, while most sociologists focus on the current ills of modern societies. Strange as this might seem, few practitioners in either discipline perceive a commonality in their fields of study.

In disciplines without analytical foundations, initiation into the discipline is craft-like, consisting of becoming acquainted with a heterogeneous set of classic writings. In anthropology we read, for instance, Franz Boaz, Margaret Mead, Clifford Geertz, Bronislaw Malinowski, Gregory Bateson, Alfred Radcliffe-Brown, and other greats, and we study a number of exemplary analyses of particular societies. In sociology, we read Max Weber, Émile Durkheim, Georg Simmel, Pierre Bourdieu, Raymond Boudon, Irwin Garfinkel, Anthony Giddens, Jeffrey Alexander, and other greats. We also study some exemplary pieces of empirical research, such as James Coleman's treatment of the culture of the American high school and William Foote Whyte's analysis of an Italian slum in Boston's North End.

With anthropology and sociology, we have the bizarre situation of two supposedly scientific disciplines studying the same thing—human society—and sharing nothing in common in terms of core theory or empirical data! In this book I have outlined an analytical foundation for both disciplines, which I treated collectively under the rubric of "sociology." Traditional anthropological research then becomes research in the sociology of small-scale societies and small-scale communities in modern societies. The analytical core of sociological theory is thus a model of human society in general.

12.2   Cross-Disciplinary Conflicts in the Behavioral Sciences

The behavioral sciences exhibit several conflicting models of decision making and strategic interaction. These include the psychological, the sociological, the biological, and the economic. These four models are not simply *different*, which is to be expected given their distinct explanatory aims, but are also *incompatible*. That is, each makes assertions taken to be key to understanding human behavior that are denied or ignored by the others. This means, of course, that they cannot all be correct. In fact all four are flawed. However, they can be modified and integrated to produce a core theoretical framework for modeling choice and strategic interaction for all of the behavioral sciences. Such a framework, of course, is rather bare-boned and must be substantially enriched in different directions to meet the particular needs of each discipline.

Roughly speaking, the received wisdom in psychology, found in virtually every first-year graduate textbook, is that humans are poor decision makers whose social interactions are plagued by biases and misperceptions. People are not *logical*, the saying goes, they are *psychological*. The received wisdom in sociology and anthropology is that human beliefs, values, and behavior are the product of socialization into the prevailing culture, which is structured to sustain a certain level of social cooperation. The received wisdom in economics is that individuals are self-regarding maximizers of personal material well-being and comforts, and efficient social cooperation in a group of self-regarding agents can be achieved through the proper set of material incentives. The received wisdom in sociobiology is that humans, like all other creatures, are inclusive fitness maximizers, which means they maximize the biological fitness of themselves and their various relatives, weighted by their degree of relatedness to these relatives. Political science draws upon these various alternatives *ad libidem*, and has no received wisdom concerning decision making and social interaction of its own.

These descriptions are broad generalizations that require careful qualification and amplification, as was presented at several points in previous chapters. Yet however carefully qualified, these received wisdoms are each and every one grossly inaccurate. Each chapter of this book has challenged one or more of these received wisdoms.

I wish I could say that all this is changing, but this is not the case. Only economics is currently in the process of systematically integrating insights from other fields, and these changes have not at all shaken economic theory's bizarre, and wholly insupportable, embrace of methodological individualism, according to which all social phenomena can be reduced to the strategic interaction of rational self-regarding agents (see my critique in Gintis 2009a). This curious attachment to methodological individualism is also shared by those gene's-eye sociobiologists, currently the vast majority in the discipline, who believe, without basis in fact or theory, that animal societies can be modeled using exclusively the principles of inclusive fitness maximization (see Chapter 9).

In recent years the value of transdisciplinary research in addressing questions of social theory has become clear, and *sociobiology* has become a major arena of scientific research (Wilson 1975). Moreover, contemporary socio-economic policy involves issues that fall in the interstices of the behavioral disciplines, including substance abuse, crime, corruption, tax compliance, social inequality, poverty, discrimination, and the cultural foundations of market economies. Incoherence is now an impediment to progress.

The first goal of unification in the behavioral sciences is to establish a *commen core framework* that is shared by all the behavioral sciences in the same way that mathematics and physics provide a set of core analytical principles that are accepted and deployed in all natural science disciplines. Such an analytical core lays a basis for future research in two ways. First, it sweeps away the interdisciplinary incompatibilities that now plague the behavioral sciences. This is because if a new theory in any discipline is incompatible with the common core, either the core or the new theory must be revised. Second, the analytical core, as the common focus of theoretical and empirical assessment for all behavioral scientists, can evolve in a cumulative manner much as the analytical core of the natural sciences has evolved.

The second goal is the development of a core analytical framework, based on the common core, for each of the behavioral disciplines. The benefits of a disciplinary analytical core are similar, and indeed mark a discipline as scientifically mature. Currently only economics and sociobiology enjoy this status in modeling social behavior. Microeconomic theory, including the rational actor model, game theory, and the general equilibrium model, is the common analytical core presented in doctoral programs around the world and with which all professional economists are conversant (Varian 1992; Mas-Colell et al. 1995). Of course this analytical core is currently considerably less powerful than the core models of physics and chemistry, and this core is currently under widespread revision, based on the laboratory and field results of behavioral game theory. Moreover, macroeconomic theory, which is the part of economic theory to which the public is most frequently exposed, lacks an acceptable analytical basis. Indeed, as we discussed in Chapter 11, economic dynamics, despite its overarching importance, generally has no secure representation in core economic theory.

Sociobiology has an analytical core based on evolutionary theory in general, and the principles of population biology in particular (Crow and Kimura 1970; Futuyma 1986; Hartl and Clark 2007). As with the case in economics, this analytical core is shared by virtually all professional biological researchers, although some of its basic principles as they relate to sociobiology have recently come under attack. We addressed these issues in Chapter 9.

The remaining behavioral disciplines, as we have seen, lack an analytical core. Cognitive psychology and neuroscience are young but quite advanced theoretically (Knill and Pouget 2004; Ma et al. 2006; Kording and Wolpert 2006; Tenenbaum et al. 2006; Gopnik and Tenenbaum 2007; Oaksford and Chater 2007). Social psychology and the general psychology of decision making, on the other hand, are a hodge-podge of experimental results with no internal unity. This situation, as I have argued throughout this book, results from their rejection of the rational actor model and game theory, the latter being incoherent without the rational actor model.

Sociology and anthropology lack coherent analytical organizing principles for the same reason, although this certainly does not prevent their practitioners from developing highly cogent and insightful analyses of particular situations, times, and places. Political science has always been happy to borrow core theory from economics, but the full development of a core theory of political life has been inhibited by critical deficiencies in the standard interpretation of the rational actor model (Green and Shapiro 1994). We touch on the possibility of a core theory of political behavior based on the concepts of *zoon politikon* and Homo ludens in Chapters 2 and 3, based on an appropriately revised version of the rational actor model.

In economics and biology, empirical findings are generally evaluated as affirming or contradicting the discipline's analytical core, and novel theoretical results are seen as building on the work of others and contributing to improving the analytical core. In other behavioral disciplines, new empirical results are treated as simply adding to the collection of empirical results, or contradicting earlier empirical results. Novel theoretical structures, rather than building on established theory, are treated as free-standing theoretical creations that vie competitively for prominence in the arena of social theory. Thus, in sociology and anthropology, viewing the contributions of the great masters is akin to marveling at the heterogeneous and *sui generis* accomplishments of the great painters and poets. One may be inspired by them, but one cannot build on their accomplishments. This lack of an analytical disciplinary core is simply the sign of immature science.

Acknowledgments

I would like to thank my coauthors Christopher Boehm, Samuel Bowles, Dirk Helbing, Antoine Mandel, and Carel van Schaik for permitting me to draw freely on our joint publications. I also want to thank Eldridge Adams, Lee Altenberg, Robert Axtell, Carl Bergstrom, Robert Boyd, Bernard Chapais, Eric Charnov, Herbert Dawid, Anthony Edwards, David Erdal, Marcus Feldman, Nancy Folbre, Duncan Foley, Steven Frank, Mauro Gallegati, Michael Ghiselin, Peter Godfrey-Smith, Robert Goldstone, David Haig, Peter Hammerstein, Marc Hauser, Michael Hechter, Geoffrey Hodgson, Alan Isaac, Carlo Jaeger, Alan Kirman, Simone Landini, David Levine, Paul Lewis, Julia Lupp, Michael Macy, John Maynard Smith, Mike Mesterton-Gibbons, James O'Connell, Samir Okasha, Paul Ormerod, Mark Pagel, David Queller, Laurent Lehmann, Martin Nowak, Peter Richerson, Barkley Rosser, Joan Roughgarden, Cosmo Shalizi, Joan Silk, Eric Alden Smith, Vernon Smith, Elliot Sober, Penny Spikins, Peter Turchin, Matthijs van Veelen, Mark Van Vugt, Andrew Whiten, Polly Wiessner, David Sloan Wilson, Edward O. Wilson, Ulrich Witt, and David Wolpert for helpful comments on previous versions of this material. Finally, I want to thank my editor Seth Ditchik for his unwavering support in guiding me through the creation and production of this book, and Princeton University Press president Peter Dougherty for guiding the Press to its currently preeminent position among academic publishers.

References

Abbot, Patrick, "Inclusive Fitness and Eusociality," *Nature* 471 (24 March 2011):E1–E4.

Abbott, R. J., J. K. James, R. I. Milne, and A. C. M. Gillies, "Plant Introductions, Hybridization and Gene Flow," *Philosophical Transactions of the Royal Society of London B* 358 (2003):1123–1132.

Abrams, Samuel, Torben Iversen, and David Soskice, "Informal social networks and rational voting," *British Journal of Political Science* 41,2 (2011):229–257.

Adams, R., "Anthropological Perspectives in Ancient Trade," *Current Anthropology* 15 (1974):239–258.

Ahlbrecht, Martin and Martin Weber, "Hyperbolic Discounting Models in Prescriptive Theory of Intertemporal Choice," *Zeitschrift für Wirtschafts- und Sozialwissenschaften* 115 (1995):535–568.

Akerlof, George A., "The Market for 'Lemons': Quality Uncertainty and the Market Mechanism," *Quarterly Journal of Economics* 84 (August 1970):488–500.

Akerlof, George A., "Labor Contracts as Partial Gift Exchange," *Quarterly Journal of Economics* 97,4 (November 1982):543–569.

Alchian, Armen and Harold Demsetz, "Production, Information Costs, and Economic Organization," *American Economic Review* 62 (December 1972):777–795.

Alcock, John, *Animal Behavior: An Evolutionary Approach* (Sunderland, MA: Sinauer, 1993).

Alexander, Jeffrey, *The Civil Sphere* (New York: Oxford University Press, 2006).

Alexander, Richard D., "The Evolution of Social Behavior," *Annual Review of Ecology and Systematics* 5 (1974):325–383.

Alexander, Richard D., *The Biology of Moral Systems* (New York: Aldine, 1987).

Ali, S. Nageeb and Charles Lin, "Why People Vote: Ethical Motives and Social Incentives," *American Economic Journal: Macroeconomics* 5,2 (2013):73–98.

Allais, Maurice, "Le comportement de l'homme rationnel devant le risque, critique des postulats et axiomes de l'école Américaine," *Econometrica* 21 (1953):503–546.

Allman, J., A. Hakeem, and K. Watson, "Two Phylogenetic Specializations in the Human Brain," *Neuroscientist* 8 (2002):335–346.

Alperson-Afil, Nira, "Continual Fire-Making by Hominins at Gesher Benot Ya'aqov, Israel," *Quaternary Science Reviews* 27 (2008):1733–1739.

Ambrose, Stanley H., "Small Things Remembered: Origins of Early Microlithic Industries in Sub-Saharan Africa," *Archeological Papers of the American Anthropological Association* 12,1 (2008):9–29.

Andreghetto, Giulia, Jordi Brandts, Rosaria Conte, Jordi Sabater-Mir, and Hector Solaz, "Punish and Voice: Punishment Enhances Cooperation when Combined with Norm-Signaling," *PLOSOne* 8,6 (June 2013):1–7.

Andreoni, James, "Warm-Glow versus Cold-Prickle: The Effects of Positive and Negative Framing on Cooperation in Experiments," *Quarterly Journal of Economics* 110,1 (February 1995):1–21.

Andreoni, James and John H. Miller, "Giving According to GARP: An Experimental Test of the Consistency of Preferences for Altruism," *Econometrica* 70,2 (2002):737–753.

Andreski, Stanislav, *Military Organization and Society* (University of California Press, 1968).

Anscombe, F. and Robert J. Aumann, "A Definition of Subjective Probability," *Annals of Mathematical Statistics* 34 (1963):199–205.

Aoki, Masahiko, *Corporations in Evolving Diversity: Cognition, Governance, and Institutions* (Oxford: Oxford University Press, 2010).

Ardrey, Robert, *The Territorial Imperative: A Personal Inquiry Into the Animals Origins of Property and Nations* (Kodansha America, 1997[1966]).

Ariely, Dan, *Predictibly Irrational: The Hidden Forces That Shape Our Decisions* (New York: Harper, 2010).

Aristotle, *Nicomachean Ethics* (Newburyport, MA: Focus Publishing, 2002[350BC]).

Arkes, Hal R. and Peter Ayton, "The Sunk Cost and *Concorde* Effects: Are Humans Less Rational Than Lower Animals?," *Psychological Bulletin* 125,5 (1999):591–600.

Arrow, Kenneth J., "The Organization of Economic Activity: Issues Pertinent to the Choice of Market versus Non-market Allocations," in Subcommittee Economy in Government of the Joint Economic on Committee (ed.) *The Analysis and Evaluation of Public Expenditures* 1969) pp. 47–64.

Arrow, Kenneth J., "Political and Economic Evaluation of Social Effects and Externalities," in M. D. Intriligator (ed.) *Frontiers of Quantitative Economics* (Amsterdam: North Holland, 1971) pp. 3–23.

Arrow, Kenneth J. and Frank Hahn, *General Competitive Analysis* (San Francisco: Holden-Day, 1971).

Arrow, Kenneth J. and Gérard Debreu, "Existence of an Equilibrium for a Competitive Economy," *Econometrica* 22,3 (1954):265–290.

Arrow, Kenneth J. and Leonid Hurwicz, "On the Stability of the Competitive Equilibrium, I," *Econometrica* 26 (1958):522–552.

Arrow, Kenneth J. and Leonid Hurwicz, "Competitive Stability under Weak Gross Substitutability: Nonlinear Price Adjustment and Adaptive Expectations," Technical Report, Office of Naval Research Contract Nonr-255, Department of Economics, Stanford University (1959). Technical Report No. 78.

Arrow, Kenneth J. and Leonid Hurwicz, "Some Remarks on the Equilibria of Economics Systems," *Econometrica* 28 (1960):640–646.

Arrow, Kenneth J., H. D. Block, and Leonid Hurwicz, "On the Stability of the Competitive Equilibrium, II," *Econometrica* 27 (1959):82–109.

Arthur, Brian, *Increasing Returns and Path Dependence in the Economy* (Ann Arbor: University of Michigan Press, 1994).

Asch, Solomon, "Effects of Group Pressure on the Modification and Distortion of Judgments," in H. Guetzkow (ed.) *Groups, Leadership and Men* (Carnegie Press, 1951) pp. 177–190.

Atkinson, Anthony and Joseph E. Stiglitz, *Lectures on Public Economics* (McGraw-Hill, 1980).

Aumann, Robert J., "Subjectivity and Correlation in Randomizing Strategies," *Journal of Mathematical Economics* 1 (1974):67–96.

Aumann, Robert J., "Correlated Equilibrium and an Expression of Bayesian Rationality," *Econometrica* 55 (1987):1–18.

Aumann, Robert J., "Game Theory," in John Eatwell, Murray Milgate, and Peter Newman (eds.) *The New Palgrave: A Dictionary of Economics* Vol. 2 (London: Macmillan, 1987) pp. 460–482.

Axelrod, Robert and William D. Hamilton, "The Evolution of Cooperation," *Science* 211 (1981):1390–1396.

Axtell, Robert, "Zipf Distribution of U.S. Firm Sizes," *Science* 293 (2001):1818–1820.

Aytimur, R. Emre, Aristotelis Boukouras, and Robert Schwager, "Voting as a signaling device.," *Economic Theory* 55,3 (2014):753–777.

Bacharach, Michael, "A Theory of Rational Decision in Games," *Erkentness* 27,1 (July 1987):17–56.

Bacharach, Michael, "The Acquisition of Common Knowledge," in Cristina Bicchieri and Maria Luisa Dalla Chiara (eds.) *Knowledge, Belief, and Strategic Interaction* (Cambridge: Cambridge University Press, 1992) pp. 285–316.

Bacharach, Michael, *Beyond Individual Choice: Teams and Games in Game Theory* (Princeton: Princeton University Press, 2006). Natalie Gold and Robert Sugden (eds.).

Bacharach, Michael, Natalie Gold, and Robert Sugden, *Beyond Individual Choice: Teams and Frames in Game Theory* (Princeton: Princeton University Press, 2006).

Bachrach, Peter and Morton Baratz, "The Two Faces of Power," *American Political Science Review* 56 (1962):947–952.

Bachtrog, Doris, "Expression Profile of a Degenerating Neo-Y Chromosome in Drosophila," *Current Biology* 17,5 (2006):1694–1699.

Bakeman, Roger and John R. Brownlee, "Social Rules Governing Object Conflicts in Toddlers and Preschoolers," in Kenneth H. Rubin and Hildy S. Ross (eds.) *Peer Relationships and Social Skills in Childhood* (New York: Springer-Verlag, 1982) pp. 99–112.

Bala, V. and M. Majumdar, "Chaotic Tatonnement," *Economic Theory* 2 (1992):437–445.

Balasco, Yves, *The Equilibrium Manifold. Postmodern Developments in the Theory of General Economic Equilibrium* (Cambridge: MIT Press, 2009).

Barkow, Jerome H., Leda Cosmides, and John Tooby, *The Adapted Mind: Evolutionary Psychology and the Generation of Culture* (New York: Oxford University Press, 1992).

Barrickman, N. L., M. L. Bastian, Karin Isler, and Carel P. van Schaik, "Life History Costs and Benefits of Encephalization: A Comparative

Test Using Data from Long-term Studies of Primates in the Wild," *Journal of Human Evolution* 54 (2008):568–590.

Battalio, R. C., Leonard Green, and John H. Kagel, "Income-leisure Trade-offs of Animal Workers," *American Economic Review* 71 (1981):621–632.

Becker, Gary S., *The Economics of Discrimination* (Chicago: University of Chicago Press, 1957).

Becker, Gary S., "Investment in Human Capital: A Theoretical Analysis," *Journal of Political Economy* 70 (1962):9–49.

Becker, Gary S., *Human Capital* (New York: Columbia University Press, 1964). For the National Bureau of Economic Research.

Becker, Gary S., "Crime and Punishment: An Economic Approach," *Journal of Political Economy* 76,2 (1968):169–217.

Becker, Gary S., *A Treatise on the Family* (Cambridge: Harvard University Press, 1981).

Becker, Gary S. and George J. Stigler, "Law Enforcement, Malfeasance, and Compensation of Enforcers," *Journal of Legal Studies* 3,1 (1974):1–18.

Becker, Gary S. and Kevin M. Murphy, "A Theory of Rational Addiction," *Journal of Political Economy* 96,4 (August 1988):675–700.

Becker, Gary S., Michael Grossman, and Kevin M. Murphy, "An Empirical Analysis of Cigarette Addiction," *American Economic Review* 84,3 (June 1994):396–418.

Beer, J. S., E. A. Heerey, D. Keltner, D. Skabini, and R. T. Knight, "The Regulatory Function of Self-conscious Emotion: Insights from Patients with Orbitofrontal Damage," *Journal of Personality and Social Psychology* 65 (2003):594–604.

Beinhocker, Eric, *The Origins of Wealth: Evolution, Complexity, and the Radical Remaking of Economics* (Cambridge, MA: Harvard Business School Press, 2006).

Belin, P., R. J. Zatorre, P. Lafaille, P. Ahad, and B. Pike, "Voice-selective Areas in Human Auditory Cortex," *Nature* 403 (2000):309–312.

Ben-Bassat, Avi and Momi Dahan, "Social identity and voting behavior," *Public Choice* 151,1–2 (2012):193–214.

Benabou, Roland and Jean Tirole, "Self Confidence and Personal Motivation," *Quarterly Journal of Economics* 117,3 (2002):871–915.

Benaim, Michel and Jörgen W. Weibull, "Deterministic Approximation of Stochastic Evolution in Games," *Econometrica* 71,3 (2003):873–903.

Berg, Joyce E., John W. Dickhaut, and Thomas A. Rietz, "Preference Reversals: The Impact of Truth-Revealing Incentives," 2005. College of Business, University of Iowa.

Berger, Peter I. and Thomas Luckmann, *The Social Construction of Reality* (Anchor, 1967).

Berle, Adolph A. and Gardiner C. Means, *The Modern Corporation and Private Property* (New York: Macmillan, 1932).

Bernheim, B. Douglas, Andrey Fradkin, and Igor Popov, "The Welfare Economics of Default Options in 401 (k) Plans," 2011. National Bureau of Economic Research, No. w17587.

Betzig, Laura, "Delayed Reciprocity and Tolerated Theft," *Current Anthropology* 37 (1997):49–78.

Bicchieri, Cristina, *Rationality and Coordination* (Oxford: Oxford University Press, 1993).

Bicchieri, Cristina, *The Grammar of Society: The Nature and Dynamics of Social Norms* (Cambridge: Cambridge University Press, 2006).

Binder, J. R., J. A. Frost, T. A. Hammeke, R. W. Cox, S. M. Rao, and T. Prieto, "Human Brain Language Areas Identified by Functional Magnetic Resonance Imaging," *Journal of Neuroscience* 17 (1997):353–362.

Binford, Lewis, "Human Ancestors: Changing Views of their Behavior," *Journal of Anthropology and Archeology* 4 (1985):292–327.

Bingham, Paul M., "Human Uniqueness: A General Theory," *Quarterly Review of Biology* 74,2 (June 1999):133–169.

Bingham, Paul M. and Joanne Souza, *Death from a Distance and the Birth of a Humane Universe: Human Evolution, Behavior, History, and Your Future* (BookSurge Publishing , 2009).

Binmore, Kenneth G., *Game Theory and the Social Contract: Playing Fair* (Cambridge, MA: MIT Press, 1993).

Binmore, Kenneth G., *Game Theory and the Social Contract: Just Playing* (Cambridge, MA: MIT Press, 1998).

Binmore, Kenneth G., *Natural Justice* (Oxford: Oxford University Press, 2005).

Binmore, Kenneth G. and Avner Shaked, "Experimental Economics: Where Next?," *Journal of Economic Behavior and Organization* 73 (2010):87–100.

Bishop, D. T. and C. Cannings, "The Generalised War of Attrition," *Advances in Applied Probability* 10,1 (March 1978):6–7.

Bissonnette, Annie, Mathias Franz, Oliver Schülke, and Julia Ostnera, "Socioecology, but not Cognition, Predicts Male Coalitions Across Primates," *Behavioral Ecology* (2014).

Black, Duncan, "On the Rationale of Group Decision-making," *Journal of Political Economy* 56 (1948):23–34.

Blais, André, *To Vote or Not to Vote* (Pittsburgh: University of Pittsburgh Press, 2000).

Blau, Peter, *Exchange and Power in Social Life* (New York: John Wiley, 1964).

Bliege Bird, Rebecca L. and Douglas W. Bird, "Delayed Reciprocity and Tolerated Theft," *Current Anthropology* 38 (1997):49–78.

Blount, Sally, "When Social Outcomes Aren't Fair: The Effect of Causal Attributions on Preferences," *Organizational Behavior & Human Decision Processes* 63,2 (August 1995):131–144.

Blumenschine, Robert J., John A. Cavallo, and Salvatore D. Capaldo, "Competition for Carcasses and Early Hominid Behavioral Ecology: A Case Study and Conceptual Framework," *Journal of Human Evolution* 27 (1994):197–213.

Blurton-Jones, Nicholas G., "Tolerated Theft: Suggestions about the Ecology and Evolution of Sharing, Hoarding, and Scrounging," *Social Science Information* 26,1 (1987):31–54.

Bodmer, Walter F. and Joseph Felsenstein, "Linkage and Selection: Theoretical Analysis of the Deterministic Two Locus Random Mating Model," *Genetics* 57 (1967):237–265.

Boëda, E., J. M. Geneste, C. Griggo, N. Mercier, and S. Muhesen, "A Levallois Point Embedded in the Vertebra of a Wild Ass (Aquus Africanus): Hafting, Projectiles and Mousterian Hunting Weapons," *Antiquity* 73 (1999):394–402.

Boehm, Christopher, "Egalitarian Behavior and Reverse Dominance Hierarchy," *Current Anthropology* 34,3 (June 1993):227–254.

Boehm, Christopher, "Pacifying Interventions at Arnhem Zoo and Gombe," in Richard W. Wrangham, W. C. McGrew, Frans B. M. de Waal, and Paul G. Heltne (eds.) *Chimpanzee Cultures* (Cambridge: Harvard University Press, 1994) pp. 211–226.

Boehm, Christopher, "Impact of the Human Egalitarian Syndrome on Darwinian Selection," *American Naturalist* 150 (Supplement) (1997):S100–S121.

Boehm, Christopher, *Hierarchy in the Forest: The Evolution of Egalitarian Behavior* (Cambridge, MA: Harvard University Press, 1999).

Boehm, Christopher, "Conflict and the Evolution of Social Control," *Journal of Consciousness Studies* 7 (2000):79–183.

Boehm, Christopher, "What Makes Humans Economically Distinctive? A Three-species Evolutionary Comparison and Historical Analysis," *Journal of Bioeconomics* 6,2 (2004):109–135.

Boehm, Christopher, *Moral Origins: The Evolution of Virtue, Altruism, and Shame* (New York: Basic Books, 2012).

Boehm, Christopher, "The Biocultural Evolution of Conflict Resolution Between Groups," in D. P. Fry (ed.) *War, Peace, and Human Nature: the Convergence of Evolutionary and Cultural Views* (New York: Oxford University Press, 2013).

Boehm, Christopher and Jessica C. Flack, "The Emergence of Simple and Complex Power Structures through Social Niche Construction," in Ana Guinote and Theresa K. Vescio (eds.) *Social Psychology of Power* (Guilford Press, 2010).

Boesch, Christophe and Hedwige Boesch-Achermann, *The Chimpanzees of the Taï Forest: Behavioral Ecology and Evolution* (Oxford: Oxford University Press, 2000).

Boesch, Christophe, Gottfried Hohmann, and Linda Marchant, *Behavioural Diversity in Chimpanzees and Bonobos* (Cambridge: Cambridge University Press, 1998).

Boesch, Christophe, Grégoire Kohou, Honora Nene, and Linda Vigilant, "Male Competition and Paternity in Wild Chimpanzees of the Taï Forest," *American Journal of Physical Anthropology* 130 (2006):103–115.

Boles, Terry L., Rachel T. A. Croson, and J. Keith Murnighan, "Deception and Retribution in Repeated Ultimatum Bargaining," *Organizational Behavior and Human Decision Processes* 83,2 (2000):235–259.

Bonner, John Tyler, *The Evolution of Culture in Animals* (Princeton: Princeton University Press, 1984).

Boomsma, Jacobus J., "Only Full-sibling Families Evolved Eusociality," *Nature* 471 (March 2011):E4–E5.

Boorman, Scott A. and Paul Levitt, *The Genetics of Altruism* (New York: Academic Press, 1980).

Borgerhoff Mulder, Monique, "The Demographic Transition: Are we any Closer to an Evolutionary Explanation?," *Trends in Ecology and Evolution* 13,7 (July 1998):266–270.

Borgerhoff Mulder, Monique, Samuel Bowles, Tom Hertz, and Adrian Bell, "Intergenerational Wealth Transmission and the Dynamics of Inequality in Small-Scale Societies," *Science* 326 (2009):682–688.

Bourdieu, Pierre, *Outline of a Theory of Practice* (Cambridge: Cambridge University Press, 1972/1977).

Bourke, Andrew F. G., *Principles of Social Evolution* (Oxford: Oxford University Press, 2011).

Bowles, Samuel, *Microeconomics: Behavior, Institutions, and Evolution* (Princeton: Princeton University Press, 2004).

Bowles, Samuel, "Group Competition, Reproductive Leveling, and the Evolution of Human Altruism," *Science* 314,5805 (8 December 2006):1569–1572.

Bowles, Samuel, "Genetic Differentiation among Hunter Gatherer Groups," 2007. Santa Fe Institute.

Bowles, Samuel, "Did Warfare among Ancestral Hunter-Gatherer Groups Affect the Evolution of Human Social Behaviors," *Science* 324 (2009). Supporting Online Material.

Bowles, Samuel and Herbert Gintis, *Schooling in Capitalist America: Educational Reform and the Contradictions of Economic Life* (New York: Basic Books, 1976).

Bowles, Samuel and Herbert Gintis, *Democracy and Capitalism: Property, Community, and the Contradictions of Modern Social Thought* (New York: Basic Books, 1986).

Bowles, Samuel and Herbert Gintis, "Contested Exchange: Political Economy and Modern Economic Theory," *American Economic Review* 78,2 (May 1988):145–150.

Bowles, Samuel and Herbert Gintis, "Contested Exchange: New Microfoundations of the Political Economy of Capitalism," *Politics and Society* 18,2 (1990):165–222.

Bowles, Samuel and Herbert Gintis, "The Revenge of Homo Economicus: Contested Exchange and the Revival of Political Economy," *Journal of Economic Perspectives* 7,1 (Winter 1993):83–102.

Bowles, Samuel and Herbert Gintis, *Recasting Egalitarianism: New Rules for Communities, States, and Markets* (London: Verso, 1999). Erik Olin Wright (ed.).

Bowles, Samuel and Herbert Gintis, "Persistent Parochialism: Trust and Exclusion in Ethnic Networks," *Journal of Economic Behavior and Organization* 55,1 (2004):1–23.

Bowles, Samuel and Herbert Gintis, *A Cooperative Species: Human Reciprocity and its Evolution* (Princeton: Princeton University Press, 2011).

Bowles, Samuel, Jung-kyoo Choi, and Astrid Hopfensitz, "The Co-evolution of Individual Behaviors and Social Institutions," *Journal of Theoretical Biology* 223 (2003):135–147.

Boyd, Robert and Joan B. Silk, *How Humans Evolved (Third Edition)* (New York: W. W. Norton, 2002).

Boyd, Robert and Peter J. Richerson, *Culture and the Evolutionary Process* (Chicago: University of Chicago Press, 1985).

Boyd, Robert and Peter J. Richerson, "An Evolutionary Model of Social Learning: the Effects of Spatial and Temporal Variation," in T. R. Zentall and G. Galef Jr. (eds.) *Social Learning: Psychological and Biological Perspectives* (Hillsdale NY: Erlbaum, 1988) pp. 29–48.

Boyd, Robert and Peter J. Richerson, "Group Selection among Alternative Evolutionarily Stable Strategies," *Journal of Theoretical Biology* 145 (1990):331–342.

Boyd, Robert and Peter J. Richerson, "Punishment Allows the Evolution of Cooperation (or Anything Else) in Sizeable Groups," *Ethology and Sociobiology* 113 (1992):171–195.

Boyd, Robert and Peter J. Richerson, *The Nature of Cultures* (Chicago, IL: University of Chicago Press, 2004).

Boyd, Robert and Peter J. Richerson, *The Origin and Evolution of Cultures* (Oxford: Oxford University Press, 2005).

Boyd, Robert, Herbert Gintis, and Samuel Bowles, "Coordinated Punishment of Defectors Sustains Cooperation and Can Proliferate When Rare," *Science* 328 (30 April 2010):617–620.

Brainard, D. H. and W. T. Freeman, "Bayesian Color Constancy," *Journal of the Optical Society of America A* 14 (1997):1393–1411.

Bramble, Dennis M. and Daniel E. Lieberman, "Endurance Running and the Evolution of Homo," *Nature* 432,7015 (2004):345–352.

Bratman, Michael E., "Shared Intention," *Ethics* 104,1 (October 1993):97–113.

Brennan, Geoffrey and Loren Lomasky, *Democracy and Decision* (Cambridge: Cambridge University Press, 1993).

Brewer, Marilyn B. and Roderick M. Kramer, "Choice Behavior in Social Dilemmas: Effects of Social Identity, Group Size, and Decision Framing," *Journal of Personality and Social Psychology* 50,543 (1986):543–549.

Brock, William and Steven N. Durlauf, "Discrete Choice With Social Interactions," *Review of Economic Studies* 68,235 (April 2001):235–260.

Brown, Donald E., *Human Universals* (New York: McGraw-Hill, 1991).

Brown, J. H. and M. V. Lomolino, *Biogeography* (Sunderland, MA: Sinauer, 1998).

Brown, J. L., "Alternate Routes to Sociality in Jays—With a Theory for the Evolution of Altruism and Communal Breeding," *American Zoologist* 14,1 (1974):63–80.

Buchanan, James, *The Limits of Liberty* (Chicago: University of Chicago Press, 1975).

Burkart, Judith M. and Carel P. van Schaik, "Cognitive Consequences of Cooperative Breeding in Primates," *Animal Cognition* 13 (2010):1–19.

Burkart, Judith M., O. Allon, F. Amici, C. Fichtel, C. Finkenwirth, A. Heschl, and E. J. Meulman, "The evolutionary origin of human hypercooperation," *Nature Communications* 5 (2014).

Burkart, Judith M., Sarah Blaffer Hrdy, and Carel P. van Schaik, "Cooperative Breeding and Human Cognitive Evolution," *Evolutionary Anthropology* 18,5 (September/October 2009):175–186.

Burks, Stephen V., Jeffrey P. Carpenter, and Eric Verhoogen, "Playing Both Roles in the Trust Game," *Journal of Economic Behavior and Organization* 51 (2003):195–216.

Burrows, Anne M., "The Facial Expression Musculature in Primates and Its Evolutionary Significance," *BioEssays* 30,3 (2008):212–225.

Burt, A. and Robert L. Trivers, *Genes in Conflict: The Biology of Selfish Genetic Elements* (Cambridge: Harvard University Press, 2006).

Buss, David M., *Evolutionary Psychology: The New Science of the Mind* (Boston: Allyn & Bacon, 1999).

Buss, Leo W., *The Evolution of Individuality* (Princeton: Princeton University Press, 1987).

Byrne, Richard and Andrew Whiten, *Machiavellian Intelligence: Social Expertise and the Evolution of Intellect in Monkeys, Apes, and Humans* (Oxford: Clarendon Press, 1988).

Calvin, William H., "A Stone's Throw and Its Launch Window: Timing Precision and Its Implications for Language and Hominid Brains," *Journal of Theoretical Biology* 104 (1983):121–135.

Camerer, Colin F., *Behavioral Game Theory: Experiments in Strategic Interaction* (Princeton: Princeton University Press, 2003).

Camerer, Colin F. and Ernst Fehr, "Measuring Social Norms and Preferences Using Experimental Games: A Guide for Social Scientists," in Joseph Henrich, Robert Boyd, Samuel Bowles, Colin F. Camerer, Ernst Fehr, and Herbert Gintis (eds.) *Foundations of Human Sociality: Economic Experiments and Ethnographic Evidence from Fifteen Small-Scale Societies* (Oxford: Oxford University Press, 2004) pp. 55–95.

Camerer, Colin F. and Richard H. Thaler, "Ultimatums, Dictators, and Manners," *Journal of Economic Perspectives* 9,2 (1995):209–219.

Cameron, Lisa A., "Raising the Stakes in the Ultimatum Game: Experimental Evidence from Indonesia," *Economic Inquiry* 37,1 (January 1999):47–59.

Cameron, Steven and James Heckman, "The Non-equivalence of the High School Equivalents," *Journal of Labor Economics* 11,1 (1993):1–47.

Camille, N., "The Involvement of the Orbitofrontal Cortex in the Experience of Regret," *Science* 304 (2004):1167–1170.

Carey, Susan, *Conceptual Change in Childhood* (Cambridge: MIT Press, 1985).

Carpenter, Jeffrey P., G. Harrison, and John List, *Field Experiments in Economics* (Greenwich and London: JAI Press, 2005).

Carpenter, Jeffrey P., Stephen V. Burks, and Eric Verhoogen, "Comparing Student Workers: The Effects of Social Framing on Behavior in Distribution Games," *Research in Experimental Economics* 1 (2005):261–290.

Cavalli-Sforza, Luigi Luca and Marcus W. Feldman, "Models for Cultural Inheritance: Within Group Variation," *Theoretical Population Biology* 42,4 (1973):42–55.

Cavalli-Sforza, Luigi Luca and Marcus W. Feldman, *Cultural Transmission and Evolution* (Princeton: Princeton University Press, 1981).

Cavalli-Sforza, Luigi Luca and Marcus W. Feldman, "Theory and Observation in Cultural Transmission," *Science* 218 (1982):19–27.

Cawley, John, James Heckman, Lance Lochner, and Edward Vytlacil, "Understanding the role of cognitive ability in accounting for the recent rise in the return to education," in Kenneth J. Arrow, Samuel Bowles, and Steven Durlauf (eds.) *Meritocracy and Economic Inequality* (Princeton: Princeton University Press, 2000) pp. 230–266.

Chamberlain, Gary and Michael Rothschild, "A Note on the Probability of Casting a Decisive Vote," *Journal of Economic Theory* 25,1 (1981):152–162.

Chapais, Bernard, *Primeval Kinship: How Pair-Bonding Gave Birth to Human Society* (Cambridge: Harvard University Press, 2008).

Charness, Gary and Martin Dufwenberg, "Promises and Partnership," *Econometrica* 74,6 (November 2006):1579–1601.

Childe, Vere Gordon, *Man Makes Himself* (London: Watts, 1936).

Churchill, Steven E. and J. A. Rhodes, "The Evolution of the Human Capacity for "Killing at a Distance": The Human Fossil Evidence for the Evolution of Projectile Weaponry," in J. J. Hublin and M. P. Richards (eds.) *The Evolution of Hominin Diets: Integrating Approaches to the Study of Paleolithic Subsistence* (Berlin: Springer, 2009).

Clarke, Harold D., David Sanders, Marianne C. Stewart, and Paul Whitely, *Political Choice in Britain* (Oxford: Oxford University Press, 2004).

Clutton-Brock, T. H., "Primate Social Organization and Ecology," *Nature* 250 (1974):539–542.

Clutton-Brock, T. H., "Cooperation between Non-kin in Animal Societies," *Nature* 462 (2009):51–57.

Coleman, James S., *Equality of Educational Opportunity* (Arno Press, 1966).

Coleman, James S., *Youth: Transition to Adulthood. Report of the Panel on Youth of the President's Science Advisory Committee* (Chicago: University of Chicago Press, 1973).

Coleman, James S., "Free Riders and Zealots: The Role of Social Networks," *Sociological Theory* 6 (Spring 1988):52–57.

Coleman, James S., *Foundations of Social Theory* (Cambridge, MA: Belknap, 1990).

Coleman, James S., John W. C. Johnstone, and Kurt Jonassohn, *The Adolescent Society: The Social Life of the Teenager and its Impact on Education* (Greenwook Press, 1981[1961]).

Coleman, James S., Martin Trow, and Seymour Martin Lipset, *Union Democracy* (Free Press, 1977).

Coleman, James S., Thomas Hoffer, and Sally Kilgore, *High School Achievement* (New York: Basic Books, 1982).

Colman, Andrew M., Briony D. Pulford, and Jo Rose, "Collective Rationality in Interactive Decisions: Evidence for Team Reasoning," *Acta Psychologica* 128,2 (2008):387–297.

Conn, Paul H., David B. Meltz, and Charles Press, "The Concept of Political Rationality," *Polity* 6,2 (Winter 1973):223–239.

Conte, Rosaria and Cristiano Castelfranchi, "From Conventions to Prescriptions. Towards an Integrated View of Norms," *Artificial Intelligence and Law* 7 (1999):323–340.

Cookson, R., "Framing Effects in Public Goods Experiments," *Experimental Economics* 3 (2000):55–79.

Coser, Lewis, *The Functions of Social Conflict* (New York: The Free Press, 1956).

Cosmides, Leda, John Tooby, and Jerome H. Barkow, "Introduction: Evolutionary Psychology and Conceptual Integration," in Jerome H. Barkow, Leda Cosmides and John Tooby (eds.) *The Adapted Mind: Evolutionary Psychology and the Generation of Culture* (New York: Oxford University Press, 1992) pp. 3–15.

Cox, Gary W., "Strategic Voting Equilibria Under the Single Nontransferable Vote," *American Political Science Review* 88,3 (September 1994):608–621.

Cox, Gary W., *Making Votes Count* (Cambridge: Cambridge University Press, 1997).

Coyne, Jerry, "The Demise of Group Selection," *Why Evolution is True* (June 2012).

Crow, James F., "Breeding Structure of Populations. II. Effective Population Number," in O. Kempthorne, T. A. Bancroft, J. W. Gowen, and J. L Lush (eds.) *Statistics and Mathematics in Biology* (Ames, Iowa: Iowa State University Press, 1954) pp. 543–556.

Crow, James F. and Motoo Kimura, *An Introduction to Population Genetic Theory* (New York: Harper & Row, 1970).

Dahl, Robert A., "The Concept of Power," *Behavioral Science* 2 (1957):201–215.

Dalton, Russell J., *The Good Citizen: How a Younger Generation is Reshaping American Politics* (New York: Sage, 2008).

Darlington, P. J., "Group Selection, Altruism, Reinforcement and Throwing in Human Evolution," *Proceedings of the National Academy of Sciences* 72 (1975):3748–3752.

Dart, Raymond A., "Australopithecus africanus: The Man-Ape of South Africa," *Nature* 115,2884 (1925):195–199.

Darwin, Charles, *The Origin of Species by Means of Natural Selection, 6th Edition* (London: John Murray, 1859).

Darwin, Charles, *The Descent of Man, and Selection in Relation to Sex* (London: Murray, 1871).

Davies, Nicholas B., "Territorial Defence in the Speckled Wood Butterfly (*Pararge Aegeria*): The Resident Always Wins," *Animal Behaviour* 26 (1978):138–147.

Dawkins, Richard, *The Selfish Gene* (Oxford: Oxford University Press, 1976).

Dawkins, Richard, "Twelve Misunderstandings of Kin Selection," *Zeitschrift für Tierpsychologie* 51 (1979):184–200.

Dawkins, Richard, *The Extended Phenotype: The Gene as the Unit of Selection* (Oxford: Freeman, 1982).

Dawkins, Richard, "Replicators and Vehicles," in King's College Sociobiology Group (ed.) *Current Problems in Sociobiology* (Cambridge: Cambridge University Press, 1982) pp. 45–64.

Dawkins, Richard, *The Blind Watchmaker: Why the Evidence of Evolution Reveals a Universe without Design* (New York: W. W. Norton, 1996).

Dawkins, Richard, "Extended Phenotype-but not too Extended. A reply to Laland, Turner and Jablonka," *Biology and Philosophy* 19 (2004):377–396.

Dawkins, Richard, "The Descent of Edward Wilson," *Prospect* (May 24 2012).

Dawkins, Richard and H. J. Brockmann, "Do Digger Wasps Commit the *Concorde* Fallacy?," *Animal Behaviour* 28 (1980):892–896.

de Waal, Frans, *Good Natured: The Origins of Right and Wrong in Humans and Other Animals* (Cambridge, MA: Harvard University Press, 1996).

de Waal, Frans, *Bonobo: The Forgotten Ape* (Berkeley: University of California Press, 1997).

de Waal, Frans, *Chimpanzee Politics: Sex and Power among the Apes* (Baltimore: Johns Hopkins University Press, 1998).

Deacon, Terrence W., *Symbolic Species: The Co-Evolution of Language and the Brain* (New York: Norton, 1998).

Debreu, Gérard, "A Social Equilibrium Existence Theorem," *Proceedings of the National Academy of Sciences* 38 (1952):886–893.

Debreu, Gérard, "Representation of a Preference Ordering by a Numerical Function," in R. M. Thrall, C. H. Coombs, and R. L. Davis (eds.) *Decision Processes* (New York: John Wiley & Sons, 1954) pp. 159–165.

Debreu, Gérard, "Excess Demand Function," *Journal of Mathematical Economics* 1 (1974):15–23.

Dediu, Dan and Stephen C. Levinson, "On the Antiquity of Language: the Reinterpretation of Neandertal Linguistic Capacities and its Consequences," *Frontiers in Psychology* 4 (2013).

deMenocal, Peter, "Climate and Human Evolution," *Science* 331 (4 February 2011):540–542.

d'Errico et al., F., "Early Evidence of San Material Culture Represented by Organic Artifacts from Border Cave, South Africa," *Proceedings of the National Academy of Sciences* 109,33 (2012):13214–13219.

Dhami, Sanjit, *Foundations of Behavioral Economic Analysis* (Oxford: Oxford University Press, 2016).

di Finetti, Benedetto, *Theory of Probability* (Chichester: John Wiley & Sons, 1974).

Di Guilmi, Corrado, Mauro Gallegati, Simone Landini, and Joseph E. Stiglitz, "Towards an Analytical Solution for Agent Based Models: An Application to a Credit Network Economy," *SSRN* (2011).

DiMaggio, Paul, "Culture and Economy," in Neil Smelser and Richard Swedberg (eds.) *The Handbook of Economic Sociology* (Princeton: Princeton University Press, 1994) pp. 27–57.

DiMaggio, Paul, "The New Institutionalisms : Avenues of Collaboration," *Journal of Institutional and Theoretical Economics* 154,4 (December 1998):696–705.

Dobzhansky, Theodosius, "A Review of Some Fundamental Concepts and Problems of Population Genetics," *Cold Springs Harbor Symposium* (1953):1–15.

Dobzhansky, Theodosius, "Cultural Direction of Human Evolution," *Human Biology* 35 (1963):311–316.

Dominguez-Rodrigoa, M. and R. Barba, "New Estimates of Tooth Mark and Percussion Mark Frequencies at the FLK Zinj Site: the Carnivore-hominid-carnivore Hypothesis Falsified," *Journal of Human Evolution* 50,2 (February 2006):170–194.

Downs, Anthony, *An Economic Theory of Democracy* (Boston: Harper & Row, 1957).

Downs, Anthony, "An Economic Theory of Political Action in a Democracy," *Journal of Political Economy* 65,2 (April 1957):135–150.

Dugatkin, Lee Alan, *Cooperation among Animals* (New York: Oxford University Press, 1997).

Dugatkin, Lee Alan and Hudson Kern Reeve, "Behavioral Ecology and the Levels of Selection: Dissolving the Group Selection Controversy," *Advances in the Study of Behavior* 23 (1994):101–133.

Dugatkin, Lee Alan and Hudson Kern Reeve, *Game Theory and Animal Behavior* (Oxford: Oxford University Press, 1998).

Dunbar, Robin M., "Coevolution of Neocortical Size, Group Size and Language in Humans," *Behavioral and Brain Sciences* 16,4 (1993):681–735.

Dunbar, Robin M., *Grooming, Gossip, and the Evolution of Language* (Cambridge, MA: Harvard University Press, 1996).

Dunbar, Robin M., *The Human Story* (New York: Faber & Faber, 2005).

Dunbar, Robin M., Clive Gamble, and John A. J. Gowlett, *Social Brain, Distributed Mind* (Proceedings of the British Academy: Oxford University Press, 2010).

Durham, William H., *Coevolution: Genes, Culture, and Human Diversity* (Stanford: Stanford University Press, 1991).

Durkheim, Émile, *The Division of Labor in Society* (New York: The Free Press, 1902).

Durrett, Richard and Simon A. Levin, "Can Stable Social Groups be Maintained by Homophilous Imitation Alone?," *Journal of Economic Behavior and Organization* 57,3 (2005):267–286.

Duverger, Maurice, "Factors in a Two-Party and Multiparty System," in Maurice Duverger (ed.) *Party Politics and Pressure Groups* (New York: Thomas Y. Crowell, 1972) pp. 23–32.

Eason, P. K., G. A. Cobbs, and K. G. Trinca, "The Use of Landmarks to Define Territorial Boundaries," *Animal Behaviour* 58 (1999):85–91.

Eckel, Catherine and Herbert Gintis, "Blaming the Messenger: Notes on the Current State of Experimental Economics," *Journal of Economic Behavior and Organization* 73,1 (January 2010):109–119.

Edgerton, Robert B., *Sick Societies: Challenging the Myth of Primitive Harmony* (New York: Free Press, 1992).

Edlin, Aaron, Andrew Gelman, and Noah Kaplan, "Voting as a Rational Choice: Why and How People Vote to Improve the Well-being of Others," *Rationality and Society* 19,3 (2007):293–314.

Edwards, A. W. F., "The Fundamental Theorem of Natural Selection," *Biological Review* 69 (1994):443–474.

Eibl-Eibesfeldt, Irenäus, *Human Ethology* (New York: Aldine de Gruyter, 1989).

Eisenberg, J. F., N. A. Muckenhirn, and R. Rudran, "The Relation Between Ecology and Social Structure in Primates," *Science* 176 (1972):863–874.

Ellis, Lee, "On the Rudiments of Possessions and Property," *Social Science Information* 24,1 (1985):113–143.

Ellsberg, Daniel, "Risk, Ambiguity, and the Savage Axioms," *Quarterly Journal of Economics* 75 (1961):643–649.

Elster, Jon, *Sour Grapes: Studies in the Subversion of Rationality* (Cambridge: Cambridge University Press, 1985).

Enos, Ryan and Anthony Fowler, "Why Did You Vote," *Daily Dish* (Nov. 11 2010).

Eshel, Ilan, "On the Neighbor Effect and the Evolution of Altruistic Traits," *Theoretical Population Biology* 3 (1972):258–277.

Eswaran, Mukesh and Ashok Kotwal, "Access to Capital and Agrarian Production Organization," *Economic Journal* 96 (June 1986):482–498.

Evren, Özgür, "Altruism and Voting: A Large-turnout Result that Does not Rely on Civic Duty or Cooperative Behavior," *Journal of Economic Theory* 147,6 (2012):2124–2157.

Ewens, W. J., "A Generalized Fundamental Theorem of Natural Selection," *Genetics* 63 (October 1969):531–537.

Faravelli, Marco and Randall Walsh, "Smooth Politicians and Paternalistic Voters: A Theory of Large Elections," *National Bureau of Economic Research* (2011):2124–2157.

Farmer, J. Doyne and F. Lillo, "On the Origin of Power Law Tails in Price Fluctuations," *Quantitative Finance* 4,1 (2004):7–11.

Fedderson, Timothy, "Rational choice theory and the paradox of not voting," *Journal of Economic Perspectives* 18 (2004):99–112.

Fedderson, Timothy and Alvaro Sandroni, "A Theory of Participation in Elections," *American Economic Review* 96,4 (September 2006):1271–1282.

Fehr, Ernst and Herbert Gintis, "Human Motivation and Social Cooperation: Experimental and Analytical Foundations," *Annual Review of Sociology* 33 (August 2007):43–64.

Fehr, Ernst and Simon Gächter, "How Effective Are Trust- and Reciprocity-Based Incentives?," in Louis Putterman and Avner Ben-Ner (eds.) *Economics, Values and Organizations* (New York: Cambridge University Press, 1998) pp. 337–363.

Fehr, Ernst and Simon Gächter, "Cooperation and Punishment," *American Economic Review* 90,4 (September 2000):980–994.

Fehr, Ernst and Simon Gächter, "Altruistic Punishment in Humans," *Nature* 415 (10 January 2002):137–140.

Fehr, Ernst and Urs Fischbacher, "Third Party Punishment and Social Norms," *Evolution & Human Behavior* 25 (2004):63–87.

Fehr, Ernst, E. Kirchler, A. Weichbold, and Simon Gächter, "When social norms overpower competition: Gift exchange in experimental labor markets," *Journal of Labor Economics* 16,2 (1998):324–351.

Fehr, Ernst, Georg Kirchsteiger, and Arno Riedl, "Does Fairness Prevent Market Clearing?," *Quarterly Journal of Economics* 108,2 (1993):437–459.

Fehr, Ernst, Simon Gächter, and Georg Kirchsteiger, "Reciprocity as a Contract Enforcement Device: Experimental Evidence," *Econometrica* 65,4 (July 1997):833–860.

Feldman, Marcus W. and Lev A. Zhivotovsky, "Gene-Culture Coevolution: Toward a General Theory of Vertical Transmission," *Proceedings of the National Academy of Sciences* 89 (December 1992):11935–11938.

Feldman, Marcus W. and Luigi Luca Cavalli-Sforza, "Cultural and Biological Evolutionary Processes, Selection for a Trait under Complex Transmission," *Theoretical Population Biology* 9,2 (April 1976):238–259.

Fifer, F. C., "The Adoption of Bipedalism by the Hominids: A New Hypothesis," *Human Evolution* 2 (1987):135–47.

Fiorina, Morris P., "Information and Rationality in Elections," in John Ferejohn and James Kuklinski (eds.) *Information and Democratic Processes* (Urbana, IL: University of Illinois Press, 1990) pp. 329–342.

Fischer, A. J., "The Probability of Being Decisive," *Public Choice* 101 (1999):267–283.

Fischer, Ilan, Alex Frid, Sebastian J. Goerg, Simon A. Levin, Daniel I. Rubenstein, and Reinhard Selten, "Fusing Enacted and Expected Mimicry Generates a Winning Strategy that Promotes the Evolution of Cooperation," *Proceedings of the National Academy of Sciences* 110,25 (June 18 2013):10229–10233.

Fishburn, Peter C., *Utility Theory for Decision Making* (New York: John Wiley & Sons, 1970).

Fisher, Franklin M., *Disequilibrium Foundations of Equilibrium Economics* (Cambridge: Cambridge University Press, 1983).

Fisher, Ronald A., *The Genetical Theory of Natural Selection* (Oxford: Clarendon Press, 1930).

Fiske, P., P. T. Rintamaki, and E. Karvonen, "Mating Success in Lekking Males: a Meta-analysis," *Behavioral Ecology* 9 (1998):328–338.

Flannery, Kent and Joyce Marcus, *The Creation of Inequality: How Our Prehistoric Ancestors Set the Stage for Monarchy, Slavery, and Empire* (Cambridge: Harvard University Press, 2012).

Fletcher, Jeffrey A. and Michael Doebili, "A Simple and General Explanation for the Evolution of Altruism," *Proceedings of the Royal Society B* 276 (2009):13–19.

Florenzano, M., *General Equilibrium Analysis: Existence and Optimality Properties of Equilibria* (Berlin: Springer, 2005).

Foley, Robert A., "An Evolutionary and Chronological Framework for Human Social Behavior," *Proceedings of the British Academy* 88 (1996):95–117.

Forsythe, Robert, Joel Horowitz, N. E. Savin, and Martin Sefton, "Replicability, Fairness and Pay in Experiments with Simple Bargaining Games," *Games and Economic Behavior* 6,3 (May 1994):347–369.

Fosco, Constanza, Annick Laruelle, and Angel Sánchez, "Turnout Intentions and Random Social Networks," *Advances in Complex Systems* 14,1 (2011):31–53.

Foster, Kevin R., Tom Wenseleers, and Francis L. W. Ratnieks, "Spite: Hamilton's Unproven Theory," *Ann. Zool. Fennici* 38 (2001):229–238.

Foucault, Michel, *The Archaeology of Knowledge & The Discourse on Language* (New York: Vintage, 1982).

Fowler, James H., "Altruism and Turnout," *Journal of Politics* 68,3 (2006):674–683.

Fowler, James H. and Cindy D. Kam, "Beyond the Self: Social Identity, Altruism, and Political Participation," *Journal of Politics* 69,3 (2007):813–827.

Frank, Steven A., "Models of Parasite Virulence," *Quarterly Review of Biology* 71 (1996):37–78.

Frank, Steven A., "The Price Equation, Fisher's Fundamental Theorem, Kin Selection, and Causal Analysis," *Evolution* 51,6 (1997):1712–1729.

Frank, Steven A., *Foundations of Social Evolution* (Princeton: Princeton University Press, 1998).

Frank, Steven A., "Repression of Competition and the Evolution of Cooperation," *Evolution* 57 (2003):693–705.

Frank, Steven A. and Montgomery Slatkin, "Fisher's Fundamental Theorem of Natural Selection," *Trends in Ecology and Evolution* 7,3 (1992):92–95.

Franklin, Mark N., Richard G. Niemi, and Guy Whitten, "The Two Faces of Tactical Voting," *British Journal of Political Science* 24,4 (1994):549–557.

Freud, Sigmund, *New Introductory Lectures on Psychoanalysis* (Liveright, 1933).

Fried, Morton H., *The Evolution of Political Society: An Essay in Political Anthropology* (New York: Random House, 1967).

Friedman, Daniel, "Evolutionary Games in Economics," *Econometrica* 59,3 (May 1991):637–666.

Fry, Douglas P., *War, Peace, and Human Nature: The Convergence of Evolutionary and Cultural Views* (Oxford University Press, 2013).

Fudenberg, Drew, David K. Levine, and Eric Maskin, "The Folk Theorem with Imperfect Public Information," *Econometrica* 62 (1994):997–1039.

Furby, Lita, "The Origins and Early Development of Possessive Behavior," *Political Psychology* 2,1 (1980):30–42.

Furuichi, T., "Sexual Swelling, Receptivity, and Grouping of Wild Pygmy Chimpanzee Females at Wamba, Zaire," *Primates* 28 (1987):309–318.

Furuichi, T., "Social Interactions and the Life History of Female (Pan paniscus) at Wamba, Republic of Zaire," *International Journal of Primatology* 10 (1989):173–198.

Furuichi, T., "Agonistic interactions and matrifocal dominance rank of wild bonobos (Pan paniscus) at Wamba," *International Journal of Primatology* 18 (1997):855–875.

Futuyma, Douglas J., *Evolutionary Biology* (Sunderland, MA: Sinauer, 1986).

Gadagkar, Raghavendra, "On Testing the Role of Genetic Asymmetries Created by Haplodiploidy in the Evolution of Eusociality in the Hymenoptera," *Journal of Genetics* 70,1 (April 1991):1–31.

Gale, David, "The Law of Supply and Demand," *Math. Scand.* 30 (1955):155–169.

Gale, David, "A Note on Global Instability of Competitive Equilibrium," *Naval Research Logistics Quarterly* 10 (March 1963):81–87.

Galef, Bennett G. and Kevin N. Laland, "Social Learning in Animals: Empirical Studies and Theoretical Models," *BioScience* 55,6 (2005):489–499.

Gardner, Andy and J. J. Welsh, "A Formal Theory of the Selfish Gene," *Journal of Evolutionary Biology* 24 (2011):1801–1813.

Gardner, Andy, Stuart A. West, and Angus Buckling, "Bacteriocins, Spite and Virulence," *Proceedings of the Royal Society London B* 271 (2004):1529–1535.

Gardner, Andy, Stuart A. West, and G. Wild, "The Genetical Theory of Kin Selection," *Journal of Evolutionary Biology* 24 (2011):1020–1043.

Gauthier, David, *Morals by Agreement* (Oxford: Clarendon Press, 1986).

Gayer, G., Itzhak Gilboa, and O. Lieberman, "Rule-based and Case-based Reasoning in Housing Prices," *The BE Journal of Theoretical Economics* 7,1 (2007).

Geertz, Clifford, *The Growth of Culture and the Evolution of Mind* (Glencoe: Free Press, 1962).

Gelman, Andrew, Gary King, and John Boscardin, "Estimating the Probability of Events That Have Never Occurred: When Is Your Vote Decisive?," *Journal of the American Statistical Association* 93,441 (1998):1–9.

Gerbault, Pascale, Anke Liebert, Yuval Itan, Adam Powell, Mathias Currat, Joachim Burger, Dallas M. Swallow, and Mark G. Thomas, "Evolution of Lactase Persistence: an Example of Human Niche Construction," *Philosophical Transactions of the Royal Society of London B, Biological Sciences* 366,1566 (March 2011):863–877.

Gerber, Alan S. and Todd Rogers, "Descriptive Social Norms and Motivation to Vote: Everybody's Voting and So Should You," *The Journal of Politics* 71,1 (2009):178–191.

Gerber, Alan S., Donald Philip Green, and Christopher W. Larimer, "Social Pressure and Voter Turnout: Evidence from a Large-scale Field Experiment," *American Political Science Review* 102,1 (2008):33–48.

Geys, Benny, "Rational Theories of Voter Turnout: A Review," *Political Studies Review* 4 (2006):16–35.

Gibbons, Ann, "Spear-Wielding Chimps Seen Hunting Bush Babies," *Science* 315 (2007):1063.

Giddens, Anthony, *The Constitution of Society* (University of California Press, 1986).

Gies, Frances, *The Knight in History* (New York: Harper, 1984).

Gigerenzer, Gerd, "On the Supposed Evidence for Libertarian Paternalism," *Review of Philosophical Psychology* 6 (2015):361–383.

Gigerenzer, Gerd and P. M. Todd, *Simple Heuristics That Make Us Smart* (New York: Oxford University Press, 1999).

Gilbert, Margaret, "Modeling Collective Belief," *Synthese* 73 (1987):185–204.

Gilbert, Margaret, *On Social Facts* (New York: Routledge, 1989).

Gilboa, Itzhak and David Schmeidler, *A Theory of Case-Based Decisions* (Cambridge: Cambridge University Press, 2001).

Gintis, Herbert, "Education, Technology, and the Characteristics of Worker Productivity," *American Economic Review* 61,2 (1971):266–279.

Gintis, Herbert, "Welfare Economics and Individual Development: A Reply to Talcott Parsons," *Quarterly Journal of Economics* 89,2 (February 1975):291–302.

Gintis, Herbert, "The Nature of the Labor Exchange and the Theory of Capitalist Production," *Review of Radical Political Economics* 8,2 (Summer 1976):36–54.

Gintis, Herbert, "Strong Reciprocity and Human Sociality," *Journal of Theoretical Biology* 206 (2000):169–179.

Gintis, Herbert, "Some Implications of Endogenous Contract Enforcement for General Equilibrium Theory," in Fabio Petri and Frank Hahn (eds.)

General Equilibrium: Problems and Prospects (London: Routledge, 2002) pp. 176–205.

Gintis, Herbert, "The Hitchhiker's Guide to Altruism: Genes, Culture, and the Internalization of Norms," *Journal of Theoretical Biology* 220,4 (2003):407–418.

Gintis, Herbert, "Solving the Puzzle of Human Prosociality," *Rationality and Society* 15,2 (May 2003):155–187.

Gintis, Herbert, "The Dynamics of General Equilibrium," *Economic Journal* 117 (October 2007):1289–1309.

Gintis, Herbert, "The Evolution of Private Property," *Journal of Economic Behavior and Organization* 64,1 (September 2007):1–16.

Gintis, Herbert, "A Framework for the Unification of the Behavioral Sciences," *Behavioral and Brain Sciences* 30,1 (2007):1–61.

Gintis, Herbert, *The Bounds of Reason: Game Theory and the Unification of the Behavioral Sciences* (Princeton: Princeton University Press, 2009).

Gintis, Herbert, *Game Theory Evolving* (Princeton: Princeton University Press, 2009).

Gintis, Herbert, "Gene-culture Coevolution and the Nature of Human Sociality," *Proceedings of the Royal Society B* 366 (2011):878–888.

Gintis, Herbert, "The Clash of the Titans," *BioScience* 62,11 (November 2012):987–991.

Gintis, Herbert, "The Dynamics of Pure Market Exchange," in Masahiko Aoki, Kenneth Binmore, Simon Deakin, and Herbert Gintis (eds.) *Complexity and Institutions: Norms and Corporations* (London: Palgrave, 2012).

Gintis, Herbert, "Markov Models of Social Exchange: Theory and Applications," *ACM Transactions in Intelligent Systems and Technology* 4,3,53 (June 2013).

Gintis, Herbert, "Inclusive Fitness and the Sociobiology of the Genome," *Biology & Philosophy* 29,4 (2014):477–515.

Gintis, Herbert, "Sociobiology: Altruist Together," *Nature* 517 (2015):550–551.

Gintis, Herbert, "*Homo Ludens*: Rational Choice in Public and Private Spheres," *Journal of Economic Behavior and Organization* (2016).

Gintis, Herbert and Dirk Helbing, "Homo Socialis: An Analytical Core for Sociological Theory," *Review of Behavioral Economics* 2,1–2 (2015):1–59.

Gintis, Herbert and Rakesh Khurana, "Corporate Honesty and Business Education: A Behavioral Model," in Paul J. Zak (ed.) *Moral Markets: The Critical Role of Values in the Economy* (Princeton: Princeton University Press, 2008).

Gintis, Herbert, Carel P. van Schaik, and Christopher Boehm, "*Zoon Politikon*: The Evolutionary Origins of Human Political Systems," *Current Anthropology* 56,3 (June 2015):327–353.

Gintis, Herbert, Eric Alden Smith, and Samuel Bowles, "Costly Signaling and Cooperation," *Journal of Theoretical Biology* 213 (2001):103–119.

Gintis, Herbert, Samuel Bowles, Robert Boyd, and Ernst Fehr, *Moral Sentiments and Material Interests: On the Foundations of Cooperation in Economic Life* (Cambridge, MA: MIT Press, 2005).

Giugni, Marco G., Doug McAdam, and Charles Tilly, *From Contention to Democracy* (London: Rowman & Littlefield, 1998).

Glaeser, Edward L., B. Sacerdote, and J. Scheinkman, "Crime and Social Interactions," *Quarterly Journal of Economics* 111 (May 1996).

Glymour, Alison, D. M. Sobel, L. Schultz, and C. Glymour, "Causal Learning Mechanism in Very Young Children: Two- Three- and four-year-olds Infer Causal Relations from Patterns of Variation and Covariation," *Developmental Psychology* 37,50 (2001):620–629.

Glymour, C., *The Mind's Arrows: Bayes Nets and Graphical Causal Models in Psychology* (Cambridge: MIT Press, 2001).

Gneezy, Uri, "Deception: The Role of Consequences," *American Economic Review* 95,1 (March 2005):384–394.

Gneezy, Uri and Aldo Rustichini, "A Fine Is a Price," *Journal of Legal Studies* 29 (2000):1–17.

Godfrey-Smith, Peter, "The Strategy of Model-Based Science," *Biology and Philosophy* 21 (2006):725–740.

Godfrey-Smith, Peter, "Models and Fictions in Science," *Philosophical Studies* 143 (2009):101–116.

Goffman, Erving, *The Presentation of Self in Everyday Life* (New York: Anchor, 1959).

Golosnoy, V. and Y. Okhrin, "General Uncertainty in Portfolio Selection: A Case-based Decision Approach," *Journal of Economic Behavior and Organization* 67,3 (2008):718–734.

Good, I. J. and Lawrence S. Mayer, "Estimating the efficacy of a vote," *Behavioral Science* 20,1 (1975):25–33.

Goodall, Jane, "Tool-using and Aimed Throwing in a Community of Free-Living Chimpanzees," *Nature* 201 (1964):1264–1266.

Goodall, Jane, *The Chimpanzees of Gombe: Patterns of Behavior* (Belkap Press, 1986).

Goodnight, Charles, E. Rauch, H. Sayama, M. A. De Aguiar, M. Baranger, and Y. Bar-yam, "Evolution in Spatial Predator-prey Models and the 'Prudent Predator': The inadequacy of Steady-state Organism Fitness and the Concept of Individual and Group Selection," *Complexity* 13,5 (2008):23–44.

Gopnik, Alison and Andrew Meltzoff, *Words, Thoughts, and Theories* (Cambridge: MIT Press, 1997).

Gopnik, Alison and Joshua B. Tenenbaum, "Bayesian Networks, Bayesian Learning and Cognitive Development," *Developmental Studies* 10,3 (2007):281–287.

Gopnik, Alison and L. Schultz, *Causal Learning, Psychology, Philosophy, and Computation* (Oxford: Oxford University Press, 2007).

Gould, Stephen Jay and Richard C. Lewontin, "The Spandrels of San Marco and the Panglossian Paradigm: A Critique of the Adaptationist Programme," *Proceedings of the Royal Society of London B* 205 (1979):581–598.

Gouldner, Alvin W., "The Norm of Reciprocity: A Preliminary Statement," *American Sociological Review* 25 (1960):161–178.

Gowlett, John A. J. and Richard W. Wrangham, "Earliest Fire in Africa: Towards the Convergence of Archaeological Evidence and the Cooking Hypothesis," *Azania: Archaeological Research in Africa* 48,1 (2013):5–30.

Grafen, Alan, "Natural Selection, Kin Selection, and Group Selection," in John R. Krebs and Nicholas B. Davies (eds.) *Behavioural Ecology: An Evolutionary Approach* (Sunderland, MA: Sinauer, 1984).

Grafen, Alan, "A Geometric View of Relatedness," in Richard Dawkins and Mark Ridley (eds.) *Oxford Surveys in Evolutionary Biology volume 2* (Oxford: Oxford University Press, 1985) pp. 28–89.

Grafen, Alan, "The Logic of Divisively Asymmetric Contests: Respect for Ownership and the Desperado Effect," *Animal Behavior* 35 (1987):462–467.

Grafen, Alan, "Formal Darwinism, the Individual-as-Maximizing-Agent: Analogy, and Bet-hedging," *Proceedings of the Royal Society of London B* 266 (1999):799–803.

Grafen, Alan, "A First Formal Link between the Price Equation and an Optimization Program," *Journal of Theoretical Biology* 217 (2002):75–91.

Grafen, Alan, "Optimization of Inclusive Fitness," *Journal of Theoretical Biology* 238 (2006):541–563.

Grandmont, Jean Michel, "Temporary General Equilibrium Theory," *Econometrica* (1977):535–572.

Granovetter, Mark, "Economic Action and Social Structure: The Problem of Embeddedness," *American Journal of Sociology* 91,3 (November 1985):481–510.

Granovetter, Mark, "The Economic Sociology of Firms and Entrepreneurs," in A. Portes (ed.) *The Economic Sociology of Immigration: Essays on Networks, Ethnicity, and Entrepreneurshp* (New York: Russell Sage, 1995) pp. 128–165.

Green, Donald Philip and Ian Shapiro, *Pathologies of Rational Choice Theory: A Critique of Applications in Political Science* (New Haven, CT: Yale University Press, 1994).

Green, Leonard and John H. Kagel, "Intertemporal Choice Behavior: Evaluation of Economic and Psychological Models," in Leonard Green and John H. Kagel (eds.) *Advances in Behavioral Economics, Volume I* (Norwood, NJ: Ablex Publishing Company, 1987) pp. 185–215.

Grether, David and Charles R. Plott, "Economic Theory of Choice and the Preference Reversal Phenomenon," *American Economic Review* 69,4 (September 1979):623–638.

Grice, H. P., "Logic and Conversation," in Donald Davidson and Gilbert Harman (eds.) *The Logic of Grammar* (Encino, CA: Dickenson, 1975) pp. 64–75.

Grund, Thomas, Christian Waloszek, and Dirk Helbing, "How Natural Selection Can Create Both Self- and Other-Regarding Preferences, and Networked Minds," *Scientific Reports* 3 (2013):1480.

Grusec, Joan E. and Leon Kuczynski, *Parenting and Children's Internalization of Values: A Handbook of Contemporary Theory* (New York: John Wiley & Sons, 1997).

Guilfoos, Todd and Andreas Duus Pape, "Predicting Human Cooperation in the Prisoner's Dilemma Using Case-based Decision Theory," *Theory and Decision* 80 (2016):1–32.

Gunnthorsdottir, Anna, Kevin McCabe, and Vernon L. Smith, "Using the Machiavellianism Instrument to Predict Trustworthiness in a Bargaining Game," *Journal of Economic Psychology* 23 (2002):49–66.

Güth, Werner and Reinhard Tietz, "Ultimatum Bargaining Behavior: A Survey and Comparison of Experimental Results," *Journal of Economic Psychology* 11 (1990):417–449.

Güth, Werner, R. Schmittberger, and B. Schwarze, "An Experimental Analysis of Ultimatum Bargaining," *Journal of Economic Behavior and Organization* 3 (May 1982):367–388.

Habermas, Jürgen, *Theory of Communicative Action* (Boston: Beacon Press, 1984).

Habermas, Jürgen, *The Structural Transformation of the Public Sphere: An Inquiry into a Category of Bourgeois Society* (Cambridge: MIT Press, 1991).

Haidt, Jonathan, "The Emotional Dog and its Rational Tail: A Social Intuitionist Approach to Moral Judgment," *Psychological Review* 108 (2001):814–834.

Haidt, Jonathan, *The Righteous Mind: Why Good People Are Divided by Politics and Religion* (New York: Pantheon, 2012).

Haig, David, *Genomic Imprinting and Kinship* (New Brunswick: Rutgers University Press, 2002).

Haig, David and Alan Grafen, "Genetic Scrambling as a Defence Against Meiotic Drive," *Journal of Theoretical Biology* 153,4 (1991):531–558.

Hall, K. R. L., "Behaviour and Ecology of the Wild Patas Monkey, *Erythrocebus Patas*, in Uganda," *Journal of Zoology* 148,1 (1965):15–87.

Hamilton, William D., "The Evolution of Altruistic Behavior," *American Naturalist* 96 (1963):354–356.

Hamilton, William D., "The Genetical Evolution of Social Behavior, I," *Journal of Theoretical Biology* 7 (1964):1–16.

Hamilton, William D., "The Genetical Evolution of Social Behavior, II," *Journal of Theoretical Biology* 7 (1964):17–52.

Hamilton, William D., "Selfish and Spiteful Behaviour in an Evolutionary Model," *Nature* 228 (1970):1218–1220.

Hamilton, William D., "Altruism and Related Phenomena," *Annual Review of Ecological Systems* 3 (1972):199–232.

Hamilton, William D., "Innate Social Aptitudes of Man: An Approach from Evolutionary Genetics," in Robin Fox (ed.) *Biosocial Anthropology* (New York: John Wiley & Sons, 1975) pp. 115–132.

Hamlin, Alan and Colin Jennings, "Expressive Political Behavior: Foundations, Scope and Implications," *British Journal of Political Science* 41 (2011):645–670.

Hammond, Peter J., "Irreducibility, Resource Relatedness and Survival in Equilibrium with Individual Nonconvexities in General Equilibrium," in R. Becker, M. Boldrin, R. Jones, and W. Thomson (eds.) *Growth and Trade* (San Diego: Academic Press, 1993) pp. 73–115.

Harbaugh, William T., "If People Vote Because they Like to, then Why do so Many of Them Lie?," *Public Choice* 89,1–2 (1996):63–76.

Hardin, Garrett, "The Tragedy of the Commons," *Science* 162 (1968):1243–1248.

Hare, B., A. P. Melis, V. Woods, S. Hastings, and Richard W. Wrangham, "Tolerance Allows Bonobos to Outperform Chimpanzees on a Cooperative Task," *Current Biology* 17 (2007):619–623.

Harsanyi, John C., "Games with Incomplete Information Played by Bayesian Players," *Behavioral Science* 14 (1968):486–502.

Harsanyi, John C., "Morality and the Theory of Rational Behavior," *Social Research* 44,4 (Winter 1977).

Hartl, Daniel L. and Andrew G. Clark, *Principles of Population Genetics* (Sunderland, MA: Sinauer, 2007).

Hauser, Marc, *Wild Minds* (New York: Henry Holt, 2000).

Hawkes, Kristen, "Why Hunter-Gatherers Work: An Ancient Version of the Problem of Public Goods," *Current Anthropology* 34,4 (1993):341–361.

Hechter, Michael and Satoshi Kanazawa, "Sociological Rational Choice," *Annual Review of Sociology* 23 (1997):199–214.

Heckman, James, "Policies to Foster Human Capital," *Research in Economics* 54 (2000):3–56.

Heckman, James and Yona Rubinstein, "The Importance of Noncognitive Skills: Lessons from the GED Testing Program," *American Economic Review* 91,2 (May 2001):145–149.

Heckman, James and Yona Rubinstein, "Non Cognitive Aspects of the Intergenerational Transmission of Economic Status: Evidence from the GED," in Samuel Bowles, Herbert Gintis, and Melissa Osborne (eds.) *Unequal Chances: Family Background and Economic Success* (New York: Russell Sage Foundation, 2003).

Heckman, James, Anne Layne-Farrar, and Petra Todd, "Does Measured School Quality Really Matter? An Examination of the Earnings-Quality Relationship," in Gary Burtless (ed.) *Does Money Matter? The Effect of School Resources on Student Achievement and Adult Success* (Washington, DC: Brookings Institution Press, 1996) pp. 192–289.

Heckman, James, Jingjing Hsee, and Yona Rubinstein, "The GED as a Mixed Signal," 1999. University of Chicago.

Hedström, Peter and P. Bearman, *The Oxford Handbook of Analytical Sociology* (Oxford: Oxford University Press, 2009).

Helbing, Dirk, *Quantitative Sociodynamics* (Dordrecht: Kluwer Academic, 1995).

Helbing, Dirk, "A Stochastic Behavioral Model and a 'Microscopic' Foundation of Evolutionary Game Theory," *Theory and Decision* 40 (1996):149–179.

Helbing, Dirk, *Quantitative Sociodynamics, 2nd Edition* (Berlin: Springer-Verlag, 2010).

Helbing, Dirk, *Social Self-organization: Agent-Based Simulations and Experiments to Study Emergent Social Behavior* (Springer, 2012).

Helbing, Dirk, Attila Szolnoki, Matjaz Perc, and Gyorgy Szabo, "Evolutionary Establishment of Moral and Double Moral Standards through Spatial Interactions," *PLoS Computational Biology* 6 (2010):e10000758.

Henrich, Joseph and Francisco Gil-White, "The Evolution of Prestige: Freely Conferred Status as a Mechanism for Enhancing the Benefits of Cultural Transmission," *Evolution and Human Behavior* 22 (2001):165–196.

Henrich, Joseph, Robert Boyd, Samuel Bowles, Colin F. Camerer, Ernst Fehr, and Herbert Gintis, *Foundations of Human Sociality: Economic Experiments and Ethnographic Evidence from Fifteen Small-Scale Societies* (Oxford: Oxford University Press, 2004).

Henrich, Joseph, Robert Boyd, Samuel Bowles, Colin F. Camerer, Ernst Fehr, and Herbert Gintis, "'Economic Man' in Cross-Cultural Perspective: Behavioral Experiments in 15 Small-Scale societies," *Behavioral and Brain Sciences* 28,6 (2005):795–815.

Herbst, Daniel and Alexandre Mas, "Peer Effects on Worker Output in the Laboratory Generalize to the Field," *Science* 350,6260 (2015):545–549.

Hill, Kim R., R. S. Walker, M. Bozicevic, J. Eder, T. Headland, B. Hewlett, Ana Magdalena Hurtado, Frank Marlowe, Polly Wiessner, and Bernard Wood, "Co-residence Patterns in Hunter-gatherer Societies Show Unique Human Social Structure," *Science* 331 (2011):1286–1289.

Hilton, Denis J., "The Social Context of Reasoning: Conversational Inference and Rational Judgment," *Psychological Bulletin* 118,2 (1995):248–271.

Hirshleifer, Jack, "The Analytics of Continuing Conflict," *Synthése* 76 (1988):201–233.

Hobbes, Thomas, *Leviathan* (New York: Penguin, 1968[1651]). Edited by C. B. MacPherson.

Hodgson, Geoffrey M., "The Approach of Institutional Economics," *Journal of Economic Literature* 36,1 (1998):166–192.

Holden, C. J., "Bantu Language Trees Reflect the Spread of Farming across Sub-Saharan Africa: A Maximum-parsimony Analysis," *Proceedings of the Royal Society of London B* 269 (2002):793–799.

Holden, C. J. and Ruth Mace, "Spread of Cattle Led to the Loss of Matrilineal Descent in Africa: A Coevolutionary Analysis," *Proceedings of the Royal Society of London B* 270 (2003):2425–2433.

Hölldobler, Bert and Edward O. Wilson, *The Ants* (Cambridge, MA: Belknap Press, 1990).

Homans, George, "Social Behavior as Exchange," *American Journal of Sociology* 65,6 (May 1958):597–606.

Horn, Roger A. and Charles R. Johnson, *Matrix Analysis* (Cambridge: Cambridge University Press, 1985).

Hrdy, Sarah Blaffer, *Mother Nature: A History of Mothers, Infants, and Natural Selection* (New York: Pantheon Books, 1999).

Hrdy, Sarah Blaffer, *Mother Nature: Maternal Instincts and How They Shape the Human Species* (New York: Ballantine, 2000).

Hrdy, Sarah Blaffer, *Mothers and Others: The Evolutionary Origins of Mutual Understanding* (New York: Belknap, 2009).

Hrdy, Sarah Blaffer, "Estimating the Prevalence of Shared Care and Co-operative Breeding in the Order Primates," *Online at www.citrona.com* (June 23 2010).

Huffman, M. A. and M. A. Kalunde, "Tool-Assisted Predation on a Squirrel by a Female Chimpanzee in the Mahale Mountains, Tanzania," *Primates* 34 (1993):93–98.

Huizinga, Josef, *Homo Ludens: A Study of the Play Element in Culture* (Einaudi, 2002[1939]).

Humphrey, Nicholas, "The Social Function of Intellect," in P. P. G. Bateson and R. A. Hinde (eds.) *Growing Points in Ethology* (Cambridge: Cambridge University Press, 1976) pp. 303–317.

Hurley, Susan L., *Consciousness in Action* (Cambridge: Harvard University Press, 2002).

Huxley, Julian S., "Evolution, Cultural and Biological," *Yearbook of Anthropology* (1955):2–25.

Ihara, Y., "Evolution of Culture-dependent Discriminate Sociality: a Gene-Culture Coevolutionary Model," *Proceedings of the Royal Society of London B* 366 (2011).

Isaac, Barbara, "Throwing and Human Evolution," *African Archaeological Review* 5 (1987):3–17.

Isaac, Glynn, "Food Sharing and Human Evolution: Archaeological Evidence from the Plio-Pleistocene of East Africa," *The Harvey Lecture Series* (1977):311–325.

Isaac, Glynn, "The food-sharing Behavior of Protohuman Hominids," *Scientific American* 238 (1978):90–108.

Jablonka, Eva and Marion J. Lamb, *Epigenetic Inheritance and Evolution: The Lamarckian Case* (Oxford: Oxford University Press, 1995).

Jaeggi, A. V., Jeroen M. G. Stevens, and Carel P. van Schaik, "Tolerant Food Sharing and Reciprocity is Precluded by Despotism Among Bonobos but not Chimpanzees," *American Journal of Physical Anthropology* 143 (2010):41–51.

James, William, "Great Men, Great Thoughts, and the Environment," *Atlantic Monthly* 46 (1880):441–459.

Joas, Hans and Wolfgang Knöbl, *Social Theory* (Cambridge: Cambridge University Press, 2009).

Johnson, Eric J. and Daniel G. Goldstein, "Do Defaults Save Lives?," *Science* 302 (2003):1338–1339.

Jolly, Alison, *The Evolution of Primate Behavior* (New York: MacMillan, 1972).

Jones, Owen D., "Time-Shifted Rationality and the Law of Law's Leverage: Behavioral Economics Meets Behavioral Biology," *Northwestern University Law Review* 95 (2001):1141–1206.

Jurmain, Robert, Harry Nelson, Lynn Kilgore, and Wenda Travathan, *Introduction to Physical Anthropology* (Cincinnati: Wadsworth Publishing Company, 1997).

Kahneman, Daniel and Amos Tversky, "Prospect Theory: An Analysis of Decision under Risk," *Econometrica* 47 (1979):263–291.

Kahneman, Daniel, Jack L. Knetsch, and Richard H. Thaler, "The Endowment Effect, Loss Aversion, and Status Quo Bias," *Journal of Economic Perspectives* 5,1 (Winter 1991):193–206.

Kandel, Denise, "Homophily, Selection and Socialization in Adolescent Friendships," *American Journal of Sociology* 84,2 (1978):427–436.

Kandori, M. G., G. Mailath, and R. Rob, "Learning, Mutation, and Long Run Equilibria in Games," *Econometrica* 61 (1993):29–56.

Kant, Immanuel, *Groundwork of the Metaphysics of Morals* (Cambridge: Cambridge University Press, 2012[1797]).

Kaplan, Hillard and Kim R. Hill, "Food Sharing among Ache Foragers: Tests of Explanatory Hypotheses," *Current Anthropology* 26,2 (1985):223–246.

Kaplan, Hillard and Kim R. Hill, "Hunting Ability and Reproductive Success among Male Ache Foragers: Preliminary Results," *Current Anthropology* 26,1 (1985):131–133.

Kaplan, Hillard, Kim R. Hill, Jane Lancaster, and Ana Magdalena Hurtado, "A Theory of Human Life History Evolution: Diet, Intelligence, and Longevity," *Evolutionary Anthropology* 9 (2000):156–185.

Kaplan, Hillard, Kim R. Hill, Kristen Hawkes, and Ana Magdalena Hurtado, "Food Sharing among Ache Hunter-Gatherers of Eastern Paraguay," *Current Anthropology* 25,1 (1984):113–115.

Kaplan, Hillard, S. W. Gangestad, Michael Gurven, Jane Lancaster, T. Mueller, and Arthur Robson, "The Evolution of Diet, Brain and Life History among Primates and Humans," in W. Roebroeks (ed.) *Guts and Brains: An Interative Approach to the Hominin Record* (Leiden, NL: Leiden University Press, 2007) pp. 47–81.

Keeley, Lawrence H., *War Before Civilization* (New York: Oxford University Press, 1996).

Keeley, Lawrence H. and Nicholas Toth, "Microwear Polishes on Early Stone Tools from Koobi Fora, Keyna," *Nature* 293 (October 1981):464–465.

Keller, Laurent and Kenneth G. Ross, "Selfish genes: A Green Beard in the Red Fire Ant," *Science* 394 (6 August 1998):573–575.

Kelly, Raymond C., *Warless Societies and the Origin of War* (Ann Arbor: University of Michigan Press, 2000).

Kelly, Raymond C., "The Evolution of Lethal Intergroup Violence," *Proceedings of the National Academy of Sciences* 102 (2005):15294–15298.

Kelly, Robert L., *The Foraging Spectrum: Diversity in Hunter-Gatherer Lifeways* (Washington, DC: The Smithsonian Institution, 1995).

Kemeny, John G. and J. Laurie Snell, *Finite Markov Chains* (Princeton: Van Nostrand, 1960).

King, Andrew J., Caitlin Douglas, Elise Huchard, Nick Isaac, and Guy Cowlishaw, "Dominance and Affiliation Mediate Despotism in a Social Primate," *Current Biology* 18 (2008):1833–1838.

King, Andrew J., Dominio D. P. Johnson, and Mark Van Vugt, "The Origins and Evolution of Leadership," *Current Biology* 19 (2009):R911–R916.

Kirby, Kris N. and Richard J. Herrnstein, "Preference Reversals Due to Myopic Discounting of Delayed Reward," *Psychological Science* 6,2 (March 1995):83–89.

Kirman, Alan, H. Foellmer, and U. Horst, "Equilibrium in Financial Markets with Heterogeneous Agents: A New Perspective," *Journal of Mathematical Economics* 41,1–2 (February 2005):123–155.

Kiyonari, Toko, Shigehito Tanida, and Toshio Yamagishi, "Social Exchange and Reciprocity: Confusion or a Heuristic?," *Evolution and Human Behavior* 21 (2000):411–427.

Klein, Richard G., *The Human Career: Human Biological and Cultural Origins* (Chicago: University of Chicago Press, 1999).

Knack, Stephen, "Civic norms, social sanctions, and voter turnout," *Rationality and Society* 4,2 (1992):133–156.

Knauft, Bruce, "Violence and Sociality in Human Evolution," *Current Anthropology* 32,4 (August–October 1991):391–428.

Knill, D. and A. Pouget, "The Bayesian Brain: the Role of Uncertainty in Neural Coding and Computation," *Trends in Cognitive Psychology* 27,12 (2004):712–719.

Koella, Jacob C., "The Spatial Spread of Altruism versus the Evolutionary Response of Egoists," *Proceedings of the Royal Society B* 267,1456 (October 2000):1979–1985.

Kording, K. P. and D. M. Wolpert, "Bayesian Decision Theory in Sensorimotor Control," *Trends in Cognitive Sciences* 10 (2006):319–326.

Kosfeld, Michael, Markus Heinrichs, Paul J. Zak, Urs Fischbacher, and Ernst Fehr, "Oxytocin Increases Trust in Humans," *Nature* 435 (June 2005):673–676.

Krakauer, David C., "Kin Selection and Cooperative Courtship in Wild Turkeys," *Nature* 434 (3 March 2005):69–72.

Krauss, Jens, David Lusseau, and Richard James, "Animal Social Networks: An Introduction," *Behavioral Ecology and Sociobiology* 63,7 (2009):967–973.

Krebs, John R. and Nicholas B. Davies, *Behavioral Ecology: An Evolutionary Approach* fourth ed. (Oxford: Blackwell Science, 1997).

Kropotkin, Pyotr, *Mutual Aid: A Factor in Evolution* (New York: Black Rose Books, 1989[1903]).

Kuhn, Thomas, *The Nature of Scientific Revolutions* (Chicago: University of Chicago Press, 1962).

Kummer, Hans and Marina Cords, "Cues of Ownership in Long-tailed Macaques, *Macaca fascicularis*," *Animal Behavior* 42 (1991):529–549.

Lack, David, "Interrelationship in Breeding Adaptations as shown by Marine Birds," *Proceedings of the XIVth Ornithology Congress, Oxford* (1966):3–42.

Laland, Kevin N., F. John Odling-Smee, and Marcus W. Feldman, "Group Selection: A Niche Construction Perspective," *Journal of Consciousness Studies* 7,1/2 (2000):221–224.

Lambert, Patricia, "Patterns of Violence in Prehistoric Hunter-Gatherer Societies of Coastal Southern California," in Debra L. Martin and David W. Frayer (eds.) *Troubled Times: Violence and Warfare in the Past* (Amsterdam: Gordon and Breach, 1997) pp. 77–109.

Langergraber, Kevin E., "Generation Times in Wild Chimpanzees and Gorillas Suggest Earlier Divergence Times in Great Ape and Human

Evolution," *Proceedings of the National Academy of Sciences* 109,39 (2012):15716–15721.

Lasswell, Harold and Abraham Kaplan, *Power and Society: A Framework for Political Enquiry* (New Haven: Yale University Press, 1950).

Leakey, Mary, *Olduvai Gorge, Excavations in Beds I and III, 1960–1963, Vol. 3* (Cambridge: Cambridge University Press, 1971).

Ledyard, John O., "The Paradox of Voting and Candidate Competition: A General Equilibrium Analysis," in George Horwich and James Quirk (eds.) *Essays in Contemporary Fields of Economics* (West Lafayette, In: Purdue University Press, 1981) pp. 54–80.

Ledyard, John O., "Public Goods: A Survey of Experimental Research," in John H. Kagel and Alvin E. Roth (eds.) *The Handbook of Experimental Economics* (Princeton: Princeton University Press, 1995) pp. 111–194.

Lee, Richard Borshay, *The !Kung San: Men, Women and Work in a Foraging Society* (Cambridge: Cambridge University Press, 1979).

Lee, Richard Borshay and Irven DeVore, *Man the Hunter* (Transaction Publishers, 1968).

Leffler, Ellen M., "Multiple Instances of Ancient Balancing Selection Shared Between Humans and Chimpanzees," *Science* 339 (29 March 2013):1578–1582.

Leigh, Egbert G., *Adaptation and Diversity* (San Francisco, CA: Freeman, Cooper, 1971).

Leigh, Egbert G., "How does Selection Reconcile Individual Advantage with the Good of the Group?," *Proceedings of the National Academy of Sciences, USA* 74,10 (October 1977):4542–4546.

Lerner, Abba, "The Economics and Politics of Consumer Sovereignty," *American Economic Review* 62,2 (May 1972): 258–266.

Levine, David K. and Thomas R. Palfrey, "The paradox of voter participation? A laboratory study," *American Political Science Review* 101,1 (2007):143–158.

LeVine, Robert A. and Donald T. Campbell, *Ethnocentrism: Theories of Conflict, Ethnic Attitudes, and Group Behavior* (New York: Wiley, 1972).

Lewis, David, *Conventions: A Philosophical Study* (Cambridge, MA: Harvard University Press, 1969).

Lewontin, Richard C., "The Units of Selection," in Richard Johnston (ed.) *Annual Review of Ecology and Systematics* (Palo Alto: Annual Review Inc., 1970).

Lewontin, Richard C., "Sleight of Hand," *The Sciences* (July-August 1981):23–26.

Li, Ming and Dipjyoti Majumdar, "A psychologically based model of voter turnout," *Journal of Public Economic Theory* 12,5 (2010):979–1002.

Liberman, Uri and Marcus W. Feldman, "On the Evolution of Epistasis I: Diploids Under Selection," *Theoretical Population Biology* 67,3 (2005):141–160.

Lichtenstein, Sarah and Paul Slovic, "Reversals of Preferences between Bids and Choices in Gambling Decisions," *Journal of Experimental Psychology* 89 (1971):46–55.

Lijphart, Arend, "Unequal Participation: Democracy's Unresolved Dilemma," *American Political Science Review* 91 (March 1997):1–14.

Lindenberg, Sigwart, "Utility and Morality," *Kyklos* 36,3 (1983):450–468.

Lindenberg, Sigwart, "Social Rationality," in G. Ritzer (ed.) *Encyclopedia of Social Theory, Vol. II* (Thousand Oaks: Sage, 2004) pp. 759–760.

Linton, Ralph, *The Study of Man* (New York: Appleton-Century-Crofts, 1936).

Lombard, M. and L. Phillipson, "Indications of Bow and Stone-Tipped Arrow Use 64,000 Years Ago in KwaZulu-Natal, South Africa," *Antiquity* 84 (2010):1–14.

Loomes, Graham, "When Actions Speak Louder than Prospects," *American Economic Review* 78,3 (June 1988):463–470.

Lorenz, Konrad, *On Aggression* (New York: Harcourt, Brace and World, 1963).

Luce, Robert Duncan, *Individual Choice Behavior* (New York: Dover, 2005).

Luce, Robert Duncan and Patrick Suppes, "Preference, Utility, and Subjective Probability," in Robert Duncan Luce, Robert R. Bush, and Eugene Galanter (eds.) *Handbook of Mathematical Psychology, vol. III* (New York: Wiley, 1965).

Luhmann, Niklas, *Introduction to Systems Theory* (Polity, 2012[2004]).

Lukes, Stephen, *Power: A Radical View* (London: Macmillan, 1974).

Lumsden, Charles J. and Edward O. Wilson, *Genes, Mind, and Culture: The Coevolutionary Process* (Cambridge, MA: Harvard University Press, 1981).

Lupo, Karen D. and James F. O'Connell, "Cut and Tooth Mark Distributions on Large Animal Bones: Ethnoarchaeological Data from the Hadza and their Implications for Current Ideas about Early Human Carnivory," *Journal of Archaeological Science* 29 (2002):85–109.

Ma, W. J., J. M. Beck, P. E. Latham, and A. Pouget, "Bayesian Inference with Probabilistic Population Codes," *Nature Neuroscience* 9,11 (2006):1432–1438.

Macdonald, D. W. and C. Sillero-Zubiri, *Biology and Conservation of Wild Canids* (New York: Oxford University Press, 2004).

Mace, Ruth and Mark Pagel, "The Comparative Method in Anthropology," *Current Anthropology* 35 (1994):549–564.

Machina, Mark J., "Choice under Uncertainty: Problems Solved and Unsolved," *Journal of Economic Perspectives* 1,1 (Summer 1987):121–154.

Mackie, John, *Ethics: Inventing Right and Wrong* (Penguin, 1977).

Maestripieri, Dario, *Machiavellian Intelligence: How Rhesus Macaques and Humans Have Conquered the World* (Chicago: University of Chicago Press, 2007).

Majolo, B., J. Lehmann, A. Bortoli de Vizioli, and G. Schino, "Fitness-related Benefits of Dominance in Primates," *American Journal of Physical Anthropology* 147 (2012):652–660.

Malécot, Gustave, *Les Mathématiques de l'Hérédité* (Paris: Masson, 1948).

Malthus, Thomas Robert, *An Essay on the Principle of Population* (Anonymously Published, 1798).

Mandel, Antoine and Herbert Gintis, "Stochastic stability in the Scarf economy," *Mathematical Social Sciences* 67 (2014):44–49.

Mandel, Antoine and Herbert Gintis, "Decentralized Pricing and the Strategic Stability of Walrasian General Equilibrium," *Journal of Mathematical Economics* 63 (2016):84–92.

Mandel, Antoine, Simone Landini, Mauro Gallegati, and Herbert Gintis, "Price Dynamics, Financial Fragility, and Aggregate Volatility," *Journal of Economic Dynamics & Control* 51 (2015):257–277.

Mandeville, Bernard, *The Fable of the Bees: Private Vices, Publick Benefits* (Oxford: Clarendon, 1924 [1705]).

Mantel, Rolf, "On the Characterization of Aggregate Excess Demand," *Journal of Economic Theory* 7 (1974):348–53.

Mantel, Rolf, "Homothetic Preferences and Community Excess Demand Functions," *Journal of Economic Theory* 12 (1976):197–201.

Markus, Gregory, "The Impact of Personal and National Economic Conditions on the Presidential Vote: A Pooled Cross-sectional Analysis," *American Journal of Political Science* 32,1 (February 1988):137–154.

Marshall, Alfred, *Principles of Economics (Eighth Edition)* (London: MacMillan, 1930).

Martinez, I., J.-L. Arsuaga, R. Quam, J.-M. Carretero, A. Gracia, and L. Rodriguez, "Human Hyoid Bones From the Middle Pleistocene Site of the Sima de los Huesos (Sierra de Atapuerca, Spain)," *Journal of Human Evolution* 54,1 (2008):118–124.

Martinez, I., M. Rosa, J.-L. Arsuaga, P. Jarabo, R. Quam, C. Lorenzo, A. Gracia, J.-M. Carretero, J.-M. Bermu de de Castro, and E. Carbonell, "Auditory Capacities in Middle Pleistocene Humans from the Sierra de Atapuerca in Spain," *Proceedings of the National Academy of Sciences* 101 (2004):9997–9981.

Marx, Karl, *The Communist Manifesto* (New York: Signet Classics, 1998[1848]).

Marzke, Mary W., "Precision Grips, Hand Morphology, and Tools," *American Journal of Physical Anthropology* 102 (1997):91–110.

Mas-Colell, Andreu, Michael D. Whinston, and Jerry R. Green, *Microeconomic Theory* (New York: Oxford University Press, 1995).

Masclet, David, Charles Noussair, Steven Tucker, and Marie-Claire Villeval, "Monetary and Nonmonetary Punishment in the Voluntary Contributions Mechanism," *American Economic Review* 93,1 (March 2003):366–380.

Maynard Smith, John, "Group Selection and Kin Selection," *Nature* 201 (1964):1145–1147.

Maynard Smith, John, "The Theory of Games and the Evolution of Animal Conflicts," *Journal of Theoretical Biology* 47 (1974):209–221.

Maynard Smith, John, "Group Selection," *Quarterly Review of Biology* 51 (1976):277–283.

Maynard Smith, John, *Evolution and the Theory of Games* (Cambridge: Cambridge University Press, 1982).

Maynard Smith, John, "Evolutionary Progress and Levels of Selection," in M. H. Nitecki (ed.) *Evolutionary Progress* (Chicago: University of Chicago Press, 1988) pp. 219–30.

Maynard Smith, John and Eörs Szathmáry, "The Major Evolutionary Transitions," *Nature* 374 (1995):227–232.

Maynard Smith, John and G. A. Parker, "The Logic of Asymmetric Contests," *Animal Behaviour* 24 (1976):159–175.

Maynard Smith, John and George R. Price, "The Logic of Animal Conflict," *Nature* 246 (2 November 1973):15–18.

Maynard Smith, John and M. G. Ridpath, "Wife Sharing in the Tasmanian Native Hen, *Tribonyx mortierii*: A Case of Kin Selection?," *American Naturalist* 96 (1972):447.

Maynard Smith, John and N. Warren, "Models of Cultural and Genetic Change," *Evolution* 36,3 (1982):620–627.

Mayr, Ernst, *The Growth of Biological Thought: Diversity, Evolution & Inheritance* (Cambridge: Harvard University Press, 1982).

McCloskey, Deirdre N., *The Bourgeois Virtues: Ethics for an Age of Commerce* (Chicago: University of Chicago Press, 2010).

McFadden, Daniel, "Conditional Logit Analysis of Qualitative Choice Behavior," in Paul Zarembka (ed.) *Frontiers in Econometrics* (New York: Academic Press, 1973) pp. 105–142.

McGrew, William Clement, *Chimpanzee Material Culture: Implications for Human Evolution* (Cambridge: Cambridge University Press, 1992).

McGrew, William Clement, *The Cultured Chimpanzee: Reflections on Cultural Primatology* (Cambridge: Cambridge University Press, 2004).

McKenzie, C. R. M., M. J. Liersch, and S. R. Finkelstein, "Recommendations Implicit in Policy Defaults," *Psychological Science* 17 (2006):414–420.

McKenzie, L. W., "On the Existence of a General Equilibrium for a Competitive Market," *Econometrica* 28 (1959):54–71.

McKenzie, L. W., "Stability of Equilibrium and Value of Positive Excess Demand," *Econometrica* 28 (1960):606–617.

McPherron, S. P., "Evidence for Stone-tool Assisted Consumption of Animal Tissues Before 3.39 Million Years Ago at Dikika, Ethiopia," *Nature* 466,7308 (2010):857–860.

McPherson, M., L. Smith-Lovin, and J. Cook, "Birds of a Feather: Homophily in Social Networks," *Annual Review of Sociology* 27 (2001):415–444.

Mead, George Herbert, *Mind, Self, and Society* (Chicago: University of Chicago Press, 1934).

Mead, Margaret, *Sex and Temperament in Three Primitive Societies* (New York: Morrow, 1963).

Mealey, Linda, "The Sociobiology of Sociopathy," *Behavioral and Brain Sciences* 18 (1995):523–541.

Medina, Martin, *The World's Scavengers* (AltaMira Press, 2007).

Mednick, S. A., L. Kirkegaard-Sorenson, B. Hutchings, J Knop, R. Rosenberg, and F. Schulsinger, "An Example of Bio-social Interaction Research: The Interplay of Socio-environmental and Individual Factors in the Etiology of Criminal Behavior," in S. A. Mednick and K. O. Christiansen (eds.) *Biosocial Bases of Criminal Behavior* (New York: Gardner Press, 1977) pp. 9–24.

Merton, Robert, *Social Theory and Social Structure* (New York: Free Press, 1968[1957]).

Mesoudi, Alex, Andrew Whiten, and Kevin N. Laland, "Towards a Unified Science of Cultural Evolution," *Behavioral and Brain Sciences* 29 (2006):329–383.

Mesterton-Gibbons, Mike, "Ecotypic Variation in the Asymmetric Hawk-Dove Game: When Is Bourgeois an ESS?," *Evolutionary Ecology* 6 (1992):1151–1186.

Mesterton-Gibbons, Mike and Eldridge S. Adams, "Landmarks in Territory Partitioning," *American Naturalist* 161,5 (May 2003):685–697.

Metz, J. A. J., S. D. Mylius, and O. Diekmann, "When Does Evolution Optimize," *Evolutionary Ecology Research* 10 (2008):629–654.

Michod, Richard E., "Cooperation and Conflict in the Evolution of Individuality. 1. The Multilevel Selection of the Organism," *American Naturalist* 149 (April 1997):607–645.

Michod, Richard E. and William D. Hamilton, "Coefficients of Relatedness in Sociobiology," *Nature* 288 (18/25 December 1980):694–697.

Milinski, Manfred, "Byproduct Mutualism, Tit-for-Tat Reciprocity and Cooperative Predator Inspection," *Animal Behavior* 51 (1996):458–461.

Miller, B. L., A. Darby, D. F. Benson, J. L. Cummings, and M. H. Miller, "Aggressive, Socially Disruptive and Antisocial Behaviour Associated with Fronto-temporal Dementia," *British Journal of Psychiatry* 170 (1997):150–154.

Miller, Geoffrey, *The Mating Mind: How Sexual Choice Shaped the Evolution of Human Nature* (New York: Anchor, 2001).

Miller, John H. and Scott E. Page, *Complex Adaptive Systems: An Introduction to Computational Models of Social Life* (Princeton: Princeton University Press, 2007).

Milton, K., "Habitat, Diet, and Activity Patterns of Free-ranging Woolly Spider Monkeys," *International Journal of Primatology* 5,5 (1984):491–514.

Mischel, Walter and Ebbe B. Ebbeson, "Attention in delay of gratification," *Journal of Personality and Social Psychology* 16,2 (October 1970):329–337.

Moll, Jorge, Roland Zahn, Ricardo di Oliveira-Souza, Frank Krueger, and Jordan Grafman, "The Neural Basis of Human Moral Cognition," *Nature Neuroscience* 6 (October 2005):799–809.

Morowitz, Harold, *The Emergence of Everything: How the World Became Complex* (Oxford: Oxford University Press, 2002).

Morris, Desmond, *The Naked Ape: A Zoologist's Study of the Human Animal* (New York: Delta, 1999[1967]).

Morris, Stephen, "The Common Prior Assumption in Economic Theory," *Economics and Philosophy* 11 (1995):227–253.

Musgrave, Richard Abel, *The Theory of Public Finance* (McGraw-Hill, 1959).

Nash, John F., "Equilibrium Points in n-Person Games," *Proceedings of the National Academy of Sciences* 36 (1950):48–49.

Navarrete, A., Karin Eisler, and Carel P. van Schaik, "Energetics and the Evolution of Human Brain Size," *Nature* 480 (2011):91–93.

Nelson, Richard and Sidney Winter, *An Evolutionary Theory of Economic Change* (Cambridge: Harvard University Press, 1982).

Newman, Mark, Albert-Laszlo Barabasi, and Duncan J. Watts, *The Structure and Dynamics of Networks* (Princeton: Princeton University Press, 2006).

Niemi, Richard G., Guy Whitten, and Mark N. Franklin, "Constituency Characteristics, Individual Characteristics and Tactical Voting in the 1987 British General Election," *British Journal of Political Science* 22 (1992):229–240.

Nikaido, Hukukaine, "On the Classical Multilateral Exchange Problem," *MetroEconomica* 8 (1956):135–145.

Nikaido, Hukukaine, "Stability of Equilibrium by the Brown-von Neumann Differential Equation," *MetroEconomica* 27 (1959):645–671.

Nikaido, Hukukaine and Hirofumi Uzawa, "Stability and Nonnegativity in a Walrasian Tâtonnement Process," *International Economic Review* 1 (1960):50–59.

Nisbett, Richard E. and Dov Cohen, *Culture of Honor: The Psychology of Violence in the South* (Boulder, CO: Westview Press, 1996).

Nishida, Toshisada, "The Ant-Gathering Behavior by the Use of Tools Among Wild Chimpanzees of the Mahali Mountains," *Journal of Human Evolution* 2 (1973):357–370.

Nishida, Toshisada and K Hosaka, "Coalition Strategies among Adult Male Chimpanzees of the Mahale Mountains, Tanzania," in W. C. McGrew, L. F. Marchant, and T. Nishida (eds.) *Great Ape Societies* (Cambridge: Cambridge University Press, 1996) pp. 114–134.

Noble, Denis, "Neo-Darwinism, the Modern Synthesis and Selfish Genes: Are they of Use in Physiology?," *Journal of Physiology* 589,5 (2011):1007–1015.

Noe, Ronald and Peter Hammerstein, "Biological Markets: Supply and Demand Determine the Effect of Partner Choice in Cooperation, Mutualism, and Mating," *Behavioral Ecology and Sociobiology* 35 (1994):1–11.

Nowak, Martin A., "Five Rules for the Evolution of Cooperation," *Science* 314 (2006):1560–1563.

Nowak, Martin A. and Karl Sigmund, "Evolution of Indirect Reciprocity by Image Scoring," *Nature* 393 (1998):573–577.

Nowak, Martin A., Corina E. Tarnita, and Edward O. Wilson, "The Evolution of Eusociality," *Nature* 466,26 August (2010):1057–1062.

Nozick, Robert, "Coercion," in Sidney Morgenbesser, Patrick Suppes, and Morton White (eds.) *Philosophy, Science and Method* (New York: St. Martin's Press, 1969) pp. 440–472.

Oaksford, Mike and Nick Chater, *Bayesian Rationality: The Probabilistic Approach to Human Reasoning* (Oxford: Oxford University Press, 2007).

O'Brien, M. J. and R. L. Lyman, *Applying Evolutionary Archaeology* (New York: Kluwer Academic, 2000).

O'Connell, James F., Kristen Hawkes, Karen D. Lupo, and Nicholas G. Blurton-Jones, "Male Strategies and Plio-Pleistocene Archaeology," *Journal of Human Evolution* 43,6 (December 2002):831–872.

Odling-Smee, F. John, Kevin N. Laland, and Marcus W. Feldman, *Niche Construction: The Neglected Process in Evolution* (Princeton: Princeton University Press, 2003).

Ok, Efe A. and Yusufcan Masatlioglu, "A General Theory of Time Preference," 2003. Economics Department, New York University.

Okada, Daijiro and Paul M. Bingham, "Human Uniqueness Self-interest and Social Cooperation," *Journal of Theoretical Biology* 254,2 (2008):261–270.

Okasha, Samir and Kenneth G. Binmore, *Evolution and Rationality: Decisions, Co-operation, and Strategic Behavior* (Cambridge: Cambridge University Press, 2012).

Olson, Mancur, *The Logic of Collective Action: Public Goods and the Theory of Groups* (Cambridge: Harvard University Press, 1965).

O'Neil, Dennis, "Analysis of Early Hominins," anthro.palomar.edu/hom inid/australo˙2.htm 2012.

Oosterbeek, Hessel, Randolph Sloop, and Gus van de Kuilen, "Cultural Differences in Ultimatum Game Experiments: Evidence from a Meta-analysis," *Experimental Economics* 7 (2004):171–188.

Ossadnik, W., D. Wilmsmann, and B. Niemann, "Experimental Evidence on Case-based Decision Theory," *Theory and Decision* (2012):1–22.

Otterbein, Keith F., *How War Began* (Texas A&M University Press, 2004).

Pagel, Mark, *Wired for Culture:* (New York: W. W. Norton, 2012).

Paley, William, *Natural Theology* (Philadelpha: John Morgan, 1802).

Palfrey, Thomas R. and Howard Rosenthal, "Voter Turnout and Strategic Uncertainty," *American Political Science Review* 79 (1985):62–78.

Pandit, Sagar and Carel P. van Schaik, "A Model for Leveling Coalitions among Primate Males: toward a Theory of Egalitarianism," *Behavioral Ecology and Sociobiology* 55 (2003):161–168.

Pareto, Vilfredo, *Corso di Economia Politica* (Turin: Unione Tipographico-Editrice Torinese, 1896–1897).

Pareto, Vilfredo, *Manuale di Economia Politica* (Milan: EGEA–Universita Bocconi Editore, 1906).

Parsons, Talcott, *The Structure of Social Action* (New York: McGraw-Hill, 1937).

Parsons, Talcott, *Essays in Sociological Theory: Pure and Applied* (Glencoe, IL: The Free Press, 1949).

Parsons, Talcott, *The Social System* (Glencoe: The Free Press, 1951).

Parsons, Talcott, "On the Concept of Political Power," *Proceedings of the American Philosophical Society* 107,3 (1963):232–262.

Parsons, Talcott, "Evolutionary Universals in Society," *American Sociological Review* 29,3 (June 1964):339–357.

Parsons, Talcott, *Sociological Theory and Modern Society* (New York: Free Press, 1967).

Parsons, Talcott, "Commentary on Herbert Gintis, 'A Radical Analysis of Welfare Economics and Individual Development'," *Quarterly Journal of Economics* 89,2 (1975):280–290.

Parsons, Talcott and Edward Shils, *Toward a General Theory of Action* (Cambridge, MA: Harvard University Press, 1951).

Parsons, Talcott and Neil J. Smelser, *Economy and Society. A Study in the Integration of Economic and Social Theory* (Glencoe: The Free Press, 1956).

Paul, A., S. Preuschoft, and Carel P. van Schaik, "The Other Side of the Coin: Infanticide and the Evolution of Affiliative Male-infant Interactions in Old World Primates," in Carel P. van Schaik and C. H. Janson (eds.) *Infanticide by Males and its Implications* (Cambridge: Cambridge University Press, 2000) pp. 269–292.

Pearl, J., *Causality* (New York: Oxford University Press, 2000).

Pepper, John W., "Simple Models of Assortment Through Environmental Feedback," *Artificial Life* 13,1 (2007):1–9.

Pinker, Steven, *The Blank Slate: The Modern Denial of Human Nature* (New York: Viking, 2002).

Pinker, Steven, "The Cognitive Niche: Coevolution of Intelligence, Sociality and Language," *Proceedings of the National Academy of Sciences* 107,2 (2010):8993–8999.

Pinker, Steven, "The False Allure of Group Selection," *Edge Original Essay (Online)* (June 2012).

Plooij, F. X., "Tool-using during Chimpanzees' Bushpig Hunt," *Carnivore* 1 (1978):103–106.

Plott, Charles R., "The Application of Laboratory Experimental Methods to Public Choice," in Clifford S. Russell (ed.) *Collective Decision Mak-*

ing: Applications from Public Choice Theory (Baltimore, MD: Johns Hopkins University Press, 1979) pp. 137–160.

Plourde, Aimée, "Human Power and Prestige Systems," in Peter M. Kappeler and Joan B. Silk (eds.) *Mind the Gap: Tracing the Origins of Human Universals* (Berlin and Heidelberg: Springer, 2010) pp. 139–152.

Poincaré, Henri, *Science and Method* (Cosimo Classics, 1908).

Pomiankowski, Andrew N., "The Costs of Choice in Sexual Selection," *Journal of Theoretical Biology* 128 (1987):195–218.

Popper, Karl, *Objective Knowledge: An Evolutionary Approach* (Oxford: Clarendon Press, 1979).

Popper, Karl, *The Logic of Scientific Discovery* (London: Routledge Classics, 2002[1959]).

Potts, Richard, *Humanity's Descent: The Consequences of Ecological Instability* (New York: Aldine de Gruyter, 1996).

Potts, Richard, "Environmental Hypotheses of Hominin Evolution," *Yearbook of Physical Anthropology* 41 (1998):93–138.

Price, George R., "Selection and Covariance," *Nature* 227 (1970):520–521.

Price, George R., "Fisher's 'Fundamental Theorem' Made Clear," *Annals of Human Genetics* 36 (1972):129–140.

Price, Michael E. and Mark Van Vugt, "The Evolution of Leader-follower Reciprocity: The Theory of service-for-prestige," *Frontiers in Human Neuroscience* 8 (2014).

Pruetz, Jill D. and Paco Bertolani, "Savanna Chimpanzees, Pan troglodytes verus, Hunt with Tools," *Current Biology* 17 (2007):412–417.

Quattrone, George A. and Amos Tversky, "Constrasting Rational and Psychological Analyses of Political Choice," *American Political Science Review* 82,3 (September 1988):719–736.

Queller, David C., "A General Model for Kin Selection," *Evolution* 42,2 (1992):376–380.

Rand, A. S., "Ecology and Social Organization in the Iguanid Lizard *Anolis lineatopus*," *Proceedings of the U.S. National Museum* 122 (1967):1–79.

Rauch, James E., "Trade and Networks: An Application to Minority Retail Entrepreneurship," June 1996. Russell Sage Working Paper.

Rawls, John, *A Theory of Justice* (Cambridge, MA: Harvard University Press, 1971).

Relethford, John H., *The Human Species: An Introduction to Biological Anthropology* (New York: McGraw-Hill, 2007).

Richerson, Peter J. and Robert Boyd, "The Evolution of Ultrasociality," in I. Eibl-Eibesfeldt and F. K. Salter (eds.) *Indoctrinability, Ideology and Warfare* (New York: Berghahn Books, 1998) pp. 71–96.

Richerson, Peter J. and Robert Boyd, *Not by Genes Alone* (Chicago: University of Chicago Press, 2004).

Richerson, Peter J., Robert Boyd, and Robert L. Bettinger, "Was Agriculture Impossible during The Pleistocene But Mandatory during The Holocene? A Climate Change Hypothesis," *American Antiquity* 66 (2001):387–411.

Riechert, S. E., "Games Spiders Play: Behavioural Variability in Territorial Disputes," *Journal of Theoretical Biology* 84 (1978):93–101.

Rieseberg, Loren H., "Chromosomal Rearrangements and Speciation," *Trends in Ecology and Evolution* 16,7 (2001):351–358.

Riker, William H. and Peter C. Ordeshook, "A Theory of the Calculus of Voting," *American Political Science Review* 62,1 (March 1968):25–42.

Riley, Joe R., Uwe Greggers, Alan D. Smith, Don R. Reynolds, and Randolf Menzel, "The Flight Paths of Honeybees Recruited by the Waggle Dance," *Nature* 435,7039 (2005):205–207.

Riley, Margaret A. and Michelle Lizotte-Waniewski, "Population Genomics and the Bacterial Species Concept," *Methods in Molecular Biology* 532 (2009):367–377.

Risen, Jane L., "Believing what we do not Believe," *Psychological Review* (October 2015):1–27.

Rivera, M. C. and J. A. Lake, "The Ring of Life Provides Evidence for a Genome Fusion Origin of Eukaryotes," *Nature* 431 (2004):152–155.

Roach, Neil T., Madhusudhan Venkadesan, Michael J. Rainbow, and Daniel E. Lieberman, "Elastic Energy Storage in the Shoulder and the Evolution of High-speed Throwing in *Homo*," *Nature* 498 (27 June 2013).

Roebroeks, W. and P. Villa, "On the Earliest Evidence for Habitual Use of Fire in Europe," *Proceedings of the National Academy of Sciences* 108 (2011):5209–5214.

Roemer, John, *A General Theory of Exploitation and Class* (Cambridge: Harvard University Press, 1982).

Roemer, John, "Kantian Optimization: A Microfoundation for Cooperation," *Journal of Public Economics* 127 (2015):45–57.

Rogers, Alan R., "Group Selection by Selective Emigration: The Effects of Migration and Kin Structure," *American Naturalist* 135,3 (March 1990):398–413.

Rotemberg, Julio, "Attitude-dependent Altruism: Turnout and Voting," *Public Choice* 140 (2009):223–244.

Roth, Alvin E., "Bargaining Experiments," in John H. Kagel and Alvin E. Roth (eds.) *The Handbook of Experimental Economics* (Princeton: Princeton University Press, 1995).

Roth, Alvin E., Vesna Prasnikar, Masahiro Okuno-Fujiwara, and Shmuel Zamir, "Bargaining and Market Behavior in Jerusalem, Ljubljana, Pittsburgh, and Tokyo: An Experimental Study," *American Economic Review* 81,5 (December 1991):1068–1095.

Rousseau, Jean-Jacques, *Discourse on the Origin and Foundations of Inequality Among Men* (New York: Penguin, 1984).

Rousset, François and S. Billard, "A Theoretical Basis for Measures of Kin Selection in Subdivided Populations," *Proceedings of the National Academy of Science* 61 (2007):2320–2330.

Rousset, François and S. Lion, "Much Ado About Nothing: Nowak et al.'s Charge Against Inclusive Fitness," *Journal of Evolutionary Biology* 24 (2011):1386–1392.

Rozin, Paul, L. Lowery, S. Imada, and Jonathan Haidt, "The CAD Triad Hypothesis: A Mapping between Three Moral Emotions (Contempt, Anger, Disgust) and Three Moral Codes (Community, Autonomy, Divinity)," *Journal of Personality & Social Psychology* 76 (1999):574–586.

Saari, Donald G., "Iterative Price Mechanisms," *Econometrica* 53,5 (September 1985):1117–1131.

Sahle, Yonatan, W. Karl Hutchings, David R. Braun, Judith C. Sealy, Leah E. Morgan, Agazi Negash, and Balemwal Atnafu, "Earliest Stone-Tipped Projectiles from the Ethiopian Rift Date to More Than 279,000 Years Ago," *PLoS ONE* 8,11 (2013):1–9.

Sally, David, "Conversation and Cooperation in Social Dilemmas," *Rationality and Society* 7,1 (January 1995):58–92.

Samuelson, Larry, *Evolutionary Games and Equilibrium Selection* (Cambridge, MA: MIT Press, 1997).

Samuelson, Larry and Jianbo Zhang, "Evolutionary Stability in Asymmetric Games," *Journal of Economic Theory* 57,2 (1992):363–391.

Samuelson, Paul, *The Foundations of Economic Analysis* (Cambridge: Harvard University Press, 1947).

Samuelson, Paul, "Wages and Interests: A Modern Dissection of Marxian Economics," *American Economic Review* 47 (1957):884–921.

Samuelson, William and Richard Zeckhauser, "Status Quo Bias in Decision Making," *Journal of Risk and Uncertainty* 1 (1988):7–59.

Sartre, Jean-Paul, *Existentialism and Human Emotions* (Philosophical Library, 1957).

Savage, Leonard J., *The Foundations of Statistics* (New York: John Wiley & Sons, 1954).

Scarf, Herbert, "Some Examples of Global Instability of Competitive Equilibrium," *International Economic Review* 1 (1960):157–172.

Schelling, Thomas C., *The Strategy of Conflict* (Cambridge, MA: Harvard University Press, 1960).

Schlatter, Richard Bulger, *Private Property: History of an Idea* (New York: Russell & Russell, 1973).

Schuessler, Alexander A., *A Logic of Expressive Choice* (Princeton, NJ: Princeton University Press, 2000).

Schulkin, J., *Roots of Social Sensitivity and Neural Function* (Cambridge, MA: MIT Press, 2000).

Schultz, L. and Alison Gopnik, "Causal Learning across Domains," *Developmental Psychology* 40 (2004):162–176.

Schumpeter, Joseph A., *The Theory of Economic Development: An Inquiry into Profits, Capital, Credit, Interest and the Business Cycle* (Oxford: Oxford University Press, 1934[1911]).

Schumpeter, Joseph A., "The Creative Response in Economic History," *Journal of Economic History* 7 (1947):149–159.

Schumpeter, Joseph A., *Imperialism and Social Classes* (New York: Augustus M. Kelley, 1951).

Schuppli, Caroline, Karin Isler, and Carel P. van Schaik, "How to Explain the Unusually Late Age at Skill competence Among Humans," *Journal of Human Evolution* 63 (2012):843–850.

Schutz, Alfred, *Phenomenology of the Social World* (Northwestern University Press, 1932/1967).

Schwander, Tanja, Romain Libbrecht, and Laurent Keller, "Supergenes and Complex Phenotypes," *Current Biology* 24,7 (2014):R288–R294.

Searle, John, *The Construction of Social Reality* (New York: Free Press, 1995).

Sears, David O., Richard R. Lau, Tom R. Tyler, and Harris M. Allen, "Self-interest vs. symbolic politics in policy attitudes and presidential voting," *American Political Science Review* 74,3 (1980):670–684.

Seeley, Thomas D., "Honey Bee Colonies are Group-Level Adaptive Units," *American Naturalist* 150 (1997):S22–S41.

Segerstrale, Ullica, *Defenders of the Truth: The Sociobiology Debate* (Oxford: Oxford University Press, 2001).

Senar, J. C., M. Camerino, and N. B. Metcalfe, "Agonistic Interactions in Siskin Flocks: Why are Dominants Sometimes Subordinate?," *Behavioral Ecology and Sociobiology* 25 (1989):141–145.

Service, E. R., *Origin of the State and Civilization: The Process of Cultural Evolution* (New York: Norton, 1975).

Shachar, Ron and Barry Nalebuff, "Follow the leader: Theory and evidence on political participation," *American Economic Review* 89,3 (1999):525–547.

Shea, John J., "The Origins of Lithic Projectile Point Technology: Evidence from Africa, the Levant, and Europe," *Journal of Archaeological Science* 33 (2006):823–846.

Shennan, Stephen, *Quantifying Archaeology* (Edinburgh: Edinburgh University Press, 1997).

Shultz, Susanne, Christopher Opie, and Quentin D. Atkinson, "Stepwise Evolution of Stable Sociality in Primates," *Nature* 479 (November 2011):219–222.

Sigg, Hans and Jost Falett, "Experiments on Respect of Possession and Property in Hamadryas Baboons (*Papio hamadryas*)," *Animal Behaviour* 33 (1985):978–984.

Silk, Joan B., "The Path to Sociality," *Nature* 49 (November 2011):182–183.

Simmel, Georg, *Georg Simmel on Individuality and Social Forms* (Chicago: University of Chicago Press, 1972).

Simon, Herbert, "Theories of Bounded Rationality," in C. B. McGuire and Roy Radner (eds.) *Decision and Organization* (New York: American Elsevier, 1972) pp. 161–176.

Simon, Herbert, "A Mechanism for Social Selection and Successful Altruism," *Science* 250 (1990):1665–1668.

Simpson, George Gaylord, "The Role of the Individual in Evolution," *Journal of the Washington Academy of Sciences* 31 (1941):1–20.

Skibo, James M. and R. Alexander Bentley, *Complex Systems and Archaeology* (Salt Lake City: University of Utah Press, 2003).

Skutch, A. F., "Helpers Among Birds," *Condor* 63 (1961):198–226.

Skyrms, Brian, *The Stag Hunt and the Evolution of Social Structure* (Cambridge: Cambridge University Press, 2004).

Smaldino, Paul E., Jeffrey C. Schank, and Richard McElreath, "Increased Costs of Cooperation Help Cooperators in the Long Run," *American Naturalist* 181,4 (2013):451–463.

Smith, Adam, *The Theory of Moral Sentiments* (New York: Prometheus, 1759/2000).

Smith, Vernon L., "Microeconomic Systems as an Experimental Science," *American Economic Review* 72,5 (December 1982):923–955.

Smith, Vernon L. and Arlington W. Williams, "Experimental Market Economics," *Scientific American* 267,6 (December 1992):116–121.

Sobel, D. M. and N. Z. Kirkham, "Bayes Nets and Babies: Infants' Developing Statistical Reasoning Abilities and their Representations of Causal Knowledge," *Developmental Science* 10,3 (2007):298–306.

Sober, Elliot and David Sloan Wilson, *Unto Others: The Evolution and Psychology of Unselfish Behavior* (Cambridge, MA: Harvard University Press, 1998).

Solow, Robert, "Another Possible Source of Wage Stickiness," *Journal of Macroeconomics* 1 (1979):79–82.

Soltis, Joseph, Robert Boyd, and Peter J. Richerson, "Can Group-functional Behaviors Evolve by Cultural Group Selection: An Empirical Test," *Current Anthropology* 36,3 (June 1995):473–483.

Sonnenschein, Hugo, "Do Walras' Identity and Continuity Characterize the Class of Community Excess Demand Functions?," *Journal of Ecomonic Theory* 6 (1973):345–354.

Sopher, Barry and Gary Gigliotti, "Intransitive Cycles: Rational Choice or Random Error: An Answer Based on Estimation of Error Rates with Experimental Data," *Theory and Decision* 35 (1993):311–336.

Spirtes, P., C. Glymour, and R. Scheines, *Causation, Prediction, and Search* (Cambridge: MIT Press, 2001).

Stake, Jeffrey Evans, "The Property Instinct," *Philosophical Transactions of the Royal Society of London B* 359 (2004):1763–1774.

Stalnaker, Robert, "A Theory of Conditionals," in Nicholas Rescher (ed.) *Studies in Logical Theory* (London: Blackwell, 1968).

Stalnaker, Robert, "Knowledge, Belief, and Counterfactual Reasoning in Games," *Economics and Philosophy* 12 (1996):133–163.

Starmer, Chris, "Developments in Non-Expected Utility Theory: The Hunt for a Descriptive Theory of Choice under Risk," *Journal of Economic Literature* 38 (June 2000):332–382.

Sterck, E. H. M., David P. Watts, and Carel P. van Schaik, "The evolution of Female Social Relationships in Nonhuman Primates," *Behavioral Ecology and Sociobiology* 41 (1997):291–309.

Sterelny, Kim, "From Hominins to Humans: How *Sapiens* Became Behaviourally Modern," *Proceedings of the Royal Society of London B* (2011).

Stevens, Elisabeth Franke, "Contests Between Bands of Feral Horses for Access to Fresh Water: The Resident Wins," *Animal Behaviour* 36,6 (1988):1851–1853.

Stevens, Jeroen M. G., Hilde Vervaecke, Han de Vries, and Linda van Elsacker, "Sex Differences in the Steepness of Dominance Hierarchies in Captive Bonobo Groups," *International Journal of Primatology* 28,6 (December 2007):1417–1430.

Steyvers, Mark, Thomas L. Griffiths, and Simon Dennis, "Probabilistic Inference in Human Semantic Memory," *Trends in Cognitive Sciences* 10 (2006):327–334.

Stiglitz, Joseph E. and Andrew Weiss, "Credit Rationing in Markets with Imperfect Information," *American Economic Review* 71 (June 1981):393–411.

Stiner, M. C., "Carnivory, Coevolution, and the Geographic Spread of the genus *Homo*," *Journal of Archeological Research* 10 (2002):1–63.

Stiner, M. C., R. Barkai, and A. Gopher, "Cooperative Hunting and Meat Sharing," *Proceedings of the National Academy of Sciences* 106,32 (2009):13207–13212.

Strassmann, Joan E., "Kin Selection and Eusociality," *Nature* 471 (March 24 2011):E5–E6.

Strier, Karen B., "Activity Budgets of Woolly Spider Monkeys, or Muriquis," *American Journal of Primatology* 13,4 (1987):385–395.

Strier, Karen B., "Causes and Consequences of Nonaggression in the Woolly Spider Monkey, or Muriqui," in J. Silverberg and J. P. Gray (eds.)

Aggression and Peacefulness in Humans and Other Primates (New York: Oxford University Press, 1992) pp. 100–116.

Sugden, Robert, *The Economics of Rights, Co-operation and Welfare* (Oxford: Basil Blackwell, 1986).

Sugden, Robert, "Spontaneous Order," *Journal of Economic Perspectives* 3,4 (Fall 1989):85–97.

Sugden, Robert, "An Axiomatic Foundation for Regret Theory," *Journal of Economic Theory* 60,1 (June 1993):159–180.

Sugden, Robert, "The Logic of Team Reasoning," *Philosophical Explorations* 6,3 (2003):165–181.

Sugiyama, L. S. and R. Chacon, "Effects of Illness and Injury on Foraging among the Yora and Shiwiar: Pathology Risk as Adaptive Problem," in L. Cronk, N. Chagnon, and W. Irons (eds.) *Adaptation and Human Behavior: An Anthropological Perspective* (New York: Aldine de Gruyter, 2000) pp. 371–395.

Tacon, P. and C. Chippendale, "Australia's Ancient Warriors: Changing Depictions of Fighting in the Rock Art of Arnhem Land, N.T.," *Cambridge Archaeological Journal* 4 (1994):211–248.

Taylor, Michael, *Anarchy and Cooperation* (London: John Wiley & Sons, 1976).

Taylor, Michael, *Community, Anarchy, and Liberty* (Cambridge: Cambridge University Press, 1982).

Taylor, Michael, *The Possibility of Cooperation* (Cambridge: Cambridge University Press, 1987).

Taylor, Peter, "Altruism in Viscous Populations: An Inclusive Fitness Model," *Evolutionary Ecology* 6 (1992):352–356.

Taylor, Peter, "Inclusive Fitness Arguments in Genetic Models of Behavior," *Journal of Mathematical Biology* 34 (1996):654–674.

Taylor, Peter and Leo Jonker, "Evolutionarily Stable Strategies and Game Dynamics," *Mathematical Biosciences* 40 (1978):145–156.

Taylor, Scott and Leonardo Campagna, "Avian Supergenes," *Science* 351,6272 (2016):446–447.

Tenenbaum, Joshua B., Thomas L. Griffiths, and Charles Kemp, "Bayesian models of Inductive Learning and Reasoning," *Trends in Cognitive Science* 10 (2006):309–318.

Tesfatsion, Leigh and Kenneth L. Judd, *Handbook of Computational Economics II: Agent-Based Computational Economics* (Amsterdam: Elsevier/North-Holland, 2006).

Thaler, Richard H., *The Winner's Curse* (Princeton: Princeton University Press, 1992).

Thaler, Richard H. and Cass Sunstein, *Nudge: Improving Decisions about Health, Wealth, and Happiness* (New York: Penguin, 2008).

Thieme, Harmut, "Lower Palaeolithic Hunting Spears from Germany," *Nature* 385,27 (February 1997):807–810.

Thorpe, I. J. N., "Anthropology, Archaeology, and the Origin of Warfare," *World Archaeology* 35,1 (2003):145–165.

Tilly, Charles, "Charivaris, Repertoires and Urban Politics," in John M. Merriman (ed.) *French Cities in the Nineteenth Century* (New York: Holmes and Meier, 1981) pp. 73–91.

Tomasello, Michael, *A Natural History of Human Thinking* (Cambridge: Harvard University Press, 2014).

Tomasello, Michael and Malinda Carpenter, "Shared Intentionality," *Developmental Science* 10,1 (2007):121–125.

Tomasello, Michael, Malinda Carpenter, Josep Call, Tanya Behne, and Henrike Moll, "Understanding and Sharing Intentions: The Origins of Cultural Cognition," *Behavioral and Brain Sciences* 28,5 (2005):675–691.

Tönnies, Ferdinand, *Community and Civil Society* (Cambridge: Cambridge University Press, 2001[1887]).

Tooby, John and Irven DeVore, "The Reconstruction of Hominid Evolution through Strategic Modeling," in W. G. Kinzey (ed.) *The Evolution of Human Behavior: Primate Models* (Albany: SUNY Press, 1987) pp. 183–237.

Tooby, John and Leda Cosmides, "The Psychological Foundations of Culture," in Jerome H. Barkow, Leda Cosmides, and John Tooby (eds.) *The Adapted Mind: Evolutionary Psychology and the Generation of Culture* (New York: Oxford University Press, 1992) pp. 19–136.

Torii, M., "Possession by Non-human Primates," *Contemporary Primatology* (1974):310–314.

Trivers, Robert L., "The Evolution of Reciprocal Altruism," *Quarterly Review of Biology* 46 (1971):35–57.

Trivers, Robert L., "Parental Investment and Sexual Selection, 1871–1971," in B. Campbell (ed.) *Sexual Selection and the Descent of Man* (Chicago: Aldine, 1972) pp. 136–179.

Tuomela, R., *The Importance of Us: A Philsophical Study of Basic Social Emotions* (Stanford: Stanford University Press, 1995).

Turchin, Peter, *Ultrasociety: How 10,000 Years of War Made Humans the Greatest Cooperators on Earth* (Beresta Books, 2015).

Turchin, Peter and Andrey Korotayev, "Population Dynamics and Internal Warfare: A Reconsideration," *Social Evolution & History* 5,2 (2006):112–147.

Turner, Jonathan H., *Handbook of Sociological Theory* (Berlin: Springer, 2006).

Tversky, Amos and Daniel Kahneman, "Extensional versus Intuitive Reasoning: The Conjunction Fallacy in Probability Judgement," *Psychological Review* 90 (1983):293–315.

Tversky, Amos and Daniel Kahneman, "Loss Aversion in Riskless Choice: A Reference-Dependent Model," *Quarterly Journal of Economics* 106,4 (November 1991):1039–1061.

Tversky, Amos, Paul Slovic, and Daniel Kahneman, "The Causes of Preference Reversal," *American Economic Review* 80,1 (March 1990):204–217.

Ullmann-Margalit, Edna, *The Emergence of Norms* (Oxford: Clarendon Press, 1977).

Uyenoyama, Marcy K. and Marcus W. Feldman, "Theories of Kin and Group Selection: A Population Genetics Approach," *Theoretical Population Biology* 17 (1980):380–414.

van Schaik, Carel P., "Why are Diurnal Primates Living in Groups?," *Behaviour* 87 (1983):120–144.

van Schaik, Carel P., "Social Evolution in Primates: the Role of Ecological Factors and Male Behaviour," *Proceedings of the British Academy* 88 (1996):9–31.

van Schaik, Carel P., G. R. Pradhan, and M. A. van Noordwijk, "Mating Conflict in Primates: Infanticide, Sexual Harassment and Female Sexuality," in P. M. Kappeler and Carel P. van Schaik (eds.) *Sexual Selection in Primates* (Cambridge: Cambridge University Press, 2004) pp. 131–150.

van Schaik, Carel P., Sagar Pandit, and E. R. Vogel, "Toward a General Model for Male-male Coalitions in Primate Groups," in P. M. Kappeler and Carel P. van Schaik (eds.) *Cooperation in Primates and Humans: Mechanisms and Evolution* (Berlin: Springer-Verlag, 2006) pp. 151–171.

Varian, Hal R., "The Nonparametric Approach to Demand Analysis," *Econometrica* 50 (1982):945–972.

Varian, Hal R., *Microeconomic Analysis* (New York: W. W. Norton, 1992).

Veblen, Thorstein, *The Theory of the Leisure Class* (New York: Macmillan, 1899).

Vigilant, Linda, Michael Hofreiter, Heike Siedel, and Christophe Boesch, "Paternity and Relatedness in Wild Chimpanzee Communities," *Proceedings of the National Academy of Sciences* 98,23 (2001):12890–12895.

von Neumann, John and Oskar Morgenstern, *Theory of Games and Economic Behavior* (Princeton: Princeton University Press, 1944).

von Rohr, C. Rudolf, S. E. Koski, Judith M. Burkart, C. Caws, O. N. Fraser, A. Ziltener, and Carel P. van Schaik, "Impartial Third-party Interventions in Captive Chimpanzees: A Reflection of Community Concern," *PLoS One* 7,3 (2012).

von Rueden, Christopher, "The Roots and Fruits of Social Status in Small-Scale Human Societies," in J. T. Cheng (ed.) *The Psychology of Social Status* (New York: Springer, 2015).

von Rueden, Christopher, Michael Gurven, Hillard Kaplan, and Jonathan Stieglitz, "Leadership in an Egalitarian Society," *Human Nature* (September 2014).

Vrba, Elisabeth S., "The Fossil Record of African Antelopes (Mammalia, Bovidae) in Relation to Human Evolution and Paleoclimate," in Elisabeth S. Vrba, G. Denton, L. Burckle, and T. Partridge (eds.) *Paleoclimate and Evolution with Emphasis on Human Origin* (New Haven: Yale University Press, 1995) pp. 385–424.

Wadley, L., T. Hodgskiss, and M. Grant, "Implications for Complex Cognition from the Hafting of Tools with Compound Adhesives in the Middle Stone Age, South Africa," *Proceedings of the National Academy of Sciences* 106,24 (2009):9590–9594.

Wallace, Anthony F. C., *Culture and Personality* (New York: Random House, 1970).

Walras, Léon, *Elements of Pure Economics* (London: George Allen and Unwin, 1874).

Washburn, S. L., "Human Behavior and the Behavior of other Animals," *American Psychologist* 33 (1978):405–418.

Watson, John B., "Psychology as the Behaviorist Views it," *Psychological Review* 20 (1913):158–177.

Watts, David P., M. Muller, S. Amsler, G. Mbabazi, and John C. Mitani, "Lethal Intergroup Aggression by Chimpanzees in the Kibale National Park, Uganda," *American Journal of Primatology* 68 (2006):161–180.

Weber, Max, *The Protestant Ethic and the Spirit of Capitalism* (Penguin, 2002[1905]).

Weeden, James and Robert Kurzban, *The Hidden Agenda of the Political Mind: How Self-Interest Shapes Our Opinions and Why We Won't Admit It* (Princeton: Princeton University Press, 2014).

Weibull, Jörgen W., *Evolutionary Game Theory* (Cambridge, MA: MIT Press, 1995).

Weigel, Ronald M., "The Application of Evolutionary Models to the Study of Decisions Made by Children during Object Possession Conflicts," *Ethnology and Sociobiology* 5 (1984):229–238.

Weisberg, Michael, "Who is a Modeler?," *British Journal of the Philosophy of Science* 58 (2007):207–233.

Wenseleers, Tom and Francis L. W. Ratnieks, "Tragedy of the Commons in *Melipona* Bees," *Proceedings of the Royal Society of London B* 271 (2004):S310–S312.

West-Eberhard, Mary Jane, "The Evolution of Social Behavior by Kin Selection," *Quarterly Review of Biology* 50 (1975):1–33.

West-Eberhard, Mary Jane, "Sexual Selection, Social Competition, and Speciation," *Quarterly Review of Biology* 58,2 (June 1983):155–183.

West, Stuart A., *Sex Allocation* (Princeton University Press, 2009).

West, Stuart A., Ashleigh S. Griffin, and Andy Gardner, "Evolutionary Explanations for Cooperation," *Current Biology* 17 (2007):R661–R672.

West, Stuart A., Claire El Mouden, and Andy Gardner, "Sixteen Common Misconceptions about the Evolution of Cooperation in Humans," *Evolution & Human Behavior* 32,4 (2011):231–262.

West, Stuart A., Stephen P. Diggle, Angus Buckling, Andy Gardner, and Ashleigh S. Griffin, "The Social Lives of Microbes," *Annual Review of Ecology, Evolution, and Systematics* 38 (2007):53–77.

Wetherick, N. E., "Reasoning and Rationality: A Critique of Some Experimental Paradigms," *Theory & Psychology* 5,3 (1995):429–448.

Whallon, Robert, "The Human Revolution," in P. Mellars and C. Stringer (eds.) *The Human Revolution: Behavioural and Biological Perspectives on the Origins of Modern Humans* (Princeton: Princeton University Press, 1989).

Wheeler, William Morton, *The Social Insects* (New York: Harcourt, Brace, 1928).

White, Harrison, *Identity and Control: How Social Formations Emerge* (Princeton: Princeton University Press, 2008).

Whiten, Andrew and David Erdal, "The human socio-cognitive niche and its evolutionary origins," *Philosophical Transactions of the Royal Society B: Biological Sciences* 367,1599 (2012):2119–2129.

Whittaker, John C. and Grant McCall, "Handaxe-Hurling Hominids: An Unlikely Story," *Current Anthropology* 42,4 (2001):566–572.

Whyte, William F., *Money and Motivation* (New York: Harper & Row, 1955).

Wiessner, Polly, "Xaro: A Regional System of Reciprocity for Reducing Risk Among the !Kung San," 1977. Ph.D. Dissertation, Department of Anthropology, University of Michigan.

Wiessner, Polly, "From Spears to M16s: Testing the Imbalance of Power Hypothesis Among the Enga," *Journal of Anthropological Research* 62 (2006):165–191.

Wiessner, Polly, "The Power of One: The Big Man Revisited," in Kevin J. Vaugh, Jelmer W. Eerkins, and John Kantner (eds.) *The Evolution of Leadership: Transitions in Decision Making from Small-scale to Middle-Range Societies* (Santa Fe: SAR Press, 2009) pp. 195–221.

Wilkins, J., B. J. Schoville, K. S. Brown, and M. Chazan, "Evidence for Early Hafted Hunting Technology," *Science* 338 (2012):942–946.

Willems, Erik P., Barbara Hellriegel, and Carel P. van Schaik, "The Collective Action Problem in Primate Territory Economics," *Proceedings of the Royal Society B* 280 (2013):1–7.

Williams, George C., *Adaptation and Natural Selection: A Critique of Some Current Evolutionary Thought* (Princeton: Princeton University Press, 1966).

Wilson, David Sloan, "Structured Demes and the Evolution of Group-Advantageous Traits," *American Naturalist* 111 (1977):157–185.

Wilson, David Sloan, "Hunting, Sharing, and Multilevel Selection: The Tolerated Theft Model Revisited," *Current Anthropology* 39 (1998):73–97.

Wilson, David Sloan, "Social Semantics: Towards a Genuine Pluralism in the Study of Social Behaviour," *Journal of Evolutionary Biology* 21 (2008):368–373.

Wilson, David Sloan and Edward O. Wilson, "Rethinking the Theoretical Foundation of Sociobiology," *The Quarterly Review of Biology* 82,4 (December 2007):327–348.

Wilson, David Sloan, G. B. Pollock, and Lee Alan Dugatkin, "Can Altruism Evolve in Purely Viscous Populations?," *Evolutionary Ecology* 6 (1992):331–341.

Wilson, Edward O., *Sociobiology: The New Synthesis* (Cambridge, MA: Harvard University Press, 1975).

Wilson, Edward O., *Consilience: The Unity of Knowledge* (New York: Knopf, 1998).

Wilson, Edward O., *The Social Conquest of Earth* (New York: W. W. Norton, 2012).

Wilson, Edward O. and Bert Hölldobler, "Eusociality: Origin and Consequences," *Proceedings of the National Academy of Sciences* 102,38 (2005):13367–71.

Wimsatt, William C., *Re-Engineering Philosophy for Limited Beings* (Cambridge: Harvard University Press, 2007).

Winter, Fabian, Heiko Rauhut, and Dirk Helbing, "How Norms Can Generate Conflict: An Experiment on the Failure of Cooperative Micromotives on the Macro-level," *Social Forces* 90,3 (March 2012):919–948.

Winterhalder, Bruce and Eric Alden Smith, *Evolutionary Ecology and Human Behavior* (New York: Aldine de Gruyter, 1992).

Wolfram Research, *Mathematica 10.0* (Champaign, IL: Wolfram Research, 2014).

Wood, Bernard, "Reconstructing Human Evolution: Achievements, Challenges, and Opportunities," *Proceedings of the National Academy of Sciences* 107,Supplement 2 (2010):8902–8909.

Woodburn, James, "Egalitarian Societies," *Man* 17,3 (1982):431–451.

Wrangham, Richard W., "An Ecological Model of Female-bonded Primate Groups," *Behaviour* 75 (1980):262–300.

Wrangham, Richard W. and Dale Peterson, *Demonic Males: Apes and the Origins of Human Violence* (New York: Mariner Books, 1996).

Wrangham, Richard W. and Luke Glowacki, "Intergroup Aggression in Chimpanzees and War in Nomadic Hunter-Gatherers: Evaluating the Chimpanzee Model," *Human Nature* 23 (2012):5–29.

Wrangham, Richard W. and Rachel Carmody, "Human Adaptation to the Control of Fire," *Evolutionary Anthropology* 19 (2010):187–199.

Wrong, Dennis H., "The Oversocialized Conception of Man in Modern Sociology," *American Sociological Review* 26 (April 1961):183–193.

Wynn, T., "Hafted Spears and the Archaeology of Mind," *Proceedings of the National Academy of Sciences* 106,24 (2009):9544–9545.

Wynne-Edwards, V. C., *Animal Dispersion in Relation to Social Behavior* (Edinburgh, UK: Oliver and Boyd, 1962).

Yamagishi, Toshio, N. Jin, and Toko Kiyonari, "Bounded Generalized Reciprocity: In-Group Boasting and In-Group Favoritism," *Advances in Group Processes* 16 (1999):161–197.

Young, H. Peyton, "The Evolution of Conventions," *Econometrica* 61,1 (January 1993):57–84.

Young, H. Peyton, *Individual Strategy and Social Structure: An Evolutionary Theory of Institutions* (Princeton: Princeton University Press, 1998).

Young, H. Peyton and Mary Burke, "Competition and Custom in Economic Contracts: A Case Study of Illinois Agriculture," *American Economic Review* 91,3 (2001):559–573.

Zajonc, Robert B., "Feeling and Thinking: Preferences Need No Inferences," *American Psychologist* 35,2 (1980):151–175.

Zajonc, Robert B., "On the Primacy of Affect," *American Psychologist* 39 (1984):117–123.

Zak, Paul J. and Stephen Knack, "Trust and Growth," *The Economic Journal* 111,470 (2001):295–321.

Subject Index

Author Index